Advances in Physical Geochemistry

Volume 9

Advances in Physical Geochemistry

Series Editor: Surendra K. Saxena

Volume 1 R.C. Newton/A. Navrotsky/B.J. Wood (editors)
Thermodynamics of Minerals and Melts
1981. xii, 304 pp. 66 illus.
ISBN 0-387-90530-8

Volume 2 S.K. Saxena (editor)
Advances in Physical Geochemistry, Volume 2
1982. x, 353 pp. 113 illus.
ISBN 0-387-90644-4

Volume 3 S.K. Saxena (editor)
Kinetics and Equilibrium in Mineral Reactions
1983. vi, 273 pp. 99 illus.
ISBN 0-387-90865-X

Volume 4 A.B. Thompson/D.C. Rubie (editors)
Metamorphic Reactions: Kinetics, Textures, and Deformation
1985. xii, 291 pp. 81 illus.
ISBN 0-387-96077-5

Volume 5 J.V. Walther/B.J. Wood (editors)
Fluid-Rock Interactions during Metamorphism
1986. x, 211 pp. 59 illus.
ISBN 0-387-96244-1

Volume 6 S.K. Saxena (editor)
Chemistry and Physics of Terrestrial Planets
1986. x, 405 pp. 94 illus.
ISBN 0-387-96287-5

Volume 7 S. Ghose/J.M.D. Cohe/E. Salje (editors)
Structural and Magnetic Phase Transitions in Minerals
1988. xiii, 272 pp. 117 illus.
ISBN 0-387-96710-9

Volume 8 J. Ganguly (editor)
Diffusion, Atomic Ordering, and Mass Transport: Selected Topics in Geochemistry
1991. xiv, 584 pp. 170 illus.
ISBN 0-387-97287-0

Volume 9 L.L. Perchuk/I. Kushiro (editors)
Physical Chemistry of Magmas
1991. 352 pp. 137 illus.
ISBN 0-387-97500-4

Leonid L. Perchuk Ikuo Kushiro
Editors

Physical Chemistry of Magmas

With Contributions by
D.R. Baker Y. Bottinga A.S. Chekhmir C.D. Doyle
M.B. Epel'baum P.C. Hess I. Kushiro A.D. Kuznetsov
B.O. Mysen L.L. Perchuk E.S. Persikov N. Shimizu
Y. Tatsumi E.B. Watson

With 137 Illustrations

Springer-Verlag
New York Berlin Heidelberg London
Paris Tokyo Hong Kong Barcelona

Leonid L. Perchuk
Institute of Experimental Mineralogy
USSR Academy of Sciences
Moscow District
Chernogolovka 142432, USSR

Ikuo Kushiro
Geological Institute
University of Tokyo
Tokyo 113, Japan

Series Editor:
Surendra K. Saxena
Department of Geology
Brooklyn College
City University of New York
Brooklyn, NY 11210, USA

Library of Congress Cataloging-in-Publication Data
Physical chemistry of magmas / Leonid L. Perchuk, Ikuo Kushiro, editors: with contributions by D.R. Baker ... [et al.].
p. cm.—(Advances in physical geochemistry; v. 9)
Includes bibliographical references and index.
ISBN 0-387-97500-4
1. Silicates. 2. Silicate minerals. 3. Magmatism. 4. Geochemistry. I. Perchuk, L. L. (Leonid L'vovich) II. Kushiro, I. (Ikuo), 1934– . III. Baker, Donald R., 1927– . IV. Series.
QE389.62.P49 1991
552'1—dc20 90-25299

Printed on acid-free paper.

Typeset by Asco Trade Typesetting Ltd., Hong Kong.
Printed and bound by Edwards Brothers, Ann Arbor, MI.
Printed in the United States of America.

9 8 7 6 5 4 3 2 1

ISBN 0-387-97500-4 Springer-Verlag New York Berlin Heidelberg
ISBN 3-540-97500-4 Springer-Verlag Berlin Heidelberg New York

Preface

This volume includes papers on properties, structure and phase relationships that involve silicate melts. These problems are of interest to many scientists who are involved in geosciences and chemistry because of the fundamental nature of the topics. The last ten years have been marked by major achievements in this field of science. An international team of invited contributors with expertise in different aspects of the topic presents the most important results in studies of viscosity (E.S. Persikov), diffusion (A.S. Chekhmir, M.B. Epel'baum, E.B. Watson and D.R. Baker), density (I. Kushiro), influence of redox equilibria of iron on the melt structure (B.O. Mysen) and the role of high field strength cations on the structure and properties of silicate liquids (P.C. Hess). An important contribution to silicate melt thermodynamics is made by Bottinga who found an effective approach to calculate the baric dependences of molar volumes and entropies for the most important liquid silicates. These papers present data and some rules for estimating the properties and structures of melts as well as the implications of the physical chemistry of silicate liquids to igneous petrology.

Another set of articles illustrates the applicability of experimental and theoretical petrology to the solution of different geological problems. For example, the paper by Y. Tatsumi summarized much of the experimental, geochemical and petrochemical data in support of an elegant model for the origin of island arc magmas. The author also indicates a restricted role for subducted lithosphere as a source of water. The paper by C.D. Doyle provides an example of a very simple thermodynamic model for describing the phase relationships on the liquidus surface of silicate melts (involving divalent oxide cations).

In contrast to the last two mentioned papers, the final chapter of the volume contributed by A.D. Kuznetsov and M.B. Epel'baum deals with the Korshinskii potentials for thermodynamic description of liquidus relationships for a granitic melt with volatiles. This rather unusual and difficult-to-understand approach was reviewed in only a limited way.

Editing the present APG volume took an inordinate amount of time and required extensive work on the part of contributors, reviewers and staff of Springer-Verlag, New York. The editors would like to thank L.Ya. Aranovich, A.L. Boettcher, C.W. Burnham, D.M. Burt, A.S. Chekhmir, M.P. Dickenson, T. Fujii, G.B. Gill, N.S. Gorbachov, C.T. Herzberg, A. Hofmann, A.A. Kadik, A. J. Naldrett, A. Navrotsky, I. Nicholls, G.V. Novikov, E.S. Persikov, C.M. Scarfe, K.I. Shmulovich, B.Y. Varshal, E.B. Watson, E. Westrum, and H.S. Yoder for their cooperation and assistance in preparing the manuscripts for publication. We are grateful to G.G. Gonchar for editorial assistance and to the staff at Springer-Verlag for their cooperation and patience during the preparation of the volume.

It is very much hoped the book will introduce the reader to the world of physical chemistry of magmas.

July 1990
Moscow, Tokyo and Washington

LEONID L. PERCHUK
IKUO KUSHIRO

Contents

Contributors

Baker, D.R.	Department of Geology and Center for Glass Science and Technology, Rensselaer Polytechnic Institute, Troy, New York 12180-3590, USA
Bottinga, Y.	Institute de Physique du Globe Tour 14, 4 place Iussieu Tour 14, 75230 Paris Cedex 05, France
Chekhmir, A.S.	Institute of Experimental Mineralogy, USSR Academy of Sciences, Moscow Chernogolovka 142432, USSR
Doyle, C.D.	Pyrometallurgy Section, J. Roy Gordon Research Lab, Inco Ltd., Sheridan Park, Mississauga, Ontario, Canada LSK 1Z9
Epel'baum, M.B.	Institute of Experimental Mineralogy, USSR Academy of Sciences, Moscow Chernogolovka 142432, USSR
Hess, P.C.	Department of Geological Sciences, Brown University, Providence, Rhode Island 02912, USA
Kushiro, I.	Geological Institute, Faculty of Science, University of Tokyo, Tokyo 113, Japan
Kuznetsov, A.D.	Institute of Experimental Mineralogy, USSR Academy of Sciences, Moscow Chernogolovka 142432, USSR
Mysen, B.O.	Geophysical Laboratory, Carnegie Institution of Washington, District of Columbia 20015, USA

Perchuk, L.L.	Institute of Experimental Mineralogy, USSR Academy of Sciences, Moscow Chernogolovka 142432, USSR
Persikov, E.S.	Institute of Experimental Mineralogy, USSR Academy of Sciences, Moscow Chernogolovka 142432, USSR
Shimizu, N.	Department of Geology and Geophysics, Woods Hole Oceanographic Institution, Woods Hole, Massachusetts 02543, USA
Tatsumi, Y.	Department of Geology and Mineralogy, Faculty of Science, Kyoto University, Kyoto 606, Japan
Watson, E.B.	Department of Earth and Environmental Sciences, Rensselaer Polytechnic Institute, Troy, New York 12180-3590, USA

Chapter 1
The Viscosity of Magmatic Liquids: Experiment, Generalized Patterns. A Model for Calculation and Prediction. Applications.

Edward S. Persikov

Introduction

Knowledge of the viscosity of fluid magmatic systems is extremely important to the understanding of igneous processes. Such processes include the ascent of magma from depth, the dynamics of magma generation and evolution under variable temperature and pressure conditions, the relative proportions of effusive and intrusive rock masses, differentiation processes, and heat and mass transfer.

High-temperature high-pressure viscosity measurements in magmatic liquids involve complex scientific and methodological problems that have faced workers in physical geochemistry and petrology for more than 50 years. Much of the progress made in the field is due to Volarovich, Kani, Shaw, Khitarov, Lebedev, Carron, Murase, Kushiro, Scarfe, Urbain among others. Clearly, it would be an unrealistic task to try to study all possible magmatic liquids by experiment. No substantial theory has yet been offered to describe the viscosity of such complex systems. In addition, the existing empirical methods for viscosity calculation and prediction do not allow for the effect of volatiles so that the results produced so far are much less accurate than experimental data. When the author began viscosity studies (1969), the experimental evidence had been mainly on viscosity in anhydrous melts under atmospheric pressure and only limited work had been done on hydrous granitoid melts.

Newton's law of viscosity states that the shear stress in liquids undergoing laminar flow is proportional to the local velosity gradient perpendicular to the stress:

$$\tau = \eta \cdot \dot{\gamma} \tag{1}$$

where τ is the shear stress, $\dot{\gamma}$ is the velocity gradient, and η is the viscosity coefficient.

In the CGS system, the unit of viscosity is 1P (poise—dyn · sec · cm^{-2}); in the

Cl system it is 1 Pa · sec = 10 poises. The viscosity of normal (Newtonian) liquids decreases with increasing temperature at constant pressure and is independent of the velocity gradient and the magnitude of the shear stress for low values of $\dot{\gamma}$ and τ. Many liquids do not obey eq. (1), and in rheology these liquids are called anomalously viscous: non-Newtonian, viscous-plastic, pseudoplastic and viscous-elastic (Reiner, 1958).

Systematic viscosity studies on anhydrous natural and synthetic aluminosilicate melts have shown that the laminar flow they undergo in the liquidus region approximates closely the Newtonian model (e.g. Volarovich, 1940; Shaw, 1969). Viscosity measurements on hydrous obsidian (Shaw, 1963) and basalt (Persikov, 1981a) melts show that the flow in hydrous melts under near-liquidus conditions is also approximately Newtonian.

The purpose of the present work is to study the rheology of magmatic volatile-bearing melts as a function of their structure, composition, temperature and pressure, and to derive a method for calculating and predicting viscosities. The application of the results to some problems in the geochemistry and petrology of magmas also will be discussed.

Experimental Methods

In the present work, the falling-sphere method was adopted as by far the most reliable of the methods currently used under high pressures. The viscosity was calculated by Stokes' law. Corrections for wall effects must be applied for small containers, $0.02 \leq d/D \leq 0.32$ (d and D are the diameters of the sphere and container, respectively). In the present experiments, the Faxen correction was used (Shaw, 1963; Persikov, 1972):

$$\eta = \frac{2gr^2(\rho_2 - \rho_1)}{9v(1 - 3.3r/h)}\left[1 - 2.104\frac{d}{D} + 2.09\left(\frac{d}{D}\right)^3 - 0.95\left(\frac{d}{D}\right)^5\right] \qquad (2)$$

where h is the height of the cylinder containing the liquid, V is the velocity of fall of a sphere with radius r; ρ_2, ρ_1 are the densities of the sphere and the liquid and g is the gravitational constant.

Although the falling sphere method has a sound theoretical basis, its application involves considerable methodological difficulty, mainly in the observation of the sphere's passage through the liquid. In this connection it is common practice to use quenching in nearly all high-pressure viscosity studies.

The author's method, in which the sphere contains ^{60}Co, has proven superior to conventional quenching techniques. In the first place, this method is twice the accuracy and much better efficiency. Furthermore, both theory and experiment have proved that the use of weakly-radioactive sources (to $1.5 \cdot 10^{-4}$ curie) in high-pressure vessels ensures complete radioactive safety. It is noteworthy that the experimental technique allows for direct measurement of the time lag in the fall of the sphere once the temperature and pressure reached the planned values.

The bulk of experimental evidence obtained with this technique on anhydrous and volatile-bearing magmatic melts clearly indicates that these time lags are random and arise from various causes.

In some runs, despite efforts to keep surface effects low, the time lags were an order of magnitude higher than the time of fall of the sphere in the melt. With a quenching technique, therefore, scatter in the experimental points can be large. An attempt to overcome this limitation of the quenching method was made by Kushiro et al. (1976). The results from two or three runs under definite P–T conditions with spheres of similar diameter were plotted on time versus distance of sinking graphs, and the average settling velocity was obtained from the slopes of the lines. The only limitation of this otherwise correct modification is that there should be at least three experimental points if the lines obtained are to

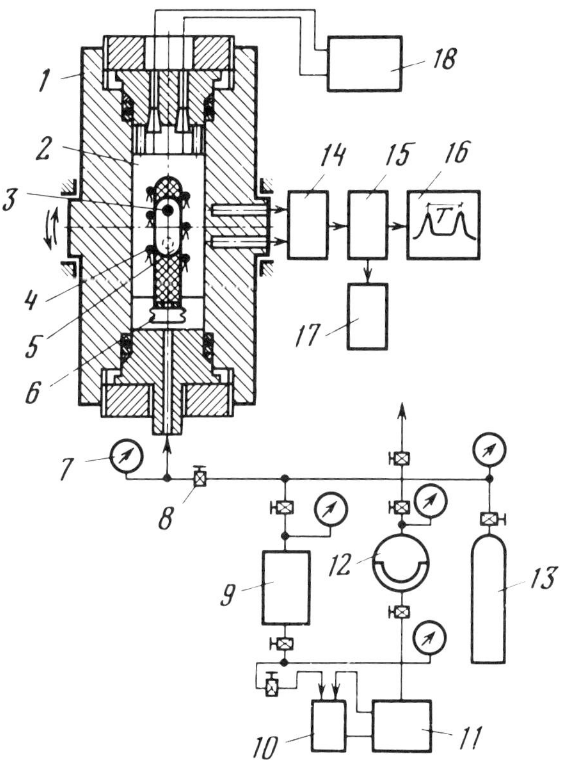

Fig. 1. Diagram of the gas pressure radiation viscosimeter. (1) Vessel. (2) Heater. (3) Sphere filled with ^{60}Co. (4) Thermocouples. (5) Capsule with experimental melt. (6) Container with the gas-fluid separator. (7) Pressure gauges. (8) Valves. (9) Booster oil-gas. (10) Oil tank. (11) High pressure oil pump. (12) Oil-gas compressor-separator. (13) Gas bottle. (14) Spectrometric gamma scintillation counter. (15) Radiometric post. (16) Electronic recorder. (17) Temperature measurement and automatic control panel.

represent the average settling velocity of the sphere adequately. Although it does not enhance the accuracy of viscosity measurements, Kushiro's modification allows erroneous results to be rejected and thus enhances the reproducibility of experiments considerably.

All the experiments reported here were carried out in a radiation high-pressure viscosimeter (Persikov, Kochkin, 1973) of a greatly improved design. The viscosimeter allows wider temperature and pressure ranges, higher accuracy, and a wider choice of fluids. Density studies can also be performed on volatile-bearing melts (Persikov, Epelbaum, 1978, Persikov, 1984).

As shown schematically in Fig. 1, the apparatus has a radiation viscosimeter-densitometer (a three-layer water-cooled container) that can move through 180° about its horizontal axis; two blind holes are drilled to collimate the γ-quantum flux emitted from the radioactive filling of the sphere.

The assembly includes an independent two-stage system for compressing cylinder gas (high purity Ar). The gas pressure is measured by a Bourdon-tube pressure gauge accurate to $\pm 1\%$. The internal heating assembly consists of a three-section heater that generates a gradient free zone symmetric about the viscosimeter rotation axis; six thermocouples (three each for regulation and control of the gradient-free zone); and a high-precision temperature regulator, type BPT-2. The error in the temperature measurement at high pressures was ± 2.5–5°C. The inner volume of the capsule with the starting material was separated from the pressure-transmitting gas by a special separator-equalizer which ensured gradient-free gas pressure on the fluid and melt. The gradient-free zone contained the reaction (platinum) capsule with the sample and sphere

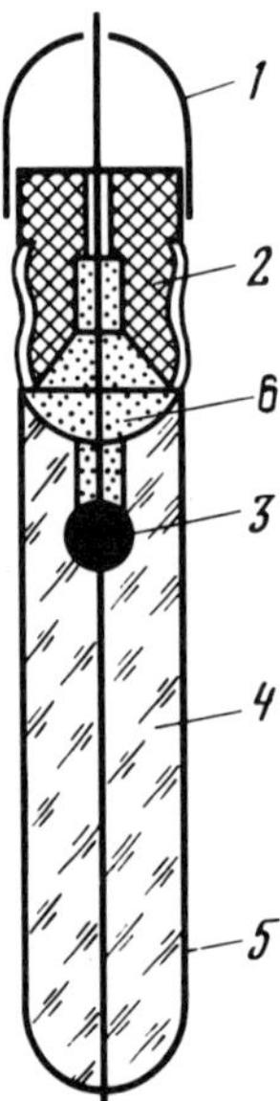

Fig. 2. Diagram of the reaction capsule. (1) Platinum cap with holes. (2) Alundum plug. (3) Sphere with radiation source. (4) Melt. (5) Platinum capsule. (6) Powdered glass (quenched melt)

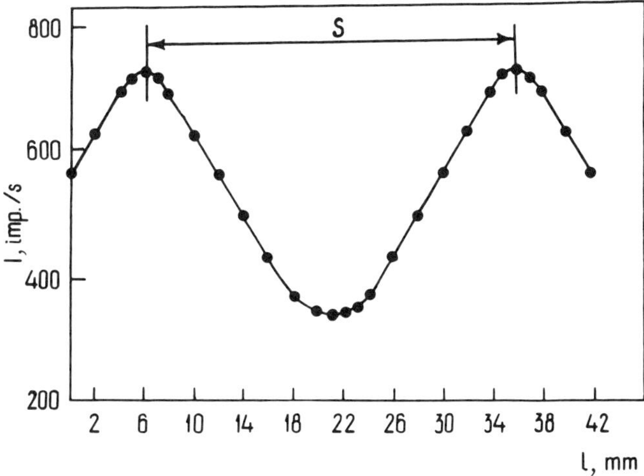

Fig. 3. Calibration curve of the radiation viscosimeter. (I) Intensity of γ-radiation, imp/sec. (S) The true distance between the collimators.

(platinum or platinum-rhodium) filled with ^{60}Co. In runs with volatile-bearing melts, the container volume was filled with H_2O, aqueous solutions of HCl, NaCl, HF, and H_2, Ar and other materials.

In the experiments the viscosity was measured by placing the charged reaction capsule (Fig. 2) in the container. The velocity of the sphere was determined from the relationship between the run time and the settling distance. The distance was obtained from the calibration curve (Fig. 3) based on a special experiment (Persikov, 1973, 1976). The time of fall was measured by changes in the γ-beam intensity. The distance between the maxima represents the time of passage between the collimators. Statistical analysis of the times of fall in calibration liquids showed that the relative uncertainty in measuring the time was not higher than $\pm 1\%$.

The viscosity of the melt at given T and P was calculated from Eq. (2) written in the form:

$$\eta = C \cdot \Delta\rho \cdot t \tag{3}$$

where t is the time taken by the sphere to cover the distance between the collimators; $\Delta\rho$ is the density difference between the sphere and the melt, and C is the apparatus constant (for the sphere used in the run). Platinum-rhodium spheres ranging in diameter from 0.13 to 0.21 cm and filled with ^{60}Co of the activity $(0.5–1.0) \cdot 10^{-4}$ curie were used in all runs.

The total relative uncertainty in measuring the viscosity on the radiation viscosimeter is within $\pm 10\%$ as shown by both experimental viscosity measurements in calibration liquids and reproducibility of experimental results on viscosities of magmatic melts. Systematic experimental studies on the viscosity of magmatic melts were carried out for a wide composition range including: acidic melts (granite, albite, potassium tetrasilicate), feldspathoid (nepheline syenite), intermediate (andesite), and basic (basalts of different composition).

Experimental Results. Discussion

Previous viscosity studies on rock and mineral melts at atmoshperic pressure show that acidic melts (SiO_2, albite, granites, obsidians) have the highest viscosity (e.g. Kani and Hosokawa, 1936; Volarovich 1940; Leontieva, 1940; Murase et al., 1962, 1973; Scarfe, 1977). However, these studies were done on granitoid melts that are virtually volatile-free and they therefore could not explain a number of structural characteristics of intrusive rocks. In particular, the homogeneity of vast granitoid batholiths, thin granitoid intrusions extending for considerable distances in dykes and sills, and the distribution of deep-seated and country rock xenoliths, were thought (Shipulin, 1969; Sharapov and Golubev, 1976; Volokhov, 1979 and oth.) to be related to the low viscosity of granitic magmas caused by the presence of volatiles.

By the time the author began his work, data on the viscosity of hydrous granitoid melts (e.g. Shaw, 1963; Burnham, 1963) was available to support the validity of such inferences. However, quantitative evaluation and analysis of any of the magmatic processes operating under various conditions in the earth's crust can only be based on systemetic investigations of the temperature, pressure and composition dependencies of the viscosity of magmatic melts.

Table 1 gives the chemical compositions for the experimental samples together with average rock compositions (from Le Maitre, 1976). The table also contains calculated values for the structural chemical characteristic, $K = 100 \cdot O_A/T$– "percentage of broken bonds between the tetrahedra of the aluminosilicate" (Carron, 1969) or $K = 100 \cdot NBO/T$—the ratio of nonbridging oxygens per tetrahedrally coordinated cations (Bockris and Reddy, 1970; Mysen et al., 1980). This ratio expresses the degree of depolymerization of aluminosilicate melts.

Viscosity of Model and Granitoid Melts

Over the 800–1450°C temperature range and at fluid pressures up to 750 MPa, temperature, composition and pressure dependences were studied for the following systems: $K_2O \cdot 4SiO_2$; $K_2O \cdot 4SiO_2 + H_2O + Ar$; $Na_2O \cdot Al_2O_3 \cdot 6SiO_2$ (albite); Ab(albite) + H_2O; Ab + Ar; granite + H_2O; granite + H_2O + HCl; granite + H_2O + NaCl; granite + H_2O + HF.

Analysis of the results, parts of which are given in Table 2 and Fig. 4–8, and comparison with other workers' results (Lebedev, Khitarov, 1979; Shaw, 1963; Burnham, 1963; Kadik and others., 1971) have made it possible to explore new facets of the rheologic characteristics (η, E) of granitoid melts.

Table 1. Chemical analyses of rock samples

Rock	SiO_2	Al_2O_3	TiO_2	FeO	Fe_2O_3	Na_2O	K_2O	MgO	CaO	Cl	F	P_2O_5	H_2O	Σ	K, (100 · NBO/T)
Granite	73.23	13.60	0.19	0.48	2,56	3.78	4.11	0.17	1.69	—	—	0.13	—	99.94	0.3
Granite[a]	71.30	14.32	0.31	1.64	1.21	3.68	4.07	0,71	1.84	—	—	0.12	0.64	99.98	10.1
Granite + H_2O	69.22	12.84	0.18	0.56	2.46	3.66	3.98	0.17	1.59	—	—	0.13	5.2	99.99	41.0
Granite + H_2O + HCl	70.49	13.56	0.17	0.17	1.81	3.18	3.51	0.19	1.42	0.14	—	0.12	5.2	99.93	36.0
Granite + H_2O + HF	67.44	12.53	0.17	0.44	2.36	3.48	3.81	0.16	1.56	—	1.6	0.12	6.27	99.94	44.0
Nepheline syenite															
Nepheline syenite[a]	60.07	20.03	0.67	1.10	1.40	8.36	6.23	0.63	0.85	—	—	0.06	—	99.4	8.3
Nepheline syenite +	54.99	20.96	0.60	2.05	2.25	8.23	5.58	0.77	2.31	—	—	0.13	0.30	99.17	22.4
H_2O	59.08	19.70	0.66	1.08	1.38	8.22	6.13	0.62	0.84	—	—	0.06	1.65	99.42	21.5
Nepheline syenite + H_2O	56.94	18.99	0.63	1.04	1.33	7.93	5.91	0.56	0.81	—	—	0.06	5.20	99.40	51.4
Andesite															
Andesite[a]	58.56	18.98	0.64	3.90	3.95	3.24	0.92	3.48	6.17	—	—	—	—	99.84	17.6
Andesite + H_2O	57.94	17.02	0.87	4.04	3.27	3.48	1.62	3.33	6.78	—	—	0.21	0.83	99.4	32.0
Andesite + H_2O	57.88	18.76	0.63	6.28	1.80	3.21	0.91	3.44	6.11	—	—	—	1.00	99.85	32.6
Basalt	55.92	18.14	0.62	6.00	1.72	3.08	0.90	3.30	5.86	—	—	—	4.60	100.3	63.3
Basalt[a]	53.54	17.29	1.05	1.11	7.46	3.59	1.64	5.46	8.32	—	—	0.2	—	99.83	31.5
Basalt + H_2O	48.91	16.44	1.63	6.09	4.90	2.70	1.41	5.88	8.78	—	—	0.44	2.61	99.79	74.0
Tholeiite*	52.20	16.85	1.02	5.52	3.03	3.50	1.60	5.32	8.11	—	—	0.2	2.55	100.20	72.0
Tholeiite + H_2O	49.58	14.79	1.98	8.03	3.38	2.37	0.43	7.30	10.36	—	—	0.24	0.91	99.37	78.9
	48.21	15.50	1.25	5.87	5.30	2.40	0.08	9.08	11.92	—	—	0.14	0.22	99.94	71.6

[a] Average rock compositions after Le Maitre, 1976

Table 2. Viscosity and activation energy of dry and hydrous model and granitoid melts

T °C	P_S MPa	P_{Fl} MPa	$N_{j,fl.}$ wt%	$\log \eta$ (η – P.)	E Kcal/mole
1	2	3	4	5	6
			$K_2O \cdot 4SiO_2$		
1000	0.1	0	0	4.34	45.2
1300	0.1	0	0	2.82	
			$K_2O \cdot 4SiO_2 + Ar + H_2O$		
1000	150	150	$H_2O = 0.2$	3.88	
			Albite, Albite + H_2O		
1400	0.1	0	0	5.25	68.0
1400	400	0	0	5.02	65.2
1100	50	50	$H_2O = 1.9$	5.26	55.0
1250	50	50	1.9	4.46	
1000	200	200	$H_2O = 4.5$	4.82	48.0
1200	200	200	4.5	3.75	
1000	400	400	$H_2O = 6.9$	3.71	42.5
1200	400	400	6.9	2.84	
1200	400	50	1.9	4.35	52.9
			Granite + H_2O		
800	50	50	$H_2O = 2.1$	6.1	48.7
1100	50	50	2.1	4.25	
800	100	100	3.3	5.8	45.5
1200	100	100	3.3	3.25	
800	200	200	5.2	4.97	41.6
1200	200	200	5.2	2.67	
850	700	700	12.3	4.2	38.8
950	700	700	12.3	3.4	
			Granite + H_2O + 0.2 M HCl		
1100	50	50	$H_2O = 2.1$, HCl = 0.1	5.04	54.6
1250	50	50	$H_2O = 2.1$, HCl = 0.1	4.36	
950	200	200	$H_2O = 5.6$, HCl = 0.1	4.65	46.3
1100	200	200	$H_2O = 5.6$, HCl = 0.1	3.85	
900	400	400	$H_2O = 8.9$, Cl = 0.1	4.1	41.0
1100	400	400	$H_2O = 8.9$, Cl = 0.1	3.5	
1100	400	50	$H_2O = 2.1$, Cl = 0.1	4.66	51.8
1200	400	50	$H_2O = 2.1$, Cl = 0.1	4.11	
			Granite + H_2O + 1 M HCl		
900	200	200	$H_2O = 5.2$, Cl = 0.14	5.2	46.4
1200	200	200	$H_2O = 5.2$, Cl = 0.14	3.58	

Table 2 (continued)

1	2	3	4	5	6
			Granite + H_2O + 0.2 M NaCl		
1000	200	200	$H_2O = 5.25$, Cl = 0.1	4.62	47.9
1250	200	200	$H_2O = 5.25$, Cl = 0.1	3.7	
900	400	400	$H_2O = 8.7$, Cl = 0.1	4.03	41.0
1100	400	400	$H_2O = 8.7$, Cl = 0.1	3.35	
			Granite + H_2O + 1 M HF		
900	200	200	$H_2O = 6.3$, F = 1.6	4.66	45.1
1200	200	200	$H_2O = 6.3$, F = 1.6	3.36	

1. *Temperature dependence* of viscosity in model and granitoid melts is exponential whatever the fluid composition or pressure (lithostatic—P_s, fluid—P_{fl}). The relationship is adequately described, within experimental error, by the well-known Arrhenius–Frenckel–Eyring equation

$$\eta = \eta_0 \cdot \exp(E_\eta / RT) \quad (4)$$

where T is the temperature in K; R is the universal gas constant, η_0 is the pre-exponential constant for the viscosity of liquids at $T \to \infty$, E_η is the activation energy of viscous flow (cal/mol).

It is very important that experimental values for the pre-exponential term in Eq. (4) is constant and approaches the theoretical value. In Fig. 4 these values are illustrated for anhydrous and hydrous albite and granite melts; and in Fig. 5 for granitoid melts, with the other workers' results extrapolated assuming $\log \eta_0 = -3.5$. On the basis of these values, comparable activation energies of viscous flow were obtained from the slopes of the temperature dependence (Figs. 4, 5), or from the simple equation:

$$E = 4.576 \cdot T \cdot (\log \eta_T + 3.5) \quad (5)$$

where η_T is the melt viscosity at T, determined from the temperature dependence. It will also be noted that for all these melts, the activation energy is independent of temperature over the entire range under study.

2. *Compositional dependence* (with respect to fluid components) of the viscosity of granitoid melts.

Systematic experimental studies of the compositional dependences on $\eta = f(N_{H_2O})$ Fig. 6 and $E = f(N_{H_2O})$, Fig. 7 in the systems granite + H_2O and albite + H_2O (Table 2), and theoretical analysis have led to the following conclusions:

1. Dissolution of H_2O considerably depresses the rheologic characteristics and thus causes profound structural changes in granitoid melts—depolymerization, or an increase in basicity. Note that the extent to which the H_2O content

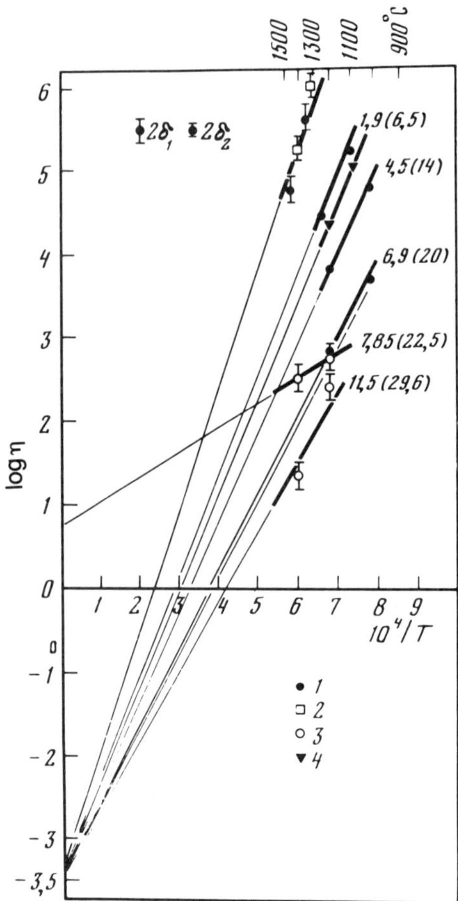

Fig. 4. Temperature dependence of viscosity of dry and hydrous albite melts (numbers at the straight lines and in parentheses are mass and molecular % H_2O, respectively, in the melts). (1) Author's experimental data. (2) Viscosity of dry albite melt from Kani et al., 1936. (3) Viscosity of hydrous albite melt from Lebedev and Khitarov, 1979. (4) Viscosity of hydrous melt at $P_{H_2O} = 50$ MPa and $P_s = 400$ MPa. δ_1 and δ_2 are the uncertainties in viscosity measurements reported in the literature and from author's experiment, respectively.

influences the rheologic properties (η, E) of the granitoid melt is controlled by the water dissolution mechanisms operating at different stages.

The complex study on the melt viscosity and water solubility in the system granitic melt-water has made it possible to determine the kinetics of the attainment of equilibrium (Persikov, 1974). These studies were combined with kinetic studies on thermal dehydration in quenched hydrous granitic glasses and investigation of their IR absorption spectra to reveal some new features in the H_2O dissolution mechanism (Persikov, 1974). These features were later confirmed for magmas over a wide compositional range—from acidic to basic (Persikov, 1981, 1984).

2. At H_2O contents up to ~4.5 and 6.4 wt% in the granite and albite melts respectively, the chemical dissolution is predominant, i.e. water occurs in the dissociated form (OH-hydroxyl). The continuing dissolution brings a change in the dissolution mechanism—the physical dissolution replaces the chemical one (molecular H_2O); and its effect on the viscosity and activation energy is

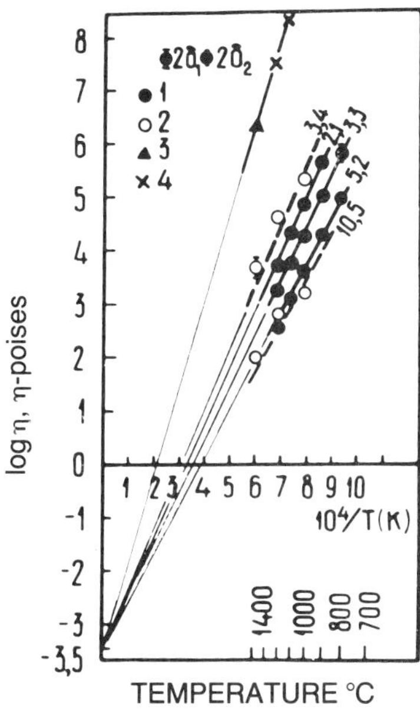

Fig. 5. Temperature dependence of viscosity in dry and hydrous granitoid melts. (1) Viscosity of hydrous melts of Barlaksky granite (Persikov, 1972, 1976). (2) Viscosity of hydrous melts of Eldzhurtinsky granite (Lebedev and Khitarov, 1979). (3, 4) Viscosity of dry granitic and obsidian melts (Kani and Hosokawa, 1936, Leontieva, 1940). δ_1; δ_2—uncertainties of the reported experimental data. Numbers at the curves are mass water contents in the melts.

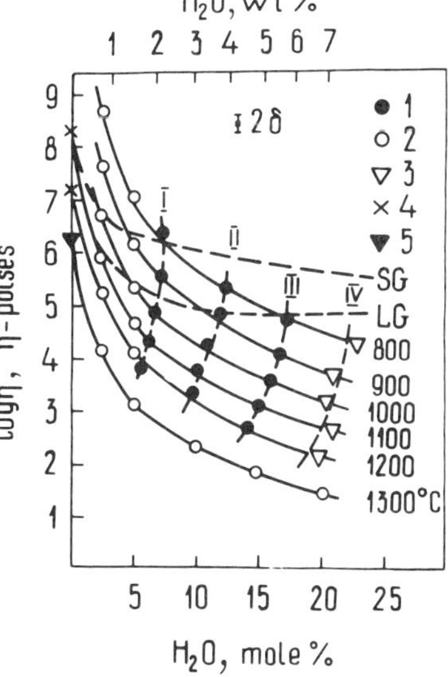

Fig. 6. Viscosity of the system Barlaksky granite-H_2O under equilibrium conditions ($P_{H_2O} = P_S$). (1, 2) Experimental and calculated values. (3) Viscosities calculated from the $P_{H_2O} = 300$ MPa. (4) Viscosity of dry granitic melt (5) Viscosity of dry obsidian melt. I, II, III, IV—are viscosity isobars; P_{H_2O} is respectively, 50, 100, 200 and 300 MPa. (SG) and (LG) are viscosities of the hydrous Barlaksky granite melt under solidus and liquidus thermodynamic parameters (T, P_{H_2O}, N_{H_2O}).

considerably weaker, leaving the melt basicity unchanged. It should be noted that the data obtained on the basis of viscosity measurements are in good agreement with those of Stolper (1982) who used spectroscopic methods.

3. Hydrous granitoid melts are rather mobile liquids with viscosities comparable to those of anhydrous basaltic melts for a wide range of temperatures and H_2O contents.

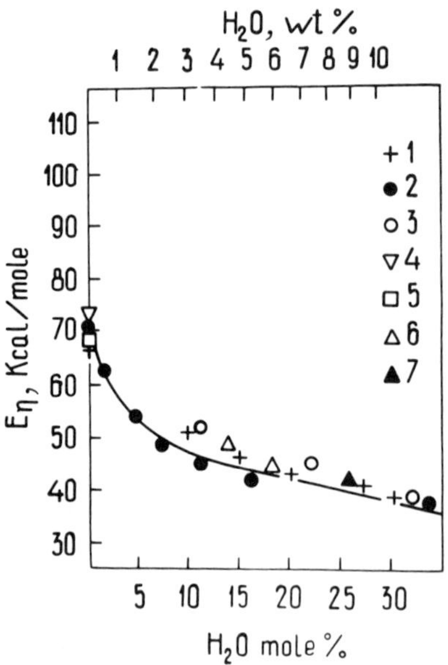

Fig. 7. Plot of the activation energy of the granitoid viscous flow versus water content in the melt. (1, 2) Albite and granitic (Barlaksky) melts. (3) Eldzhurtinsky granite–H_2O. (4, 5) Obsidian and granite dry melts. (6) System obsidian–H_2O (Shaw, 1963). (7) Pegmatite–H_2O (Burnham, 1963).

Important as it is, water is not the only component of fluid systems. Viewed in the framework of the fluid magmatic interaction in granitoid systems, other fluid components in order of significance are carbon dioxide, chlorine and fluorine (e.g. Burnham, 1979; Marakushev, 1979; Ryabchikov, 1975).

The effect of CO_2 on the rheologic properties of granitoids was not studied. However, with respect to granitoid magmatism, the effect of CO_2 on the system's rheologic behavior, i.e. a decrease in the H_2O partial pressure and a concomitant decrease in water solubility in the granitic melt, is due to mere dilution of the hydrous fluid. This conclusion is drawn reliably from the available experimental data on the low solubility of CO_2 in granitoid melts under water-carbon dioxide fluid pressure corresponding to the entire depth range in the crust (e.g. Mysen et al., 1976, Kadik and Eggler, 1976), as well as from the theoretically and experimentally established ideality of the system $H_2O + CO_2$ under magmatic conditions (e.g. Ryabchikov, 1975).

The effect of added CO_2 in hydrous granitoid fluids, as well as the polymerization effect of CO_2 on the rheologic properties of granitic melts under high pressures, can now be evaluated with reasonable accuracy using the proposed method (see below).

The effect of the added chlorine ion on the rheologic properties of granitic melts has been studied in considerable detail (Table 2). It follows from the results of these studies that, for comparable T, P, N_{H_2O}, the addition of chlorine compounds to the aqueous fluids increases the rheologic characteristics of hydrous granitic melts considerably. Notably, the magnitude of η and E depends largely on P_{H_2O} (N_{H_2O}) in the melt, decreasing with increasing pressure and is only slightly affected by the chlorine concentration or the manner in which the chlorine ion enters the solution (acid or salt).

However, an increase in the viscosity and activation energy is not related to the incorporation of chlorine into the melt structure, the Cl solubility in granitic melts being extremely low (Table 2), but rather is due to alkali and iron losses from the melt to the aqueous chloride fluid, i.e. debasification of the melt (Table 1).

The effect of the fluorine ion addition to the fluid (IM HF), on the viscosity of hydrous granitic melts was studied at $P_{fl} = 200$ MPa. It follows from Table 2 that water and fluorine have considerably different effects on the rheologic properties of hydrous granitic melts. The dissolution of the water is responsible for depolymerization of the melt structure, whereas substitution of fluorine for part of the H_2O in the melt would result in polymerization of the melt. According to the data in Table 2, equimolar substitution of fluorine for water causes a marked increase in η and E of the hydrous granitic melt over the temperature range covered. It is interesting that the effect of addition of fluorine to the hydrous fluid on the η and E of the melt correlates closely with the new conception (Kogarko a. Kriegman, 1981) that fluorine acts as a polymerizer of silicon-oxygen tetrahedra in the melt. However, the system granite + H_2O + HF that can exist only under high fluid pressures is a special case. Fluorine appears to be a polymerizer for the silicon-oxygen tetrahedra at the intermediate stage in the interaction between the aqueous fluoride fluid and the granitic melt. It has been found that addition of fluorine to the aqueous fluid produced no apparent effect on the rheologic properties of the melt: the melt did not respond to the addition. This process can be represented schematically by the set of equations:

$$2(\equiv\text{Si–O–Na}) + 2\text{HF} \rightarrow \equiv\text{Si–O–Si}\equiv + \text{H}_2\text{O} + 2\text{NaF} \rightarrow$$
$$2(\equiv\text{Si–O–H}) + 2\text{NaF} \tag{6}$$

As a result, the fluorine anions bind to alkali metal cations and the water thus produced in the melt depolymerizes once again the silicon-oxygen (aluminum-oxygen) tetrahedra. The molar amounts of bridging oxygens in the melt remain essentially the same, so the structure-sensitive characteristics of the melt (η, E) should not undergo drastic changes.

The Pressure Dependence of Viscosity of Granitoid Melts

The viscosity and activation energy of anhydrous, hydrous model and granite melts decrease with increasing total (lithostatic) and fluid pressures (Table 2). Some of the results obtained for the system granitoid melt + H_2O are given in Fig. 8.

The viscosity of granitoid melts is seen to decrease most noticeably at $P_{H_2O} = P_s$, which is the direct result of increasing water content in the melt. Comparing the results obtained for the dependence of η and E on pressure and water content in

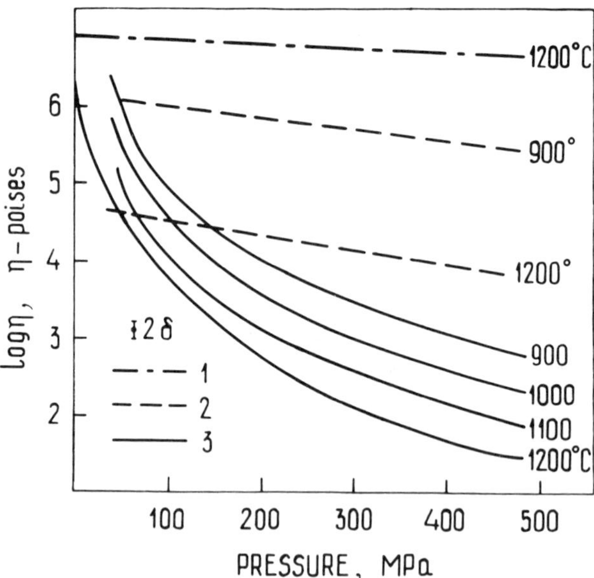

Fig. 8. Plot of the viscosity of granitoid melts versus pressure. (1, 2, 3) Respectively, $P_{H_2O} = 0$, $P_{H_2O} < P_s$ ($T = 900°C$, $H_2O = 2, 5$ wt%; $T = 1200°C$, $H_2O = 1.9$ wt%) and $P_{H_2O} = P_s$.

the melt it is found that the nature of the water pressure effect on the granite melt viscosity at constant temperature is defined by the H_2O dissolution pattern at different saturation pressures (Persikov, 1974). For instance, at $T = 1100°C$ and water pressure 100 MPa, the viscosity of the granite melt decreases by a factor of 10^4 as compared with that of the anhydrous melt, and from 100 to 500 MPa at 1100°C, the viscosity decreases by a factor of only ~30. The bulk water content in the granitic melt under these conditions is 3.3 and 11 wt%, respectively. It will be noted in this respect that the author's idea (Persikov, 1974), that the water dissolution in the system granitic melt-water proceeds mainly through a chemical mechanism up to the equilibrium pressures of $P_{H_2O} \sim 150$ MPa, correlates well with the observed inflection in the relation $E = f(N_{H_2O})$ at $N_{H_2O} \sim 4.5$ wt%. In fact, the experimental data on water solubility in granitold melts (e.g. Goranson, 1931; Burnham et al., 1958; Ostrovsky, 1963; Kadik et al., 1971; Persikov, 1972, 1974, show that the equilibrium pressure ~150 MPa (average for the 800–1200°C range) corresponds to a water content of $N_{H_2O} \sim 4.5$ wt%. The study of the lithostatic pressure effect on the viscosity of fluid-granite melts ($P_s > P_{fl}$) showed (Table 2 and Fig. 8) that the viscosity of the melt decreased with increasing P_s though by much less than under the equilibrium conditions ($P_{H_2O} = P_s$). However, this effect is more pronounced than that produced by P_s on the viscosity of anhydrous melts (Fig. 8 and Table 2).

The observed decrease in the viscosity of hydrous granite melts of constant composition with increasing total pressure at $P_{H_2O} < P_s$ and the decrease in viscosity of anhydrous albite melt under neutral gas pressure are at variance with the theoretical interpretation of the pressure dependence of viscosity of simple liquids (e.g. Frenckel, 1945; Glasstone et al., 1941). Although the mechanism of this phenomenon is not yet understood (see below) it must be related to some structural changes. In the experiments on granitoid melts, the activation energy of viscous flow was found to decrease with increasing total pressure (P_s) (Table 2 and Fig. 2–8) under isochemical conditions.

The Viscosity of Nepheline-Syenite Hydrous Melts

The starting material was an Erkhilnursky massif, Mongolia, nepheline syenite of the composition close to the nepheline syenite average composition (Table 1). The temperature dependence of the viscosity of nepheline syenite hydrous melts was determined at 900–1200°C; the viscosity dependence on water content was measured at 50 and 200 MPa water pressure. The water contents of the melt, 1.65 and 5.2 wt%, respectively, were determined on a gas-chromatographic analyzer CHN-I. The results are given in Fig. 9 and show that, whatever the water content in the melt, the temperature dependence of the viscosity of nepheline syenite melts is approximated to within $\pm 10\%$ by the exponential equation (4) with a constant pre-exponential ($\log \eta_0 = -3.5 \pm 0.1$). High-alkaline melts thus display the same behavior as that earlier determined for hydrous acidic melts.

It is noteworthy that although hydrous nepheline syenite melts are richer in alkalies, their viscosities and activation energies are close to the η and E in

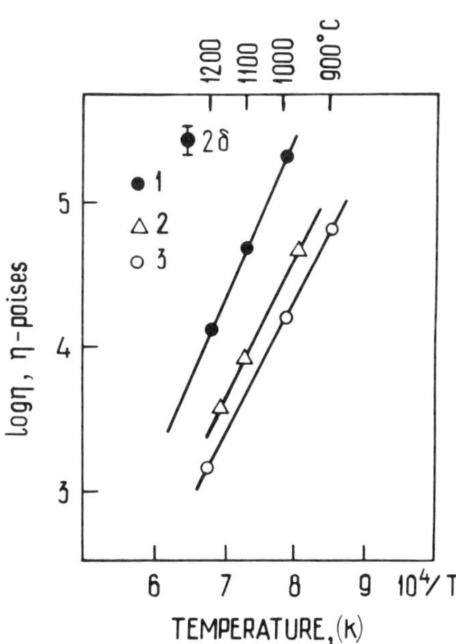

Fig. 9. The temperature dependence of viscosity of hydrous nepheline syenite melts. (1, 2) The system nepheline syenite —H_2O, N_{H_2O} is, respectively, 1.65 and 5.2 wt%. (3) The system granite—H_2O, $N_{H_2O} = 5.2$ wt%

hydrous granitic melts for comparable values of temperature, pressure and water content (lines 2, 3, Fig. 9).

Viscosities of Alkali-Earth Rock Melts of Intermediate, Basic and Ultrabasic Compositions

In Table 3 and Fig. 10, 11, 12, 13 the results of runs on dry and hydrous andesite and basalt melts obtained on the radiation viscosimeter with the relative uncertainty of ±10% are compared with the published data. In the present experi-

Table 3. Viscosity and activation energy of dry and hydrous andesite and basalt melts

T	P_S	P_{H_2O}	N_{H_2O}	$\log\eta$	E
°C	MPa		wt%	η – P.	Kcal/mol
		Andesite (Avacha)			
1200	0.1	0	0	3.75	49.1
1300	0.1	0	0	3.22	
1300	400	0	0	3.15	47.8
1350	400	0	0	2.86	
1150	25	25	1.0	3.5	45.5
1200	25	25	1.0	3.25	
1100	300	25	1.0	3.64	44.4
1200	300	25	1.0	3.01	
1100	200	200	4.6	2.8	40.1
1200	200	200	4.6	2.45	
		**Apokhonchich basalt			
1200	0.1	0	0	3.6	48.2
1300	0.1	0	0	3.16	
1200	400	0	0	3.4	47.0
1300	400	0	0	2.94	
1100	100	100	2.5	2.74	39.4
1300	100	100	2.5	1.91	
1100	375	100	2.5	2.42	37.4
1300	375	100	2.5	1.8	
1000	400	400	6.36	2.3	33.9
1300	400	400	6.36	1.27	
		Tholeiite (Hess trough Pacific)			
1100	10	10	0.22	2.86	39.9
1250	10	10	0.22	2.3	

**Approximately close values for viscosity of dry andesitic and basaltic melts are due to essentially greater degree of oxidation of the starting basalt (see Table 1)

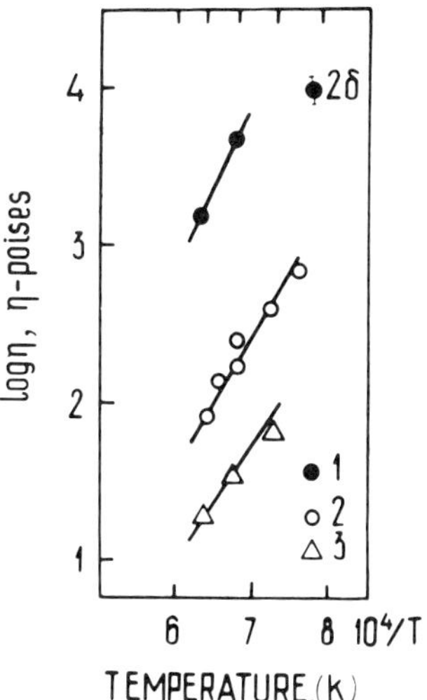

Fig. 10. The temperature dependence of viscosity of dry and hydrous melts of the Apokhonchich basalt. (1, 2, 3) The water content in the melt is, respectively, 0; 2.5; and 6.4 wt%

ments, the temperature and pressure dependences were measured at $P_{H_2O} = 0$ and $P_{H_2O} < P_s$, and the dependence of viscosities on the water content was studied to 20 mol% H_2O.

Starting materials were an Avacha andesite and basalts of different composition (Klyuchevskaya sopka, lava flow Apokhonchich, and a Hess trough tholeiite, Table 1).

Experiments with dry melts under atmospheric and high pressures were run in an argon atmosphere. The oxygen partial pressure was maintained by water in the magnetite stability field in experiments with hydrous melts under run conditions ($P_{H_2O} = 25 - 400$ MPa, $T = 1000$–$1400°C$) because the container design precluded considerable hydrogen losses from the reaction zone during the runs.

The amount of water dissolved in the andesite and basalt melts at high pressures was determined by thermally dehydrating glasses produced by the isobaric quenching of hydrous melts. The hydrous glasses were heated to $T = 1000°C$ in a helium atmosphere, and the amount of water rcleased was determined on a gas-chromatographic analyzer CHN-I (analyst S. Koshemchuk).

Temperature Dependence of Anhydrous and Hydrous Andesite and Basalt Melts

The analysis of the results (Fig. 10, 11, and Table 3) showed that this temperature dependence obeys eq. (4) closely with the constant pre-exponential ($\log \eta_0 = -3.5 \pm 0.1$), as it did with acidic melts. Therefore, Eq. (5) was used to obtain comparable values for the activation energies of viscous flow for all melts studied by experiment (Table 3). It is interesting that other workers' results on viscosity studies on dry andesitic and basaltic melts when treated on the assumption of $\eta_0 =$ constant, were also found to give a good fit (within the 10–30% uncertainty

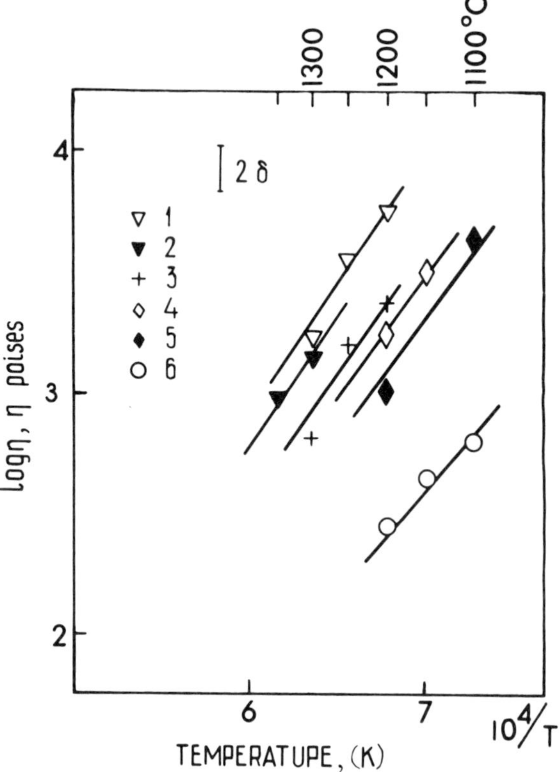

Fig. 11. The temperature dependence of viscosity of dry and hydrous melts of the Ava cha andesite (P-MPa) (1) $P_s = 0.1$. $P_{H_2O} = 0$. (2) $P_s = 400$, $P_{H_2O} = 0$. (3) $P_s = P_{H_2} = 200$. (4) $P_s = P_{H_2O} = 25$. (5) $P_s = 300$, $P_{H_2O} = 25$. (6) $P_s = P_{H_2O} = 200$.

limit) to the theoretical Eq. (4), Fig. 13. Therefore, it became possible to obtain comparable values for the activation energies of viscous flow for these melts also (Fig. 13).

Composition Dependence of Viscosity of Dry and Hydrous Andesitic and Basalt Melts

The results from the present viscosity studies on the andesite and various basalt melts (Table 3) were compared with other authors' data it was shown that anhydrous basaltic melts have a wide range of viscosity covering some 1.5 orders of magnitude. With increasing basicity of the melt, the viscosity decreases uniformly from considerably viscous high-alumina basalts to mobile olivine tholeiitic melts. Note that the rheologic characteristics (η, E) of anhydrous and hydrous andesitic melts are close to those of high-alumina basalts. Water was found to dissolve in andesitic, basaltic and granitoid melts in two forms. However, unlike granitoid melts, the change of the water dissolution mechanism (from the chemical to essentially physical) in melts of intermediate and basic compositions occurs at considerably lower bulk water contents in the melt (5–8 mol% H_2O). Indeed, the dissolution of the first 7.5 mol% H_2O in a basaltic melt causes the activation energy to decrease by about 25%, and the viscosity to decrease by nearly an order of magnitude. Subsequent addition of the same amount of water to the melt decreases E by 8% while the viscosity drops to one third. The same amount of H_2O dissolved in the granitic melt decreases its viscosity by a factor of 5000.

Pressure Dependence Studies

Viscosities in both anhydrous and hydrous andesitic and basaltic melts were found to decrease with increasing pressure (Table 3 and Fig. 12) and the same applies to acidic melts. The greatest drop occurs at $P_{H_2O} = P_s$. A smaller decrease in viscosity of andesitic and basaltic melts occurs at $P_{H_2O} < P_s$. With increasing neutral gas pressure that models the lithostatic pressure (P_s), the decrease in viscosity of anhydrous melts is even smaller: at $T = 1200°C$ and $P_{ar} = 400$ MPa, the viscosity in the basaltic melt decreased only by a factor of 1.5. The activation energies of viscous flow in both anhydrous and hydrous melts of constant composition (Table 3) also decrease with increasing P_s. This effect implies that the melts undergo structural changes with increasing total pressure.

The results obtained in this study on the P_s effect in andesitic and basaltic melts agree well with those of Scarfe (1973); Kushiro et al. (1976); Lebedev, Khitarov (1979).

The author did not study the viscosity of ultramafic melts. Ivanov and Stengelmayer (1982) reported such results from a study under atmospheric pressure.

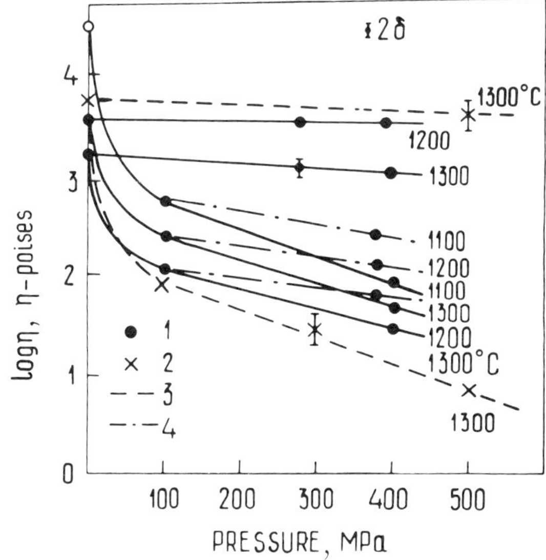

Fig. 12. The pressure dependence of viscosity of dry and hydrous basalt melts. (1) Apokhonchich basalt. (2, 3) Kirgurich basalt (the straight line is $P_{H_2O} = 0$; the curve line is $P_{H_2O} = P_s$), (Lebedev and Khitarov, 1979); (4) Apokhonchich basalt ($P_s > P_{H_2O}$, $H_2O \sim 2.5$ wt%).

However this interesting work did not include activation energies of viscous flow in ultramafic melts that are indispensable for determining rheologic properties of magmatic melts through the entire range of composition. Therefore, the results from this work on the temperature dependence of viscosity were calculated by the present author's method, These calculations show that the temperature dependence of viscosity in ultramafic melts is approximated fairly well by the Arrhenius–Frenckel relation with the constant pre-exponential ($\log \eta_0 = -3.5$). Several conclusions concerning the viscosity of magmatic melts in the alkali-earth series (intermediate, mafic, uitramafic) can be drawn from the present experimental studies:

1. The viscosity range of anhydrous melts in this series is fairly large, being some 4 orders of magnitude. With increasing depolymerization (basicity), the viscosity of melts in this series decreases uniformly. At near-liquidus temperatures, the viscosity of anhydrous andesite, basalt and dunite melts at atmospheric pressure is 5600 *P* (andesite, 1200°C), 4365 *P* (high-alumina basalt, 1200°C), 361 *P* (olivine tholeiite, 1250°C); 1.1 *P* (dunite, 1720°C).
2. The dissolution of water in intermediate and basic melts causes a decrease in viscosity although it is not so marked as in acidic melts. In this process, the proportion of the chemically bound water (OH^-) depends on the degree of depolymerization in the original (anhydrous) melts and decreases uniformly in the sequence, acidic–intermediate–basic melts.
3. The temperature dependence of viscosity in both anhydrous and hydrous melts is exponential. Depending on the structural state of the melt, the numerical values of the activation energy decrease in the sequence: andesitic (anhydrous and hydrous)—basaltic (anhydrous and hydrous)—ultrabasic melts.
4. Viscosities of anhydrous and hydrous melts (andesites and basalts) decrease under water pressure as well as neutral (Ar) gas pressure that models the lithostatic pressure. Water pressure has the greatest effect at $P_{H_2O} = P_s$.

Generalized Patterns, A Method for Calculation and Prediction of Viscosities in Magmatic Melts

Magmatic melts are regarded as polyanionic liquids of complex composition and structure (e.g. Shaw, 1972; Burnham, 1979; Bottinga and Weill, 1972; Mueller and Saxena, 1977; Mysen et al, 1980; Kushiro, 1980; Persikov, 1984). No comprehensive theory on viscosities of such complex systems has as yet been developed, so a new method for calculating and predicting rheologic characteristics of magmatic as well as compositionally simpler silicate and aluminosilicate melts has been proposed. Analytical equations for the viscosity—temperature—pressure (total and fluid)—composition (with respect to petrogenetic and volatile components) dependences were derived. The latter were based on analysis of the relationship between the structure of magmatic melts and the evolution in their chemistry viewed in the framework of the Frenckel molecular-kinetic theory of liquids. The experimental basis for the new method comprises the bulk of the author's and other scientists' results of work on viscosity of magmatic melts at atmospheric and high pressures and abundant experimental evidence on viscosities in binary, ternary and more complex silicate and aluminosilicate melts. The basic criteria for selecting the data were as follows: a) the maximum possible simplicity of analytical relations; and b) accuracy of viscosity prediction and calculation consistent with that of the experimental results.

Temperature Dependence of Viscosity of Magmatic Melts

The temperature dependence of viscosity of silicate and aluminosilicate melts is best described by the simple exponential Eq. (4). In the Frenckel (1945) and Eyring et al (1941) theories, the pre-exponential constant in Eq. (4) is independent of T, P, or liquid composition within the boundary conditions specified by the authors. This independence is very important because only with it (η_0 = const) will comparison of the activation energies in compositionally different liquids, as well as E-composition plots, have a stricit physico-chemical meaning. This problem was earlier discussed by Shaw (1972) in his model for silicate melt viscosity prediction.

Kobeko (1952) showed η_0 to be nearly constant and independent of composition in many liquids (alcohols, acids, organic and inorganic glasses etc.) and the author and co-workers have found it to remain constant in magmatic melts over a wide range of compositions. These findings formed the basis for the analysis of all experimental data available on viscosities in magmatic, silicate and aluminosilicate melts. A statistical gave the weighted mean $\eta_0 = 10^{-3.5}$, which is a good fit to the theoretical value (e.g. Frenckel, 1945).

Generalized data on the temperature dependence of viscosity in magmatic and compositionally simpler silicate and aluminosilicate melts were obtained using the determined value for the pre-exponential and the $\pm 30\%$ confidence limit for viscosity estimates. Most of these data are given in Fig. 13. Note that although the uncertainty limit is large in terms of the strict molecular theory of viscosity for compositionally simple liquids, it is commonly the highest accuracy achieved in experiments at high pressures. Results from different authors show, in places, even greater variance for compositionally similar systems, as illustrated in Fig. 13a for silica melts (lines I–I^{I}–I^{II}).

The regular arrangement of the curves for temperature dependence of viscosity in Fig. 13, allow the experimentally determined patterns to be applied with sufficient accuracy to the whole composition range of magmas—from granitoid to ultramafic. It is very important that the new activation energy values for viscous flow in these systems can be correlated with high accuracy ($\pm 1.0\%$ within the stated $\Delta\eta = \pm 30\%$ error limit), which was in principle unattainable with traditional data-handling techniques.

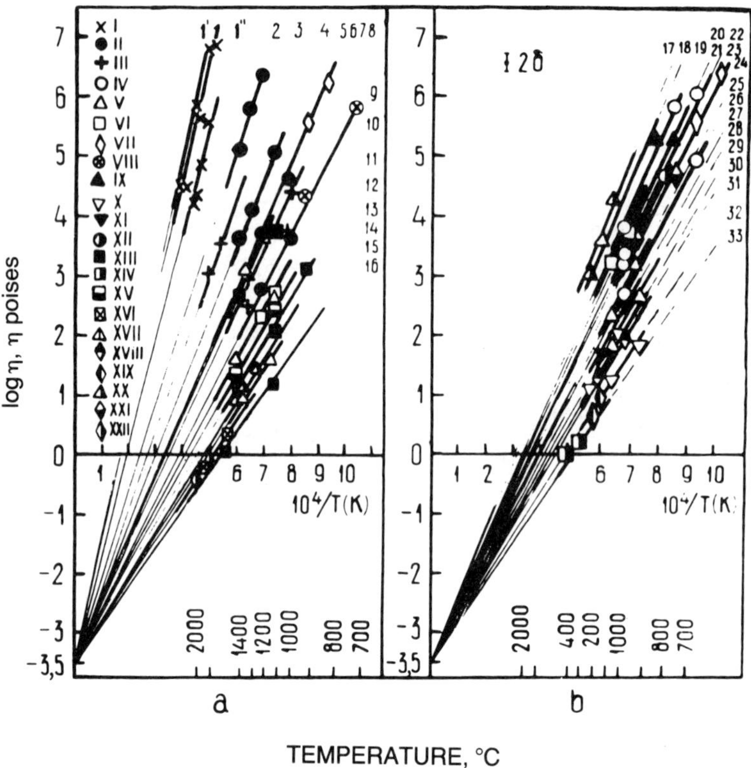

Fig. 13(a, b). The temperature dependence of viscosity of magmatic aluminosilicate and silicate melts ($\log \eta_0 = -3.5$, N_j = wt%). (I) SiO_2/1—dates of Bacon and Avray from the handbook, Mazurin et al., 1973, 1976;/. 1′—dates of Brucker from the handbook; 1″—Volarovich, 1937, Bockris et al., 1955. (II) Albite/2—dry melt, 4,6,9–hydrous melts, N_{H_2O} = 1.4; 4.5; and 6.9, Persikov, 1984, 1980a. (III) System K_2O–SiO_2/ 3,8,9–N_{K_2O} = 3.9; 22.5; 41.6—author's data and from the handbook*. (IV) System granite—H_2O/19, 22, 27–N_{H_2O} = 1.4; 2,2; 5.2—Persikov, 1972, 1976, 1984. (V) System basalt—H_2O/7,10,14,26, 29–N_{H_2O} = 0; 0.7; 1.5; 2,3; 6.4; Persikov, 1981, 1984. (VI) System andesite–H_2O/10, 20–N_{H_2O} = 0; 4.6, Persikov, 1984. (VII) System obsidian—H_2O/5, 24–N_{H_2O} = 4.3; 6.2, Shaw, 1963. (VIII) System pegmatite—H_2O/9–N_{H_2O} = 8.8, Burnham, 1963. (IX) System granite—H_2O–HCl/21–N_{H_2O} = 5.2; N_{Cl} = 0.1, Persikov, 1984. (X) Basalt/30, Murase and McBirney, 1973. (XI) Basalt/28, Euler and Winkler, 1957. (XII) System nepheline syenite—H_2O/23–N_{H_2O} = 5.2, Persikov, 1984. (XIII) System Na_2O–SiO_2/8, 12, 16–NaN_{Na_2O} = 15.4; 40; 51;—Bockris et al., 1955 and from Mazorin et al. (1973, 1977) (XIV) System MgO–SiO_2/33–N_{MgO} = 41.7, from the Mazurin et al. (1973, 1977) (1XV) System Li_2O–SiO_2/11–N_{Li_2O} = 17, from Mazurin et al. (1973, 1977). (XVI) System CaO–SiO_2/15–N_{CaO} = 54.3, from Mazurin et al. (1973, 1977). (XVII) System SiO_2–Al_2O_3–CaO/17, Shludiakov, 1980. (XVIII) System $47.1SiO_2 \cdot 5.5Al_2O_3 \cdot 41.4CaO \cdot 6.0Na_2O$/31, from the handbook. (XIX) Lunar basalt/32, Murase and McBirney, 1970. (XX) Systems granite—H_2O and basalt—H_2O/18–granite–H_2O, N_{H_2O} = 5.7, 13–basalt–H_2O, N_{H_2O} = 3.5, Lebedev and Khitarov, 1979. (XXI) System granite—H_2O–HF/25, N_{H_2O} = 6.27, N_F = 1.6, Persikov, 1984. (XXII) Dunite/16, Ivanov and Shtengelmeer, 1982/.

Pressure Dependence of Viscosity of Magmatic Melts

All present-day theories of the viscosity of liquids, whatever concepts they are based on, state that viscosity should increase with pressure. The experimental data on Al-free silicate melts (Scarfe et al., 1979) are consistent with this theoretical inference. None of these theories is able to give a plausible explanation for the experimentally established fact—that viscosity decreases with increasing total pressure in anhydrous and volatile-bearing aluminosilicate and magmatic melts.

Analysis of the generalized diagram for the pressure dependence of viscosity (Fig. 14) and the pressure effects at $P_{H_2O} < P_s$ discussed earlier, has led to a simple semi-empirical equation for the pressure and temperature dependences in magmatic melts through the range of natural compositions:

$$\log \eta = E/4.576 \cdot T - 3.5 + \alpha(P_s - P_{H_2O}) \tag{7}$$

where the viscosity piezocoefficient has two negative values: $\alpha_1 = -5.02 \cdot 10^{-4}$ MPa^{-1} for dry melts and $\alpha_2 = -1.2 \cdot 10^{-3}$ MPa^{-1} for water undersaturated melts;$^{-3.5}$ is the pre-exponential constant in Frenckel's equation; E is the activation energy of viscous flow in anhydrous melts at $P_{H_2O} = 0$, or in hydrous fluid-bearing melts at $P_s > P_{H_2O}$. According to eq. (7), the pressure correction for

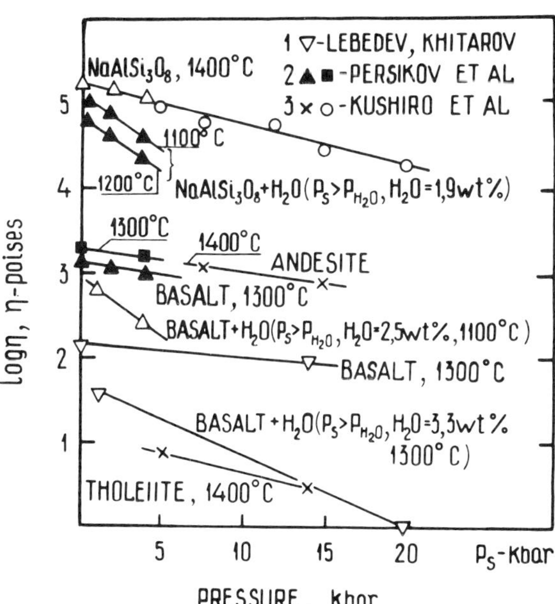

Fig. 14. A generalized diagram of the pressure dependence of viscosity in magmatic melts. (1) Basalt, basalt-H_2O, Lebedev and Khitarov, 1979. (2) Albite, Persikov, 1984; Albite–H_2O, Persikov and Epelbaum, 1980; Andesite, basalt, basalt–H_2O, Persikov, 1981, 1984. (3) Albite, andesite tholeiite-Kushiro, 1978, Kushiro et al., 1976.

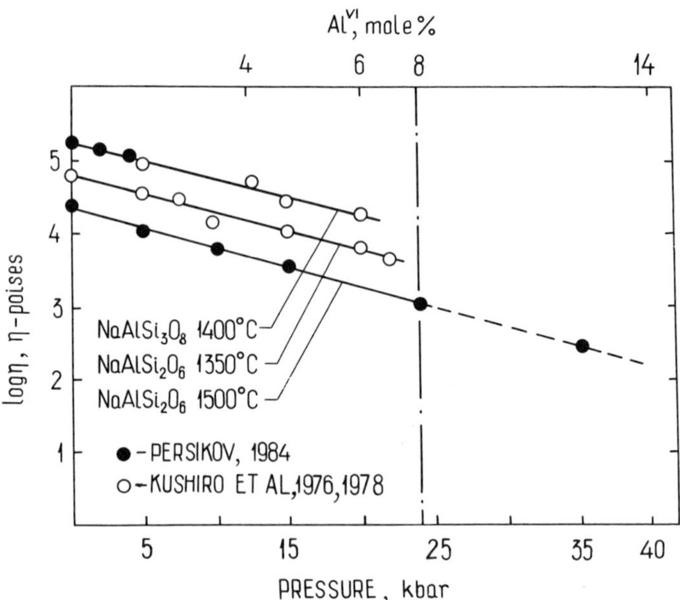

Fig. 15. The inter-relationship between viscosity, pressure and coordination transition $Al^{IV} \rightarrow Al^{VI}$ in albite and jadeite melts. (see text for details).

viscosity is zero at $P_s = P_{H_2O}$; in this instance, the P_{H_2O} effect on the melt viscosity shows up as a reduction in the activation energy owing to the dissolution of H_2O equilibrium concentrations in the melt. The viscosity piezocoefficient (α) is independent, to a first approximation, of the temperature and composition with respect to petrogenetic components because their effect is relatively small within the temperature range considered ($\sim$200°C above the liquidus) (Persikov, Epelbaum, 1979).

The more likely mechanism of this phenomenon can involve the transformation of Al from four- to six-fold oxygen coordination. However, this possibility, proposed by Waff (1975), is discarded by Mysen et al. (1980) and others. The amount of aluminum shift, which might extend to sixfold coordination in the albite and jadeite melts, is plotted in Fig. 15 as a function of pressure. This predictive diagram is based on experimental viscosity measurements and estimates obtained using the proposed method (an inverse problem was posed). It is seen that in the pressure range under experimental study (up to 24 kbar) the jadeite and albite melts seem to contain only about 8 mol% Al^{VI} and retain their albite (framework) structure (see Fig 15). Consequently, the ratio of aluminum-oxygen to silicon-oxygen tetrahedra ($\varepsilon = Al^{IV}/(Al^{IV} + Si^{IV})$) should change by approximately 3%. Clearly, such small structural changes will not be recorded adequately in the Raman spectra of quenched high-pressure albite and jadeite glasses used in the study of Mysen et al. (1980).

The results of such calculations shows that the essentially pyroxene (metasilicate) structure ($Al^{VI} > 50$ mol%) in jadeite melt could be formed at pressures higher than 65 kbar. In this example, a very narrow pressure range (from 40 to 60 kbar) can be used to determine Al^{VI} in jadeite glass by Raman spectroscopy study. The amount of Al^{VI} in the jadeite melt seems small at pressures lower than 40 kbar (see Fig. 15). On the other hand, it has proved difficult to obtain a clear jadeite glass by quenching at pressures higher than 65 kbar, because quenched crystals of jadeite form easily in quenched melt (glass).

The viscosity and activation energy of aluminosilicate and magmatic melts must show two minima and one maximum with increasing pressure. The minima would correspond to complete transformation of Al^{IV} and the Si^{IV} cations to 6-fold oxygen coordinations. The maximum may correspond to the begining of coordination change of 4-fold coordinated Si cations in such melts. The first minimum for jadeite melt must be approximately at 90 kbar, and the second at about 200 kbar.

Compositional Dependence. The Generalized Model for Calculation and Prediction of Viscosities in Magmatic Melts

As is known, among the physico-chemical properties of liquids, viscosity displays the most complex dependence on melt composition (e.g. Esin and Geld, 1966, Bottinga and Weil, 1972). Although some progress was made towards the solution to this problem by Volarovich (1937–1950), Murase (1962), Carron (1969), Bockris (1970), Bottinga and Weil (1972), Shaw (1972), Scarfe (1973), Sheludyakov (1980), Urbain (1979), Richet (1984), no reasonably accurate method has yet been devised for the whole composition range of natural magmas.

The problem reduces, in fact, to the two principal points: it is necessary to introduce some reasonably sensitive and precise criteria for assessing the structure and composition of the melt. Through the effort of many scientists this task has been fulfilled. The criterion is defined as a structural-chemical parameter of the melt (K) which reflects to some extent the dynamic equilibrium of the oxygen quasichemical forms (Persikov, 1984) or the degree of depolymerization. However, no convincing answer has yet been found to the second question—how is the viscosity of melts to be compared? Most previous models (except that used by Shaw, 1972) involved isothermal comparison, which can only be applied to a very narrow composition range. Because the liquidus temperature range in magmatic rock is very wide, the isothermal composition often has no physico-chemical meaning. It is clear that more complete information can be obtained by comparing the activation energies of viscous flow which do not depend on temperature. This comparison however has not yet been achieved.

The dependence on composition of the activation energy of viscous flow through the entire range of comparison of magmatic melts, from acidic to ultrabasic, is shown in Fig. 16. The composition is expressed through the above structural-chemical parameter K, i.e. "the basicity coefficient" or the degree of depolymerization (Carron, 1969; Mysen et al., 1980; Persikov, 1984).

Table 4. Examples of calculations of rheological properties (E, η) of hydrous granitic melts

	Chemical analyses of granite				
	Starting composition	Run product	Composition with two forms of dissolved H_2O	Network-forming cations $\times 10^3$	Oxygen gram-ions $\times 10^3$
SiO_2	73.23	64.22	69.86	1163.2	2326.4
Al_2O_3	13.60	11.93	12.97	254.6	381.8
TiO_2	0.19	0.16	0.15	—	3.9
FeO	0.48	0.42	0.46	—	6.4
Fe_2O_3	2.56	2.24	2.44	30.5	45.9
Na_2O	3.78	3.32	3.61	—	58.2
K_2O	4.11	3.60	3.92	—	41.6
MgO	0.17	0.15	0.16	—	4.0
CaO	1.69	1.48	1.61	—	28.7
P_2O_5	0.13	0.11	0.12	1.7	4.2
H_2O	—	12.3	—	—	—
$H_2O(OH^-)$	—	—	4.6	—	255.3
H_2O(Ph)	—	—	7.7[a]		
Σ	99.94	99.93	99.9	H = 1450	O = 3156.3
K	0.3	101.5	35.4		

[a] 23.1 mol% is taken into account when E is calculated with Eq. (11)

1. According to Eq. (8): $K = \dfrac{2(O - 2H)}{H} \cdot 100 = 35.4$

2. According to Eq. (9): $E_{(OH^-)} = (51\text{–}0.154K) = 45.54 \dfrac{\text{Kcal}}{\text{mol}}$

3. According to Eq. (11): $E_{Cal} = 45.55 - 0.3 \cdot 23.1 = 38.62$ Kcal/mol Experimental value of E is 38.8 Kcal/mole (see Table 2)

4. $\Delta E = [(E_{Cal} - E_{ex})/E_{ex}] \cdot 100 = -0.46\%$

5. Calculated value of η with Eq. (10) at $T = 850°C$ and $P_s = P_{H_2O}$: $\log \eta_{Cal} = E_{Cal}/4.576 \times T - 3.5 = 38620/4.576 \times 1123 - 3.5 = 4.01$ Experimental is equal to $\log \eta_{ex} = 4.2$ (see Table 2)

6. $\Delta\eta = [(\eta_{Cal} - \eta_{ex})/\eta_{ex}]$: $100 = [(10359 - 15849)/15849] \cdot 100 = -34.6\%$

7. If using the experimental value of $E_{ex.} = 38.8$ kcal/mol and $\log \eta_{ex} = 4.05$, respectively (see Eq. 5), then the calculated viscosity error will be essentially lower: $\Delta\eta = [(10359 - 11229)/11229] \cdot 100 = -7.7\%$

8. The accuracy of rheological property prediction will be essentially lower if two H_2O forms are not allowed for:

a) According to eq. (9): $E_{Cal} = (40 - 0.0405 \cdot K) = 40 - 0.0405 \cdot 101.5 = 35.9$ Kcal /mole

b) $\Delta E = [(35900 - 38800)/38800] \cdot 100 = -7.47\%$

c) Calculated value of η with eq. (10) at $T = 850°C$ and $P_s = P_{H_2O}$: $\log \eta_{Cal} = 35900/4.576 \times 1123 - 3.5 = 3.49$

d) $\Delta\eta = [(3090 - 11229)/11229] \cdot 100 = -72.5\%$

The degree of depolymerization was found from the following expression (Carron, 1969; Mysen et al., 1980):

$$K = \frac{2(\mathrm{O} - 2\mathrm{H})}{H} \cdot 100 = 100 \frac{\mathrm{NBO}}{T} \tag{8}$$

where H is the total number of network forming gram-ions (such as Si^{4+}, Al^{3+}, Fe^{3+}, P^{5+}) that are tetrahedrally coordinated with respect to oxygen and are members of the anionic part of the melt structure; 0 is the total number of oxygen gram-ions in the melt. The computation is based on the rock (melt) chemical composition taken as wt% oxides, using the simple technique of Carron (1969), Persikov (1984). (see Table 4).

In Fig. 16, the vertical lines bound a broad field of natural magma composition, $5 \le K \le 400$, which was determined in the following way. At the average water concentration in granitoid magmas (0.67 wt% in granites; 1.1% in rheolite), the basicity coefficient is greater than 5; on the other hand, in depolymerized ultramafic melts, $K = 200:400$. For example, in an anhydrous dunite melt, K will be 380.

The diagram shows that the activation energy decreases regularly in the sequence acid—ultramafic melts with increasing basicity. The kinks in the $E = f(k)$ curve divide it into four composition ranges and can be interpreted in terms of the silicate solution structure theory by Bockris et al. (1954), Esin et al. (1966), Mysen et al. (1980) in the following way. At $K = 17$, which corresponds to the

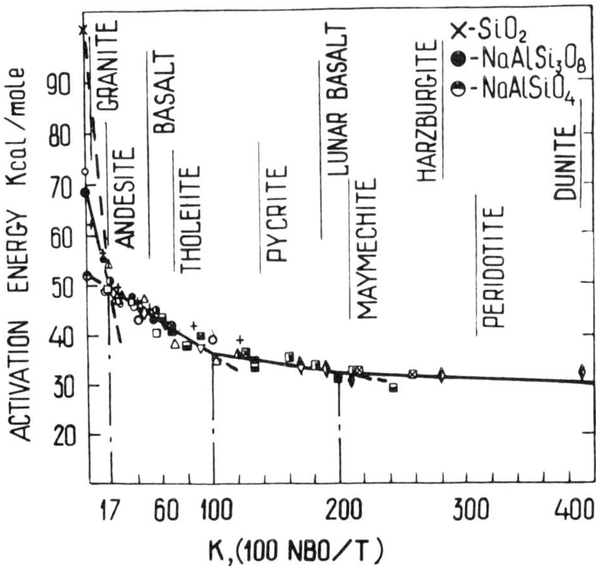

Fig. 16. A generalized diagram for the concentration dependence of the viscous flow activation energy in magmatic (aluminosilicate) and silicate melts (symobls are the same as in Fig. 13a, b, see text for details).

dry andesite average composition, the framework structure breaks down; at $K = 100$ and 200 corresponding to the average compositions of olivine tholeiite ($N_{H_2O} \sim 3$ wt%) and pyroxenite, respectively, the formation process of the di- and metasilicate structures is being completed. The results show that the differences between aluminosilicate and silicate melts show up only within the first composition range ($K < 17$). In the proposed qualitative model (Persikov, 1984), the structure of magmatic melts within this range is identical to the continuous three-dimensional framework of molten silica in which Al substitutes part of Si in the tetrahedra. Because of a considerable energy difference between the Al–O and Si–O bonds (80 and ~109 kcal/mol (e.g. Zharikov, 1969), the quantitative ratios of the activation energies of viscous flow will depend on the molar ratio $\varepsilon = Al^{IV}/(Al^{IV} + Si^{IV})$ on transition from silicate to aluminosilicate melts. This dependence accounts for the appearance of the $E = f(k)$ dependence fan for different values of ε in the composition range considered ($K < 17$). For example, in Fig. 16, the $E = f(k)$ curves are drawn for quartz ($\varepsilon = 0$), albite (granitoid) ($\varepsilon = 0.25$), and nepheline ($\varepsilon = 0.5$) melts. The composition ranges of magmatic melts this determined are largely consistent with interpretation of magmatic melt structures by Mysen et al. (1980).

The breakdown of the framework structure ($K > 17$), i.e. for the other three composition ranges, the structures of silicate and magmatic (alumnosilicate) melts become indistinguishable with respect to their energies because their activation energies are equal at the same K.

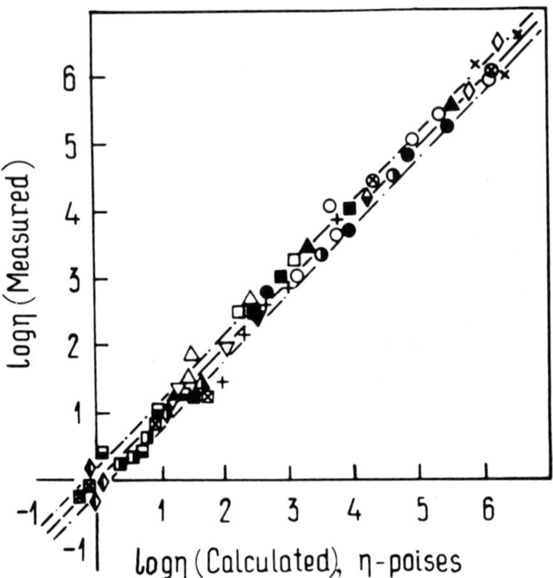

Fig. 17. Comparison of calculated and experimental viscosities of silicate, aluminosilicate and magmatic melts (symbols are the same as in fig. 13a, b). The dashed lines show the confidence limits $-\Delta\eta = \pm 30\%$.

Linear approximation of the $E = f(k)$ dependence for magmatic and aluminosilicate melts in each of the composition ranges has led to the following simple equations for E:

$$\begin{aligned} &1.\ 0 \le K \le 17 && E_1 = (70 - 1.27 \cdot K) \cdot 10^3 \text{ cal/mol} \\ &2.\ 17 \le K \le 100 && E_2 = (51 - 0.154 \cdot K) \cdot 10^3 \text{ cal/mol} \\ &3.\ 100 \le K \le 200 && E_3 = (40 - 0.0405 \cdot K) \cdot 10^3 \text{ cal/mol} \\ &4.\ 200 \le K \le 400 && E_4 = (35 - 0.015 \cdot K) \cdot 10^3 \text{ cal/mol} \end{aligned} \quad (9)$$

The correlation coefficients for the obtained dependences are, respectively: 0.83, 0.92, 0.97, 0.94.

Simultaneous solution of these equation with Eq. (7) produced the generalized equation for the concentration–temperature–pressure dependence of viscosity in magmatic melts:

$$\log \eta_{1,2,3,4} = E_{1,2,3,4}/4.576T - 3.5 + \alpha(P_s - P_{H_2O}) \quad (10)$$

Consequently, the operations required to apply the proposed method are very simple.

1. The basicity coefficient (K) is calculated from the chemical analysis, accounting for the effect of the volatiles. Water (dissociated) is calculated from the equation: $H_2O_g + O^0_m \rightleftarrows 2(OH)^-_m$; fluorine: $HF_g + O^{2-}_m \rightleftarrows OH^-_m + F^-_m$; carbon dioxide: $CO_{2g} + O^{2-}_m \rightleftarrows CO^{2-}_{3m}$, where g is the gas, and m is the melt (see Table 1).

2. Depending on the K value, the activation energy of viscous flow is calculated from one of eq. (9). At $K < 17$, ε should be allowed for.

3. The viscosity of the melt for different T, P_s, $P_{H_2O}(P_{fl})$ is found from Eq. (10). To clarify the steps performed, an example for calculating the rheologic properties of fluid-bearing granitic melts is given in Table 4.

A direct test of the method was made by way of experimental viscosity studies on anhydrous and hydrous andesite and nepheline syenite melts, for which predicted values were first obtained. The calculated and measured values were found to agree to within 15 + 30%.

A comprehensive comparison of the calculated and experimental viscosity data for magmatic, silicate and aluminosilicate melts (Fig. 17) shows good agreement, well within the stated uncertainty limits ($\Delta E = \pm 1.0\%$, $\Delta\eta = \pm 30\%$). A relatively small number of values that fall outside these limits are readily accounted for and arise from the variations in the Fe^{3+}/Fe^{2+} ratio in experiments with melts containing variable-valence elements.

Some Applications of the Above Results to the Geochemistry of Magmatic Melts and Rocks

The results from this study have so many important applications to geochemistry and petrology of natural magmas that they warrant a separate publication. Some of the uses to which the data might be put are listed below. It has now become possible to estimate and predict a number of other physico-chemical properties of magmatic melts known, in theory and by experiment, and to relate them to the rheological properties (η, E) of the melts. The more important properties are the mobility of petrogentic and volatile components, electrical conductivity, and crystal settling.

A great deal of attention has been given to analytical and numerical methods of calculating kinetics and dynamics of magmatic processes (e.g. Sharapov, Golubev, 1976; Mueller and Saxena, 1977) in the past few years. The proposed model in which the viscosity is calculated from $\eta = f(N_{fl}, T, P_s, P_{fl})$ forms a reliable quantitative basis for such calculations. Moreover, it has become possible to find solutions to inverse problems bearing on structural positions of different ions in magmatic melts, and the dissolution mechanisms and regimes of water, oxygen and other volatiles.

The Mechanism of H_2O Solubility

As was noted earlier, in the system granite (albite)-H_2O, andesite-H_2O, basalt-H_2O, the water solubility mechanism involves two stages: essentially chemical (dissociated water, OH^- hydroxyl) at the first stage and essentially physical (molecular H_2O) at subsequent stages.

Fig. 18 presents the generalized E-composition diagram for the pseudobinary system anhydrous melt-H_2O in the compositional range silica melt – basaltic melt. To construct the diagrams, experimental data on $E = f(N_{H_2O})$ dependence has been used for melts of differing composition (for example, see Fig. 7). Linear sections of these diagrams have been interpolated up to the H_2O activation energy of viscous flow. The inflection points in the curves were obtained by the linear transformation of the coordinates and although, in the regular solution theory, they mark the change in the H_2O solubility mechanism, they are seen to shift towards lower H_2O contents with increasing basicity of the anhydrous melt. The absolute values ΔE at these points decrease by nearly an order of magnitude in the series silica-basalt. It is also seen that the amount of the chemically bound H_2O in pyroxenite melts tends to zero (shown by the arrow).

In keeping with the fundamental principles of the acid-basic interaction (Korzhinskii, 1966) and the maximum polarity of chemical bonds (Kogarko, 1980), the present results clearly indicate that the chemically dissolving water is a base, relative to melts in the specified composition range. Consequently, the amphoteric nature of H_2O is manifested on dissolution in ultrabasic melts, in which the concentration of free oxygen (O^{2-}) becomes considerable (Mysen et al., 1980a). This conclusion is fully supported by the depolymerization effect H_2O has on the structure of aluminosilicate melts, which is similar to the effect produced by bases (e.g. Na_2O, K_2O) added to a silicate melt. The proposed characteristic features of the mechanism for H_2O dissolution in silicate and aluminosilicate melts (details can be found in Persikov, 1972, 1974, 1976, 1984) also show a good fit to a large set of the published data. Some of them are:

1. the viscosity and electrical conductivity patterns determined for hydrous melts (e.g. Shaw, 1963; Scarfe, 1973; Lebedev, Khitarov, 1979);

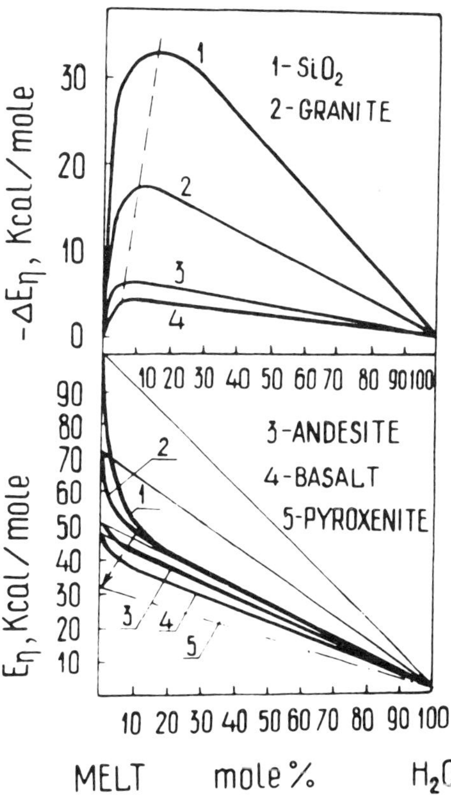

Fig. 18. A generalized diagram for the E-composition of pseudobinary aluminosilicate (silicate)—H_2O systems (see text for details).

2. an increase in the H_2O solubility with increasing acidity of the melt caused by the dissolution of fluorine (Persikov, for granitic melts, 1984; Kovalenko, for ongonitic melts, 1979);
3. an increase in the CO_2 solubility in albite melts in the presence of water (Kadik et al., 1976; Mysen et al., 1976);
4. kinetic studies of thc thermal dehydration of hydrous granitic glasses (Persikov, 1974);
5. IR and Raman spectroscopy of hydrous glasses of various compositions (Orlova, 1967; Persikov, 1972, 1974 Mysen et al., 1980a; Stolper, 1982).

Several previous works have studied in considerable detail possible reactions of dissolution of dissociated water in magmatic melts (Zharikov, 1969; Kadik, 1971; Epelbaum, 1980; Mysen et al., 1980a; Burnham 1979; Stolper, 1982). According to the present author's views, the reaction which predominates in melts with normative quartz involves the breakdown of Si–O–Si bonds. The two-stage dissolution of H_2O in magmatic melts should be taken into consideration in part in viscosity calculations with the proposed model. This consideration may readily be achieved if Eq. (10) is re-arranged in the form

$$E = E_{(OH^-)} - a \cdot N_{H_2O}^{Ph} \tag{11}$$

where E_{OH^-} is the input of the dissociated water content into the activation energy of viscous flow; $N_{H_2O}^{Ph}$ is the physical (molecular H_2O) molar concentration; and a is the empirical constant. The numerical value of a is 300 cal/mol% H_2O for acid, intermediate and basic melts.

The Relationship Between the Viscosity of Magmatic Melts and Relative Occurrence of Effusive and Plutonic Igneous Rocks

This relationship has been determined from the calculated viscosities for anhydrous and hydrous melts at the liquidus T and P using the proposed method. Fig. 19 gives some results for anhydrous and hydrous granitoid and basalt melts. The thermodynamic melting parameters of rocks are from Tuttle and Bowen, 1958; Kadik et al., 1971; Perchuk, 1973; Burnham, 1979.

The results show that through the wide length range in the crust (~2 to 20 km), where most granitoid massifs are believed to be formed, the viscosity of near-liquidus granitic and basaltic magmas is virtually independent of H_2O content within the range 2–8 wt%. Also, hydrous granitic magmas (H_2O = 2–5 wt%) have liquidity comparable with those of basaltic and andesitic effusives.

Totally different viscosity patterns found in near-liquidus granitoid and andesitic and basaltic melts at H_2O contents below 2 wt% are responsible for the contrasting relative occurrences of effusive and plutonic varieties of acid, on the one hand, and intermediate and basic rocks, on the other. Indeed, completely

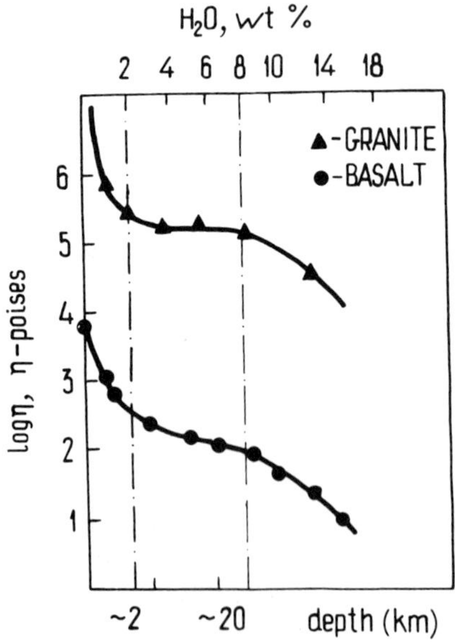

Fig. 19. Viscosity of dry and hydrous granitoid and basalt melts at the liquidus thermodynamic parameters (T, P_{H_2O}, N_{H_2O}) (see explanation in the text).

outgassed near-liquidus granitoid melts with viscosities of 10^7–10^9 poise are unable to outpour in lava flows and only occasionally erupt as extrusions or explosive ejections. The bulk of granitoid melts do not reach the surface because their formation and crystallization occurs in the plutonic facies.

Unlike granitoid melts, near-liquidus andesitic and basaltic melts retain their high mobility even when completely outgassed, as both experiments and field observations show. (Shaw et al., 1968; Andreev et al., 1978). This factor seems to account well for the presence of abundant basalts and andesites relative to gabbros and diorities in the earth's crust.

It is emphasized that the viscosity patterns determined in hydrous magmatic melts under their liquidus parameters (T, P) clearly indicate that the dissolution of the first ~ 2 wt% H_2O has a decisive effect on most processes in andesitic and basaltic magmatism.

Relation Between Viscosity and Acid-Basic Indicators of Magmatic Melts and Rocks

Petrology and geochemistry studies of magmatic melts and igneous rocks are faced with a complex problem of the quantitative evaluation of acidity-basicity and its classification (e.g. Zharikov, 1967, 1982; Marakushev, 1979a, 1979). The latter implies mainly the need of reliable criteria for analyzing the chemical evolution of multicomponent magmatic melts and rocks. The calculated viscosities of melts at their liquidus parameters agree well with the acid-basic indicators of rock-forming minerals and magmatic rocks (Table 4) in the complete series: quartz-larnite, granite-dunite. It will be noted that the viscosity as a physico-chemical parameter of the melt reflects, at the liquidus parameters (T, P_s, P_{fl}, N_{fl}), various aspects of component interaction in the melt. Viscosity is the function that depends not only on the composition but also on the ratio of the oxygens, i.e. acidity–basicity, the temperature and pressure on the liquidus.

Analysis of the results in Table 4 and Fig. 16 shows that the degree of depolymerization (K) can be used both to compare the chemistry of multicomponent magmatic rocks and their melts (e.g. Carron, 1969; Mysen, et al., 1980) and to estimate, to a first approximation, their bulk basicity. Consequently, the term "coefficient of the bulk basicity of magmatic rocks" seems appropriate. In particular, changes in the basicity of hydrous melts under the effect of dissolving water were evaluated with this parameter. Thus, at 2 wt% H_2O in the granite melt, (K) changes by a factor of 500 whereas in the tholeiite melt it changes by only a factor of 1.2. However, this value (K) can not be applied to simpler silicate and aluminosilicate systems, because, even when used with the $\varepsilon = Al^{IV}/(Al^{IV} + Si^{IV})$ ratio, it fails to reflect fully their chemistry in terms of the acid-basic interaction (Table 4).

Moreover, no experimentally determined criterion is presently available to give a strict quantitative characterization of the basicity of multicomponent

Table 5. Relationship between viscosity at T of liquidus and basicity indicators for melts of rock-forming minerals and magmatic rocks. (average rock compositions after Le Maitre, 1976)

Substance	Melting temperature (liquidus) T_1, °C	Activation energy of viscous flow, E Kcal/mol	Viscosity at T_1 η-poises $\log \eta_{T_L}$	Thermodynamic basicity indicator, Kcal (Marakushev, 1979) $\Delta Z^{H_2O}_{298,K}$	Conventional ionisation potential, Kcal/mole (Zharikov, 1967, 1982)	$\left(\frac{NBO^K}{T} \cdot 100\right)$	$\varepsilon = \frac{Al^{IV}}{Al^{IV} + Si^{IV}}$
1	2	3	4	5	6	7	8
SiO_2(Quartz)	1680	100	7.7	−1.886	226.7	0.0	0.0
$NaAlSi_3O_8$ (Albite)	1120	68	7.16	−0.23	204.2	0.0	0.25
$KAlSi_3O_8$ (Orthoclase)	1150	68	6.94	−0.476	202.8	0.0	0.25
$NaAlSi_2O_6$ (P = 0.1 MPa)	1140	62	6.1	0.5	197.5	0.0	0.33
$KAlSi_2O_6$ (Leucite)	1686	62	3.42	0.537	195.7	0.0	0.33
$MaAlSiO_4$ (nepheline)	1526	49	2.45	2.055	185.2	0.0	0.5
$CaAl_2Si_2O_8$ (Anorthite)	1550	49	2.38	1.458	197.0	0.0	0.5
$KAlSiO_4$ (Kalsilite)	1750	49	1.79	2.463	182.4	0.0	0.5
$CaMgSi_2O_6$ (Diopside)	1391	32	0.702	1.548	191.7	200	0.0
$NaAlSi_2O_6$ (Jadeite P_s = 35Kbars)	1450	32	0.56	—	—	200	0.0

$MgSiO_3$ (Enstatite)	1540	32	0.358	1.858	197.1	200	0.0
$CaSiO_3$ (Wollastonite)	1546	32	0.344	2.262	186.3	200	0.0
Fe_2SiO_4 (Fayalite)	1205	24	0.049	1.872	189.0	400	0.0
Mg_2SiO_4 (Forsterite)	1890	24	−1.076	4.006	184.5	400	0.0
Ca_2SiO_4 (Larnite)	2403	24	−1.319	5.818	169	400	0.0
Granite ($N_{H_2O} = 0$)	1090	69.6	7.66	1.02	—	0.3	—
Granite ($N_{H_2O} = 2.0$ wt%)	900	50.8	6.0	—	—	15.3	—
Andesite ($N_{H_2O} = 0$)	1100	50.0	4.46	1.88	—	17.2	—
Andesite ($N_{H_2O} = 2.0$ wt,%)	1000	45.8	4.36	—	—	33.3	—
Thleiite	1170	40.2	2.59	2.71	—	70.2	—
Tholeiite ($N_{H_2O} = 2.0$ wt.%)	1100	37.9	2.53	—	—	85.2	—
Lherzolite ($N_{H_2O} = 0$)	1350	31.3	0.714	4.07	—	215	—
Lherzolite ($N_{H_2O} = 2.0$ wt.%)	1300	30.27	0.705	—	—	240.1	—
Dunite ($N_{H_2O} = 0.0$)	1700	24.93	−0.744	4.84	—	372.0	—
Dunite ($H_{H_2O} = 2.0$ wt.%)	1670	24.0	−0.796	—	—	395	—

silicate and aluminosilicate (magmatic) melts (Zharikov, 1982). Therefore, it would seem reasonable to compare the basicities of these melts using the essentially experimental value ($\log \eta_{T_1}$), which can be termed the "conventional basicity of the melt" within the stated error limit. This value was used to obtain the continuous basicity series for melts of rock-forming minerals from quartz to larnite, and for anhydrous and hydrous magmatic melts in the series granite-dunite (Table 4).

In sum, the present generalized patterns and the new method for calculating rheologic properties (η, E) of near-liquidus magmatic melts form a sound physicochemical basis for the quantitative evaluation of heat and mass transfer as well as the dynamics of magmatism. However, more detailed work needs to be done on the rheology of magmatic melts at T and P of upper mantle and of sub-liquidus magmatic melts.

Acknowledgments

The author is indebted to V.A. Zharikov, A.A. Marakushev and L.L. Perchuk for invaluable discussion and support of the paper.

Thanks are also due to M.B. Epel'baum, P.G. Bukhtiyarov and T.V. Kalinicheva who are co-authors of some my works. Sincere gratefulness also is expressed to G.B. Lakoza and G.G Gonchar for translating the manuscript from Russian into English. The manuscript was greatly improved by critical reviews of Ch.M. Scarfe and A.A. Kadik to whom the author is particularly indebted for their constructive suggestions.

References

Andreev, V.I., Gusev, N.A., Kovalev, G.N., and Slezin, Yu.B. (1978) The dynamics of lava flows in the South outburst of the Great Tolbachik fissure eruption. *Bul. Volcanol.*, No. 5, 18–26 (in Russian).

Bockris, J.O'M., Mackenzie, J.D., and Kitchner, J.A. (1955). Viscous flow in silica and binary liquid silicates. *Trans. Faraday Soc.*, **51**, 1734–1748.

Bockris, J.O'M., and Lowe, D.L. (1954) Viscosity and structure of molten silicates. *Proc. Roy. Soc. Lond.* **A226**, 423–435.

Bockris, J.O'M., and Reddy, A.K.N. (1970). *Modern Electrochemistry, Vol.* **1**, chap. 6. Plenum Press, New York.

Bottinga, Y., and Weill, D.F. (1972) The viscosity of magmatic silicate liquids; a model for calculation. *Amer. J. Sci.*, **272**, 438–475.

Burnham, C.W. (1963) Viscosity of water-rich pegmatite melt at high pressure (Abstr.). *Geol. Soc. Amer. Spec.*, **76**, 26.

Burnham, C.W. (1979) The importance of volatile constituents. In: *The Evolution of the Igneous Rocks* (*Fiftieth Anniversary Perspectives*) edited by Yoder, H.S., Chap. 16, pp. 439–482. Princeton University Press, Princeton, New Jersy.

Carron, J.P. (1969) Vue d'ensemble sur la rheologie des magmas silicates naturels. *Bul. Soc. Franc. Miner. Cristallogr.*, **92**, 435–446.

Esin, O.A., and Geld, P.V. (1966) *Physical Chemistry of Pyrometallurgical Processes.* pp. 2, 703. Metallurgia, Moscow (in Russian).

Epelbaum, M.B. (1980) *Silicate Melts Containing Volatiles*, p. 255. Nauka, Moscow (in Russian).

Euler, R., and Winkler, H.G. (1957) Über die Viskositäten von Gestein- und Silikatschmelzen. *Glastech. Ber.*, No. 8, S. 325–332.

Frenckel, Ya.I. (1945) *The Kinetic Theory of Liquids*, p. 401, Akad. Nauk, Moscow (in Russian).

Glasstone, S., Laidler, K.J., and Eyring, H. (1941) *The Theory of Rate Processes*, p. 460. McGraw-Hill, New York.

Goranson, R.W. (1931) The solubility of water in granite magmas. *Amer. J. Sci.*, **22**, 481–502.

Ivanov, I.P., and Shtengelmaer, S.V. (1982) Viscosity and temperature of crystallization of ultramafic melts. *Geochemistry*, No. 3, 330–337 (in Russian)

Kadik, A.A., Lebedev, E.B., and Khitarov, N.I. (1971) *Water in Magmatic Melts*, p. 267. Nauka, Moscow (in Russian).

Kadik, A.A., and Eggler, D.H. (1976) The regimes of H_2O and CO_2 during formation and outgassing of acidic magmas. *Geochemistry*, No. 8, 1167–1175 (in Russian).

Kani, K., and Hosokawa, K. (1936) On the viscosity of silicate rock-forming minerals and igneous rocks. *Rev. Electrotechn. Lab.*, No. 391, 1–105.

Kobeko, P.P. (1952) *Amorphous Substances*, p. 431. ONTI, Moscow-Leningrad (in Russian).

Kogarko, L.N. (1980) Principle of polarity of chemical bond and its significance in magmatic geochemistry. *Geochemistry*, No. 9, 1286–1297 (in Russian).

Kogarko, L.N., and Krigman, L.D. (1981) *Fluorine in Silicate Melts and Magmas*, p. 125. Nauka, Moscow (in Russian).

Korzhinskii, D.S. (1966) Acid-basic interaction of components in melts. In: *Investigation of Natural and Technical Mineral Formation*, pp. 5–9. Nauka, Moscow (in Russian).

Kushiro, I. (1976) Changes in viscosity and structure of melt of $NaAlSi_2O_6$ composition at high pressures. *J. Geophys. Res.*, **81**, No. 35, 6347–6350.

Kushiro, I. (1978) Viscosity and structural changes of albite ($NaAlSi_3O_8$) melt at high pressures. *Earth Planet. Sci. Lett.*, **41**, 87–90.

Kushiro, I., Yoder, H.S., Jr., and Mysen, B.O. (1976) Viscosity of basalt and andesite melts at high pressures. *J. Geophys. Res.*, No. 35, 6351–6359.

Kushiro, I. (1980) Viscosity, density, and structure of silicate melts at high pressures, and their petrological applications. In: *Physics of Magmatic Processes*, Chap. 3, pp. 93–120, edited by Hargraves R.B., Princeton University Press, Princeton, New Jersey.

Kovalenko, N.I. (1979) *Experimental Study of the Formation of Rare Metal Lithium–Fluoride Granites*, p. 150. Nauka, Moscow (in Russian).

Lebedev, E.B., and Khitarov, N.I. (1979) *Physical Properties of Magmatic Melts*, p. 200. Nauka, Moscow (in Russian).

Le Maitre, R.W. (1976) The chemical variability of some common igneous rocks. *J. Petrol.*, **17**, No. 4, 589–637.

Leontieva, A.A. (1940) Experimental investigation of the viscosity of obsidian and hydrous glasses. *Izvest. Akad. Nauk SSSR Ser. Geol.*, No. 2, 44–55 (in Russian).

Marakushev, A.A. (1979) *Petrogenesis and Ore Formation*, p. 281. Nauka, Moscow (in Russian).

Marakushev, A.A. (1979a) A method for thermodynamic calculation of basicity indicators of rocks and minerals. *J. Petrol.*, **20**, No. 4, 821–845.

Mueller, R.F., and Saxena, S.K. (1977) *Chemical Petrology*, p. 512. Springer-Verlag, New York.

*Mazurin, O.V., Strelzina, M.V., and Shvaiko-Shvaikovskaja, T.P. (1973, 1977) *The Properties of Glasses and Glass-Forming Melts, Parts I and III. Handbook**, Nauka, Leningrad, pp. 444 and 586. (in Russian)

Murase, T. (1962) Viscosity and related properties of volcanic rocks. *J. Fas. Sci. Hokkaido Univ., Ser. VII*, No. 6, 121–125.

Murase, T., and McBirney, A.R. (1970) Viscosity of lunar lavas. *Science*, **161**, 1491–1493.

Murase, T., and McBirney, A.R. (1973) Properties of some common igneous rocks and their melts at high temperatures. *Geol. Soc. Amer. Bull.*, **84**, No. 11, 3563–3592.

Mysen, B.O., Eggler, D.H., Leits, M.G., and Hollway, J.R. (1976) Carbon dioxide in silicate melts and crystals. Part I. Solubility measurements. *Amer. J. Sci.*, **276**, No. 4, 455–479.

Mysen, B.O., Virgo, D., and Scarfe, C. (1980) Relation between the anionic structure and viscosity of silicate melts: A Raman spectroscopic study. *Amer. Minerlog.*, **65**, 680–710.

Mysen, B.O., Virgo, D., Harrison, W., and Scarfe, C. (1980a) Solubility mechanism of H_2O in silicate melts at high pressures and temperatures: A Raman spectroscopic study. *Amer. Mineralog.*, **65**, 900–914.

Orlova, G.P., and Rudnitskaya, E.S. (1965) On the water-silicate melt interaction under pressure. In: *Vitreous State*, pp. 282–284. Nauka, Moscow-Leningrad (in Russian).

Ostrovsky, I.A. (1963) On the general pattern of water dissolution in the silicate melts, *Geologiya Rudnykh Mestorozhd.*, No. 1, 103–106 (in Russian).

Persikov, E.S. (1972) The viscosity of the granitic melt at 800–1200°C and water pressure 2000atm. In: *Experimental Studies in Mineralogy* (1970–1971), pp. 93–98. Nauka, Novosibirsk (in Russian).

Persikov, E.S. (1974) Experimental studies of solubility of water in granitic melt and kinetic of the melt-water equilibrated at high pressures. *Int. Geol. Rev.*, **16**, No. 9, 1062–1067.

Persikov, E.S. (1976) Experimental investigation of the viscosity of hydrous granitic melts at high temperatures and pressures. In: *Problems of Physics of Magmatic and Ore Formation Processes*. 92–123. Nauka, Novosibirsk (in Russian).

Persikov, E.S. (1981) Relation between the viscosity of magmatic melts and some regularities of the acid and basic magmatism. *Dokl. Akad. Nauk SSSR*, **260**, No. 2, 426–429 (in Russian).

Persikov, E.S. (1981a) Experimental investigation of the viscosity of basaltic melts. *Vulkanologia Seismologia*, No. 2, 70–77 (in Russian).

Persikov, E.S. (1984) *The Viscosity of Magmatic Melts*, p. 160. Nauka, Moscow (in Russian).

Persikov, E.S., and Kochkin, Yu.N. (1973) A radiation high-pressure falling sphere viscosimeter. *Geolog. Geophis.* No. 8, 138. (in Russian).

Persikov, E.S., and Epelbaum, M.B. (1978) An apparatus for viscosity and density measurements in magmatic melts at high pressures. In: *Experiment and Methods of High Gas and Solid-Media Pressures*, pp. 94–98. Nauka, Moscow (in Russian).

Persikov, E.S., and Epelbaum, M.B. (1979) The effect of pressure on the viscosity of hydrous magmatic melts. *Dokl. Akad. Nauk SSSR*, **245**, No. 5, 1198–2000 (in Russian).

Persikov, E.S., and Epelbaum, M.B. (1980) A study on the viscosity of hydrous albite melts at high pressures. In: *Contributions to Physicochemical Petrology*, No. 9, 111–118. Nauka, Moscow (in Russian).

Persikov, E.S., and Epelbaum, M.B. (1980a) Experimental study of the pressure effect on the viscosity of hydrous magmatic melts. *High Pressure Sci. Technol. Proc.* (*Oxford*), **2**, 868–870.

Persikov, E.S., and Kalinicheva, T.V. (1982) Concentration and temperature dependences of the viscosity of magmatic melts (A method for calculation and prediction). *Dokl. Akad. Nauk SSSR*, **266**, No. 6, 1467–1471 (in Russian).

Perchuk, L.L. (1973) *Thermodynamic Regime of Depth Petrogenesis*, p. 318. Nauka, Moscow (in Russian).

Reiner, M. (1958) *Rheology*, p. 224. Springer-Verlag, Berlin.

Richet, P. (1984) Viscosity and configurational entropy of silicate melts *Geochim. Cosmochim. Acta*, **48**, 471–483.

Ryabchikom, I.D. (1975) *Thermodynamics of a Fluid Phase in Granitic Magmas*, p. 230. Nauka, Moscow (in Russian).

Scarfe, C.M. (1977) Viscosity of a pantellerite melt at one atm. *Can. Mineralog.* **15**, 185–189.

Scarfe, C.M. (1973) Viscosity of basaltic magmas at varying pressure. *Nature*, **241**, 101–109.

Scarfe, C.M., Mysen, B.O., and Virgo D. (1979) Changes in viscosity and density of melts of sodium disilicate, sodium metasilicate and diopside composition with pressure. *Carnegie. Inst. Wash. Year Book*, **78**, 547–551.

Shaw, H.R. (1963) Obsidian–H_2O viscosities at 1000 and 2000 bars in temperature range 700° to 900°C. *J. Geophys. Res.*, **68**, No. 23, 6337–6343.

Shaw, H.R. (1969) Rheology of basalt in the melting range. *J. Petrol.* **10**, No. 3, 510–535.

Shaw, H.R. (1972) Viscosities of magmatic silicate liquids: An empirical method of prediction. *Amer. J. Sci.*, **272**, No. 11, 870–893.

Shaw, H.R., Wright, T.L., Peck, D.L., and Okamura, R. (1968) The viscosity of basaltic magma: An analysis of field measurements in Makaopuhi lava lake, *Hawaii Amer. J. Sci.*, **266**, 225–264.

Sharapov, V.N., and Golubem, V.S. (1976) *The Dynamics of Magma-Rock Interaction*, p. 231. Nauka, Novosibirsk (in Russian).

Sheeludyakov, L.N. (1980) *Composition, Structure and Viscosity of Homogeneous Silicate and Aluminosilicate Melts*, p. 157. Nauka, Alma-Ata (in Russian).

Shipulin, F.K. (1969) On the energy of intrusion processes. In: *Problems of Petrology and Genetic Mineralogy*, pp. 80–93. Nauka, Moscow (in Russian).

Stolper E. (1982) Water in silicate glasses: an infrared spectroscopic study. *Contrib. Miner. Petrol.*, **81**, 1–17.

Tuttle, O.F., and Bowen, N.L. (1958) Origin of granite in the light of experimental studies of the system $NaAlSi_3O_8$–$KAlSi_3O_8$ SiO_2–H_2O. *Geol. Soc. Amer. Mem.*, **74**, 153.

Urbain, G., Carron, J.P., and Calas, G. (1979) Estimation de la viscosité de certains aluminosilicates liquides d'intérêt géologique. In: *Hautes Températures et Sciences de la Terre.* Editions du C.N.R.S., 73–87.

Volarovich, M.P. (1940) A study of viscosity of molten rocks. *Zap. Vseros. Mineral. Obchestva*, **69**, No. 2/3, 310–313 (in Russian).

Volarovich, M.P., and Korchemkin, L.I. (1937) Relationship between the viscosity of molten rocks and acidity after Levinson-Lessing. *Dokl. Akad. Nauk SSSR*, **17**, No. 8, 413–418 (in Russian).

Volarovich, M.P., Tolstoy, M.D., and Korchemkin, L.I. (1936) A study of viscosities of Alaghez molten lavas. *Dokl. Akad. Nauk SSSR*, **1**, No. 8, 321–324 (in Russian).

Volokhov, I.M. (1979) *Magmas, Intratelluric Solutions and Magmatic Formations*, p. 166. Nauka, Novosibirsk (in Russian).

Waff, H.S. (1975) Pressure-induced coordination changes in magmatic liquids. *Geophys. Res. Lett.*, **2**, No. 5, 193–196.

Yoder, H.S., Jr. (1976) Generation of basaltic magma. *Proc. Natl. Acad. Sci. USA*, p. 215.

Zharikov, V.A. (1967) Indicators of acidity-basicity of minerals. *Geolog. Rudnykh Mestorozhd.*, No. 5, 75–89 (in Russian).

Zharikov, V.A. (1969) Component regime in melts and magmatic replacement. In: *Problems of Petrology and Genetic Mineralogy*, Vol. 1, pp. 62–79. Nauka, Moscow (in Russian).

Zharikov, V.A. (1982) Criteria of acidity of mineral formation processes, In: *Acid–Basic Properties of Chemical Elements, Minerals, Rocks and Natural Solutions*, pp. 63–91. Nauka, Moscow (in Russian).

Zharikov, V.A., Persikov, E.S., Epelbaum, M.B., and Chekmir, A.S. (1978) Magmatic melt properties study in connection with transport of component. In: *International Geodynamics Conferencs "Western pacific" and magma genesis.* Abstracts of papers, Tokyo.

Chapter 2
Relations Between Structure, Redox Equilibria of Iron, and Properties of Magmatic Liquids

Bjorn O. Mysen

Introduction

An understanding of how intensive and extensive properties control the properties of magmatic liquids (needed to describe and understand static and dynamic processes in magmatic systems) requires characterization of how the structure of silicate liquids affect their properties. To answer this need, recent experimental and theoretical petrological studies have been focused on structure and the relationships between structure and properties of silicate liquids as a function of bulk composition, temperature, pressure, and (for aliovalent cations) oxygen fugacity (e.g., Kushiro, 1975, 1980; Hess, 1980; Stebbins, 1989; Mo et al., 1982; Mysen and Virgo, 1985; Dingwell and Virgo, 1988; Mysen et al., 1982, 1983, 1984, 1985a,b,c; Bottinga et al., 1981, 1982; Murdoch et al., 1985; Kirkpatrick et al., 1986).

A common and useful basis for description of silicate melt structure is based on the concept that the cations may be subdivided into network formers and network modifiers (Bottinga and Weill, 1972). Network formers occupy the central location in interconnected tetrahedra linked with bridging oxygen atoms to form more complex anionic units (Fig. 1). Network-modifying cations occupy central positions of non-tetrahedral oxygen polyhedra. The oxygens at the corners of these polyhedra are shared with neighboring tetrahedra and with oxygen in other polyhedra. These latter oxygens are termed nonbridging oxygen (Fig. 1). The ratio of nonbridging oxygens to tetrahedrally coordinated cations (NBO/T) expresses the degree of polymerization of silicate melts. The value of NBO/T ranges from 0 for fully polymerized melts (tectosilicate melt) to 4 for melts with

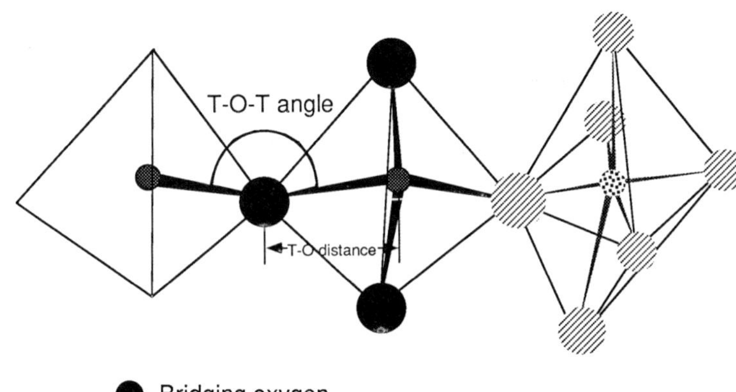

Fig. 1. Schematic representation of the geometric relationships of bridging and nonbridging oxygen and network-forming and network-modifying cations.

no bridging oxygen (orthosilicate melt)[1]. The values for most magmatic liquids are between 0 and 1 with an increase in NBO/T as the magma becomes more mafic (Fig. 2).

In order to calculate the value of NBO/T for a melt, it is necessary to determine which cations, and in what proportions, are network formers and network modifiers. This information is available from experimental studies of melt structure in a variety of chemical systems (see Mysen, 1988, for review). Iron is particularly interesting in this respect because Fe^{3+} may occur as a network-former and as a network-modifier depending on the $Fe^{3+}/\Sigma Fe$ (Virgo and Mysen, 1985), whereas Fe^{2+} generally is considered to be a network modifier (but see also Cooney et al. 1987; and Waychunas et al. 1988, for alternative views). Inasmuch as both ferric and ferrous iron are abundant in natural magmatic liquids, changes of the redox ratio of iron ($Fe^{3+}/\Sigma Fe$) can affect their degree of polymerization significantly. Any melt property that depends on NBO/T therefore, will also depend on the redox ratio of iron.

In addition to the relationships between $Fe^{3+}/\Sigma Fe$ and melt structure, the

[1] In this chapter, the degree of polymerization of species or units in the melts is described in terms of their individual degree of polymerization ($NBO/T = 0, 1, 2, 3$ and 4) and stoichiometric expressions. TO_2, T_2O_5, TO_3, T_2O_7, and TO_4 (T = tetrahedrally-coordinated cation,—most frequently Al and Si), are used. For simple silicates, Si replaces T. Another terminology, frequently used in the NMR literature, is Q^4, Q^3, Q^2, Q^1, and Q^0, respectively. The superscript in the Q-notations refers to the number of *bridging* oxygen in the species in question.

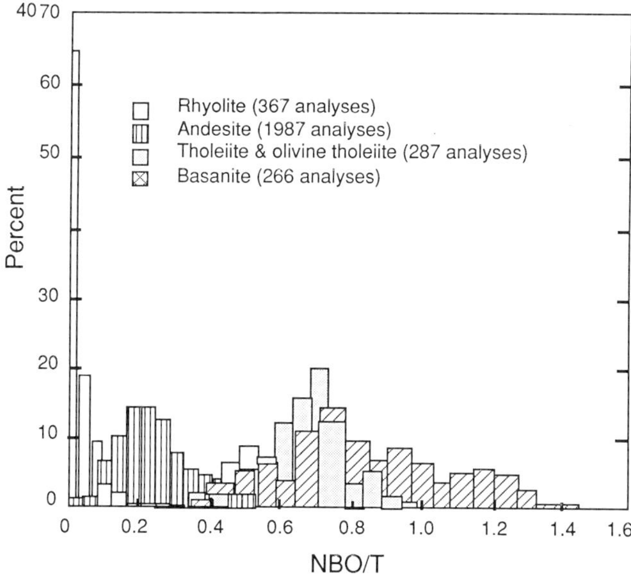

Fig. 2. *NBO/T* ranges of extrusive rocks (data from Chayes, 1975a,b, 1985)

redox ratio of iron is important in elucidating petrogenetic histories of magmatic rocks as $Fe^{3+}/\Sigma Fe$ as a function of temperature (e.g., Kennedy, 1948; Sack et al., 1980; Thornber et al., 1980), oxygen fugacity (e.g., Fudali, 1965), and pressure (Mysen and Virgo, 1978, 1983, 1985; Mo et al., 1982). For example, temperature–oxygen fugacity relationships of natural basalts have been deduced from laboratory-calibration of redox ratios of silicate melts (Carmichael and Ghiorso, 1986). In this chapter, the relationships between redox equilibria, melt structure and intensive variables will be discussed.

Principal Features of the Melt Structure

In order to appreciate better the role of ferric and ferrous iron in silicate melts and the interaction between Fe^{3+} and Fe^{2+} and the overall melt structure, a brief summary of the principal facets of the structure of silicate liquids is useful.

Although current understanding of structure is based on spectroscopic information, many of the principles commonly accepted were originally deduced from the properties of the liquids. The rheological behavior of binary alkaline earth and alkali silicate melts led Bockris et al. (1955, 1956) to suggest that the anionic structure of these liquids may be treated in terms of a small number of coexisting, relatively simple anionic complexes (see also Bockris and Reddy, 1970, for summary of this information and interpretations). On the basis of a theoreti-

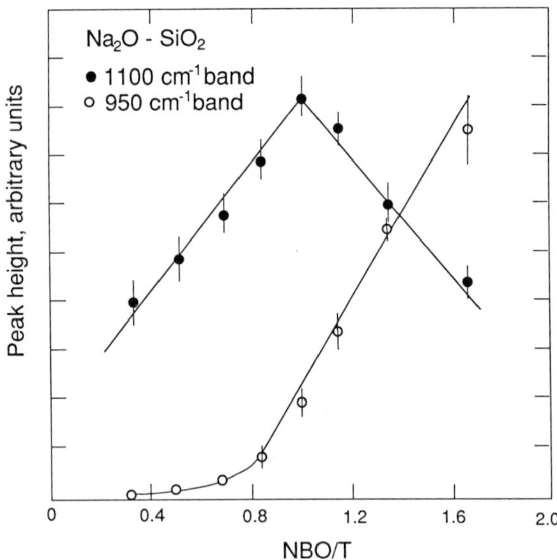

Fig. 3. Relationship between peak intensity and bulk composition from Raman spectra of quenched melts on the join Na_2O–SiO_2. The 1100-cm^{-1} band is an Si–O$^-$ stretch band typically diagnostic of the existence of $Si_2O_5{}^{2-}$ units in the melt, and the 950-cm^{-1} band is from Si-O$^-$ stretching diagnostic of $SiO_3{}^{2-}$ structural units (Brawer and White, 1975, 1977; Furukawa et al., 1981; Mysen et al., 1982; McMillan and Piriou, 1983).

cal discussion of Raman spectra of molecular compounds (Brawer, 1975), Brawer and White (1975, 1977) reported Raman spectra and suggested that liquids formed by melting of simple binary compounds retained many of the essential structural features of the crystalline analogs. For example, sodium and potassium disilicate melts contain a large proportion of disilicate ($Si_2O_5{}^{2-}$) structural units. To maintain mass balance, smaller proportions of units that are less polymerized, and of units that are more polymerized than disilicate, are also present.

An example of the results from the early Raman spectroscopic studies is shown in Fig. 3. The 1100- and 950-cm^{-1} bands (Fig. 3) result from the existence of structural units with, on the average, 1 and 2 nonbridging oxygens per silicon, respectively. This treatment was expanded in a summary (Mysen et al., 1982; Mysen, 1988) of Raman spectroscopic data on several alkali and alkaline earth silicate melts (Fig. 4). Recent unit abundance data obtained by ^{29}Si NMR (Schneider et al., 1987; Stebbins, 1987; marked SS87 and S87, respectively) are shown in Fig. 4. The NMR data corroborate the original Raman data. The mol fractions of structural units obtained by both NMR and Raman spectroscopy for a given bulk composition are in accord within analytical uncertainty. These and other results (e.g., Virgo et al., 1980; Furukawa et al., 1981; Matson et al., 1983; Murdoch et al., 1985; Kirkpatrick et al., 1986; Mysen et al., 1985c; Brandriss

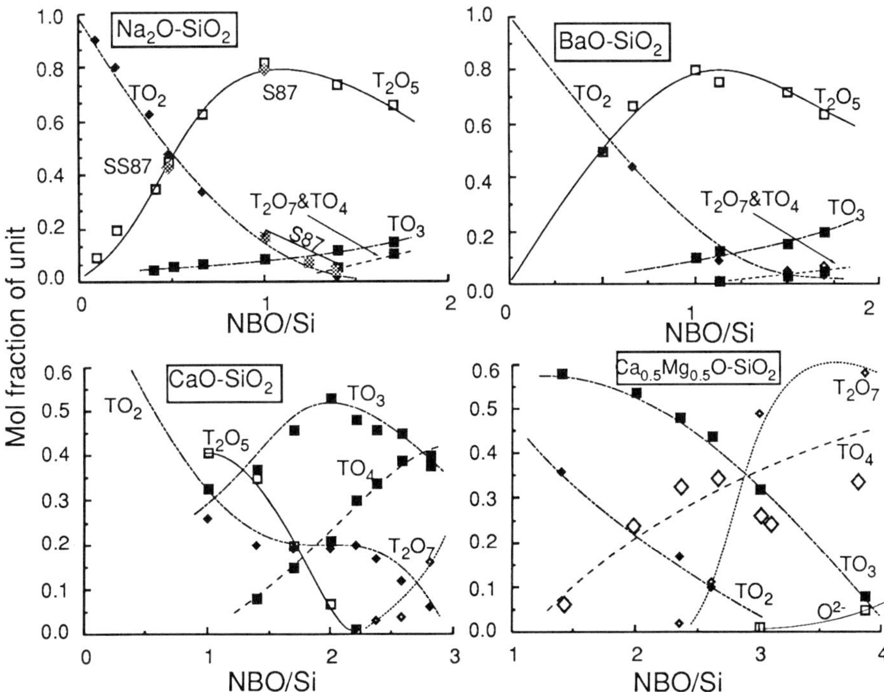

Fig. 4. Relative abundance of structural units in quenched melts on the joins $CaO-SiO_2$, $BaO-SiO_2$, $(Ca_{0.5}Mg_{0.5})O-SiO_2$, and Na_2O-SiO_2 (data from Mysen et al., 1982).

and Stebbins, 1988; Stebbins, 1989) led to two important conclusions in regard to the anionic structure of silicate melts. (1) The melts consist of a relatively small number of anionic units with comparatively simple stoichiometries (Table 1). (2) These stoichiometries do not vary as continuous functions of bulk melt NBO/Si (or metal/silicon). Rather, the proportions of the units vary (Fig. 4) as a function of the composition of the melt.

Aluminum is in tetrahedral coordination in silicate melts provided mono- or divalent cations (e.g., alkalies or alkaline earths) for electrical charge-balance are available (Taylor and Brown, 1979a,b; Seifert et al., 1982; Navrotsky et al., 1982, 1985; McMillan et al., 1982; McKeown et al., 1984; Mysen et al., 1980a,b, 1981a, 1982, 1985c). However, significant differences exist between Al^{3+} charge-balanced with alkaline earths and Al^{3+} charge-balanced with alkali metals. In the former, there is significant (Al, Si) ordering, with Al^{3+} forming primarily $Al_2Si_2O_8^{2-}$ units (Seifert et al., 1982). In systems where the bulk Al/(Al + Si) differs from 0.5, combinations of Al-free silica and pure aluminate units will also be present (Seifert et al., 1982; see also Fig. 5). The proportions of alkaline earth charge-balanced fully polymerized aluminosilicate structural units vary as systematic functions of bulk melt Al/(Al + Si) (Fig. 5).

Table 1. Compositional ranges (expressed as bulk melt *NBO/Si*[a]) of coexisting anionic units in binary[b] metal oxide silicate melts.

NBO/Si Range	Anionic Units
	Na_2O-SiO_2
> 0–0.05	SiO_2, $Si_2O_5{}^{2-}$
0.05–1.0	SiO_2, $Si_2O_5{}^{2-}$, $SiO_3{}^{2-}$
1.0–1.4	$Si_2O_5{}^{2-}$, $SiO_3{}^{2-}$, $SiO_4{}^{4-}$
1.4–2.0	$Si_2O_5{}^{2-}$, $SiO_3{}^{2-}$, $Si_2O_7{}^{6-}$, $SiO_4{}^{4-}$
> 2.0	Not available
	$BaO-SiO_2$
> 0–0.2	SiO_2, $Si_2O_5{}^{2-}$
0.2–1.0	SiO_2, $Si_2O_5{}^{2-}$, $SiO_3{}^{2-}$
1.0–1.1	SiO_2, $Si_2O_5{}^{2-}$, $SiO_3{}^{2-}$, $SiO_4{}^{4-}$
1.1–2.4	$Si_2O_5{}^{2-}$, $SiO_3{}^{2-}$, $Si_2O_7{}^{6-}$, $SiO_4{}^{4-}$
> 2.4	$SiO_3{}^{2-}$, $Si_2O_7{}^{6-}$, $SiO_4{}^{4-}$
	$CaO-SiO_2$
> 0–0.3	SiO_2, $Si_2O_5{}^{2-}$
0.3–1.0	SiO_2, $Si_2O_5{}^{2-}$, $SiO_3{}^{2-}$
1.0–2.2	$Si_2O_5{}^{2-}$, $SiO_3{}^{2-}$, $SiO_4{}^{4-}$
2.2–3.0	SiO_3, $Si_2O_7{}^{6-}$, $SiO_4{}^{4-}$
> 3.0	$SiO_3{}^{2-}$, $Si_2O_7{}^{6-}$, O^{2-}[c]
	$(Ca_{0.5}Mg_{0.5})O-SiO_2$
> 0–1.2	SiO_2, $Si_2O_5{}^{2-}$, $SiO_3{}^{2-}$
1.2–2.0	SiO_2, $Si_2O_5{}^{2-}$, $SiO_3{}^{2-}$, $SiO_4{}^{4-}$
2.0–2.4	$Si_2O_5{}^{2-}$, $SiO_3{}^{2-}$, $SiO_4{}^{4-}$
2.4–3.0	$SiO_3{}^{2-}$, $Si_2O_7{}^{6-}$, $SiO_4{}^{4-}$
> 3.0	$SiO_3{}^{2-}$, $Si_2O_7{}^{6-}$, $SiO_4{}^{4-}$, O^{2-}

[a] For M_2O-SiO_2 melts, the NBO/Si = M/Si, whereas for $MO-SiO_2$ melts, NBO/Si = 2.0M/Si.
[b] For ternary melts where the third component is aluminum, Al^{3+} occurs principally as aluminate component and Fe^{3+} as ferrite component.
[c] O^{2-} denotes free oxygen.
Data summary by Mysen et al. (1982)

From the ^{29}NMR spectra of such melts the bands broaden systematically (Fig. 6) with increasing Z/r^2 of the charge-balancing cation (Oestrike et al., 1987). This broadening may reflect the changes in shielding of Si in the various structural units depending on the local (Si, Al)-environment. With Al^{3+} charge-balanced with alkali metals there is random substitution of Al^{3+} for Si^{4+}. The Al/(Al + Si) of alkali charge-balanced fully polymerized units is positively correlated with the

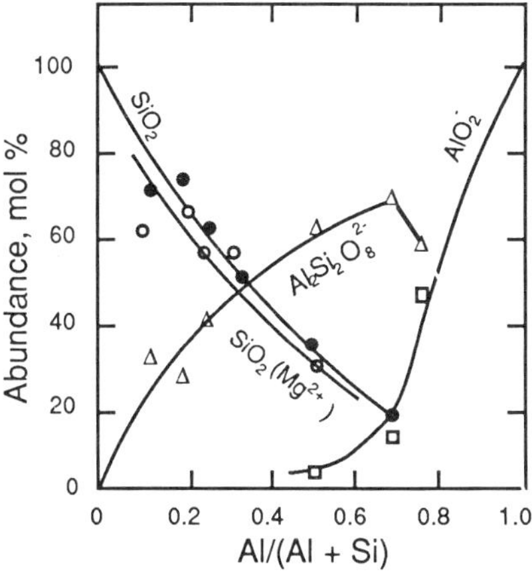

Fig. 5. Abundance of fully polymerized structural units in melts in the joins $CaAl_2O_4$–SiO_2 and $MgAl_2O_4$–SiO_2 (data from Seifert et al., 1982).

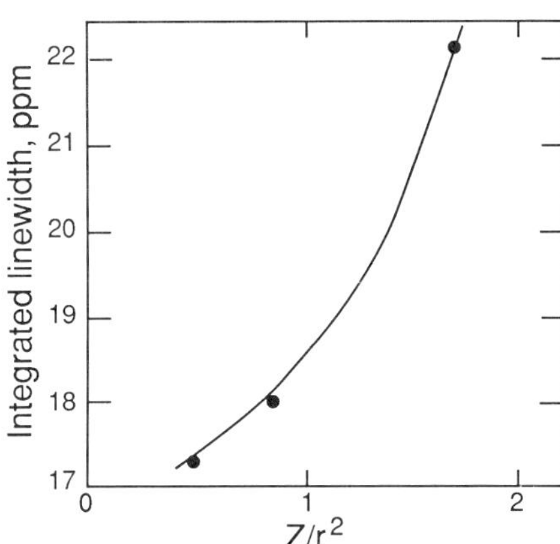

Fig. 6. ^{29Si}NMR line widths from spectra of glasses on the joins $CaAl_2O_4$–SiO_2, $NaAlO_2$–SiO_2, and $MgAl_2O_4$–SiO_2 (data from Oestrike et al., 1987).

Al/(Al + Si) of the melt itself [bulk melt Al/(Al + Si)]. As a result, the (Al, Si)–O bond distance increases and the intertetrahedral angle [(Si, Al)–O–(Si, Al)] decreases as a systematic function of the proportion of Al^{3+} substituted for Si^{4+} (Fig. 7). These structural changes result in a weakening of the bridging (Si, Al)–O bonds. Any melt property of a fully polymerized alkali aluminosilicate melt that depends on the bridging oxygen bond strength (e.g., diffusion, conductivity, and viscosity) will, thus be affected.

In alkaline earth aluminosilicate melts, the properties of the individual structural units are not affected by the Al/(Al + Si) of the melt. However, the proportions of these units vary with bulk melt Al/(Al + Si). Properties of the melt, being functions of the proportion of each unit will also be affected. These differences in the structural behavior of Al^{3+}, depending on the electronic properties of the charge-balancing cation(s), are also reflected in different thermochemical (Navrotsky et al., 1982) and rheological (Kushiro, 1980, 1981; Seifert et al., 1982) behavior of aluminosilicate melts.

In melts with bulk melt $NBO/T > 0$, Al^{3+} is distributed between the coexisting structural units and exhibits a pronounced preference for those units with the smallest NBO/T (Mysen, 1990a; Mysen et al., 1981a, 1985c; Domine and Piriou, 1986; Oestrike et al., 1987). As a result, in a melt with constant bulk melt NBO/T (where T includes both Al^{3+} and Si^{4+}), increasing Al/(Al + Si) at the same bulk melt NBO/T results in increased abundance of a highly polymerized unit such as TO_2 (analogous to SiO_2). As a consequence of this increase in abundance, increasing bulk melt Al/(Al + Si) is also associated with an increase in abundance of highly depolymerized units to maintain mass balance (Fig. 8).

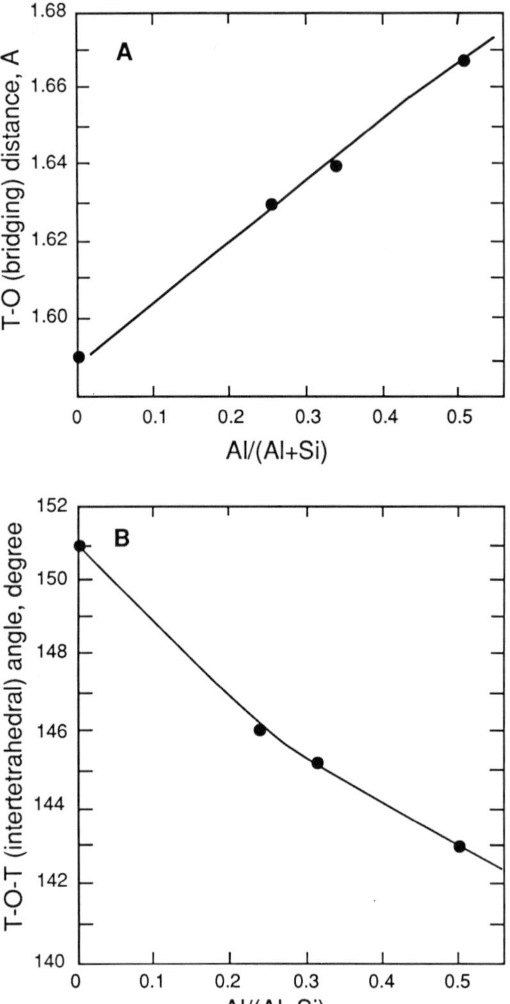

Fig. 7. Bond angle and bond length in alkali aluminate-silica (data from Konnert and Karle, 1973; Taylor and Brown, 1979a, b).

This distribution of Al^{3+} is both pressure- and temperature-dependent (Mysen, 1990a, b; Mysen et al., 1985c).

To provide a firm basis for establishment of the relationship between redox equilibria and melt composition (and, therefore, melt structure as summarized above), the relationships between the redox equilibria and each possible structural factor must be identified and quantified. With this goal in mind, the structural roles of Fe^{2+} and Fe^{3+} and the ferric/ferrous ratio will be evaluated in the succeeding sections, first as a function of individual compositional (extensive) variables and then as a function of intensive variables (temperature, oxygen fugacity, and pressure).

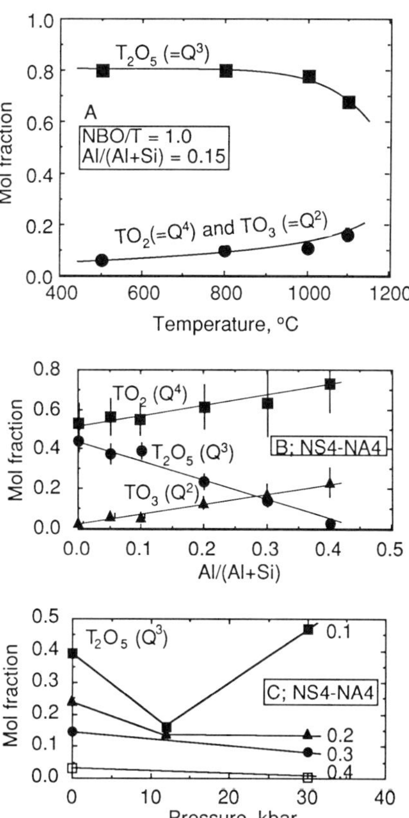

Fig. 8. Relative abundance of structural units in melts on the composition joins $Na_2Si_4O_9$–$Na_6Al_4O_9$ and $Na_2Si_2O_5$–$Na_4Al_2O_5$ as a function of Al/(Al + Si), temperature, and pressure. The symbol T denotes $Al^{3+} + Si^{4+}$. Note in particular the decreasing abundance of T_2O_5 and the increasing abundance of TO_2 with increasing Al/(Al + Si). The top panel shows unit distribution from *in-situ*, high-temperature Raman spectroscopy of a melt composition $Na_2Si_2O_5$ (85 mol%)–$Na_4Al_2O_5$ (15 mole%). The two lower panels are from Raman spectra of quenched melts on the join $Na_2Si_4O_9$ (NS4)–$Na_6Al_4O_9$ (NA4) (data from Mysen, 1989a; Seifert et al., 1981 and Mysen et al., 1985b).

Structural Positions of Ferric and Ferrous Iron in Silicate Melts

The structure of silicate melts in general and the structural positions of ferric and ferrous iron in particular have been studied by various spectroscopic methods. The redox ratio of iron can also be obtained with the aid of some of the same methods, although wet-chemical methods are also commonly employed for this purpose.

The local geometric environments in silicate melts have been studied by Raman, NMR, and X-ray absorption K-edge spectroscopy (Brawer and White, 1975, 1977; Furukawa et al., 1981; Mysen et al., 1980a–c, 1982, 1985a–d; McMillan, 1984; McMillan and Piriou, 1983; Matson et al., 1983; Calas and Petiau, 1983; Brown et al., 1978; Kirkpatrick et al., 1982, 1986; Stebbins, 1987, 1989; Domine and Piriou, 1986; Waychunas et al., 1988). For iron-bearing materials, luminescence, absorption, EPR, and Mössbauer spectroscopy have also been employed (Mao et al., 1973; Nolet et al., 1979; Fox et al., 1982; Calas and Petiau, 1983; Dyar and Burns, 1983; Dyar, 1985; Mysen et al., 1980a, 1984, 1985a,b; Mysen and Virgo, 1989). Examples of such spectra are shown in Fig. 9.

Luminescence spectra of simple metal oxide–silica–ferric iron-bearing glasses (Fox et al., 1982; see also Fig. 9) reveal two peaks (near 14,600 and 16,200 cm^{-1}). These two peaks have been ascribed to the existence of two distinct ferric iron–oxygen tetrahedra, with all or a significant portion of the tetrahedral position occupied by Fe^{3+} (Fox et al., 1982). The corresponding Raman spectra of such materials reveal two stretch bands associated with Fe^{3+}–oxygen bonds near 900 and 990 cm^{-1}, respectively (Mysen et al., 1980a, 1984; Fox et al., 1982). Thus, luminescence and Raman spectroscopy appear able to identify the existence of two Fe^{3+}–O polyhedra at least in oxidized alkali silicate melts. The proportion of the two types of ferric–oxygen tetrahedra is a systematic function of the metal/silicon ratio of the melt (see also Fox et al., 1982; Virgo et al., 1983, for detailed discussion). In the K-edge absorption spectra, the presence of Fe^{3+} (IV)–O bonds in iron-bearing glasses results in a distinct energy change compared with octahedrally coordinated iron (Waychunas et al., 1983; Calas and Petiau, 1983), although the method seems insufficiently sensitive to discern the existence of two types of tetrahedra. Similarly, the EPR spectrum of ferric iron-bearing glasses gives a distinct signal for tetrahedrally coordinated Fe^{3+} (Goldman, 1983; Calas and Petiau, 1983; see also Fig. 9).

Mössbauer spectroscopy however, has been the method of choice for determining the structural roles of ferric and ferrous iron in silicate melts and glasses (e.g., Dyar, 1985; Levy et al., 1976; Mysen and Virgo, 1978; Dyar and Burns, 1981; Seifert et al., 1979; Mysen et al., 1980a, 1984, 1985a; see Fig. 9b, for example). In ^{57}Fe Mössbauer spectroscopy each iron–oxygen polyhedron gives rise to a distinct quadrupole-split doublet.[2] The values (in mm/s) of these quadrupole splits and the isomer shifts depend on the oxidation state of iron, the number of oxygen ligands around ferric or ferrous iron, and the distortion of the oxygen polyhedron around the central iron cation (e.g., Waychunas and Rossman, 1983; Mao et al., 1973; Seifert et al., 1979; Calas and Petiau, 1983; Annersten and Halenius, 1976; Annersten and Olesch, 1978). The isomer shift generally increases as the number of oxygen ligands around Fe^{2+} and Fe^{3+} increases.

The quadrupole split values tend to increase with increasing distortion of the iron–oxygen polyhedron. In a spectrum such as that shown in Fig. 9b, the innermost quadrupole split doublet (shaded grey) results from tetrahedrally coordinated ferric iron, whereas the outermost split doublet (black) is due to octahedrally coordinated ferrous iron. The Mössbauer spectra are generally insufficiently sensitive to resolve the two distinct ferric iron–oxygen tetrahedra

[2] The quadrupole split (referred to as *QS* with appropriate subscript referring to either Fe^{3+} or Fe^{2+}) is the velocity difference between the two absorption bands in the doublet (see Fig. 9). The isomer shift (denoted *IS* with the appropriate subscript) is the position (velocity) of the center of gravity of the doublet relative to a reference such as metallic iron. Together, the quadrupole splits and the isomer shifts are referred to as hyperfine parameters.

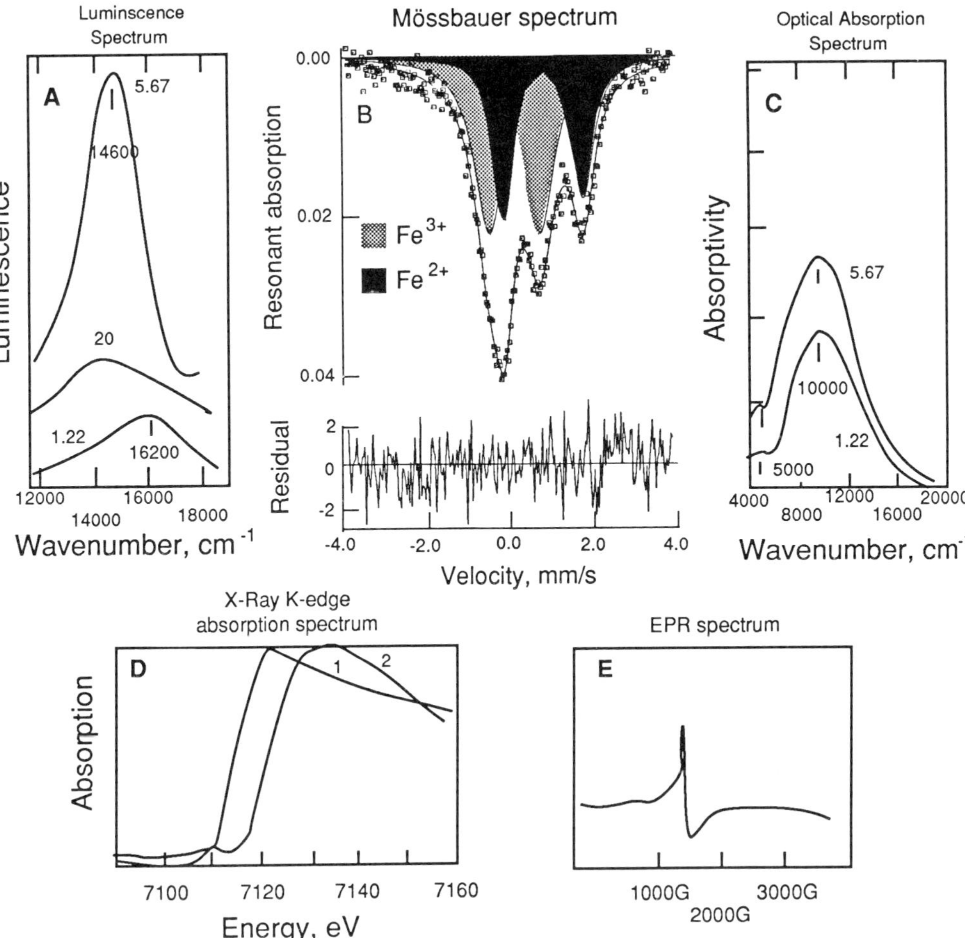

Fig. 9. Examples of spectra of iron-bearing glasses. Luminescence and absorption spectra, simplified after Fox et al. (1982), of glasses in the system Na_2O–SiO_2. Sample $SiO_2/Na_2O = 5.67$, $\Sigma Fe_2O_3 = 0.49$ wt%, $Fe^{3+}/\Sigma Fe = 0.75$; $SiO_2/Na_2O = 2.0$, $\Sigma Fe_2O_3 = 0.53$ wt%, $Fe^{3+}/\Sigma Fe = 0.73$; and $SiO_2/Na_2O = 1.22$, $\Sigma Fe_2O_3 = 0.50$ wt%, $Fe^{3+}/\Sigma Fe = 0.78$. K-edge absorption and X-band EPR spectrum after Calas and Petiau (1983). In K-edge spectra, sample 1, $Na_2Si_2O_5$ glass with 6 wt% Fe_2O_3 and with essentially all iron as Fe^{3+}; sample 2, $CaMgSi_2O_6$ glass with 20% of (Ca + Mg) replaced by Fe^{2+}. The EPR spectrum is from $CaMgSi_2O_6$ glass with 1 wt% Fe_2O_3 added and equilibrated in air. The Mössbauer spectrum is from a quenched melt in the the system CaO–SiO_2–P_2O_5–Fe–O with $NBO/Si = 1.7$, with 16 wt% P_2O_5 and 5 wt% iron oxide added as Fe_2O_3 with $Fe^{3+}/\Sigma Fe = 0.63$ (unpublished data from Mysen, 1989).

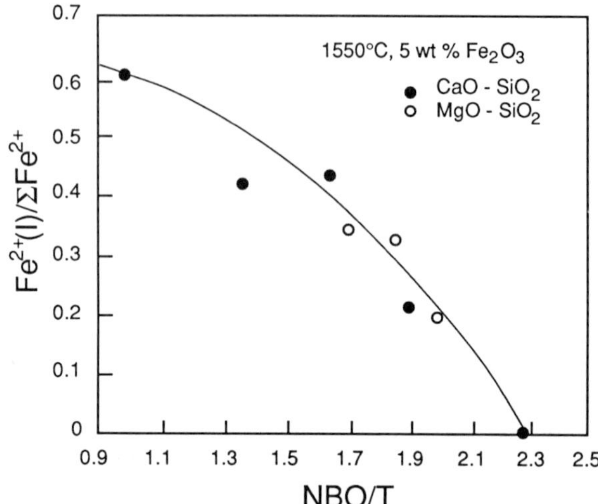

Fig. 10. Proportion of Fe^{2+}-oxygen polyhedra in glasses on the joins $CaO-SiO_2$ and $MgO-SiO_2$ with 5 wt% Fe_2O_3 added and equilibrated in air at 1550°C as a function of polymerization (*NBO/T*) of the glass. The Fe^{2+}(I) polyhedron is that with the highest value of $IS_{Fe_{2+}}$ and the lowest value of $QS_{Fe_{2+}}$ (data from Virgo et al., 1981; Mysen et al., 1984).

inferred from luminescence (Fox et al., 1982) and Raman spectroscopy (Fox et al., 1982; Virgo et al., 1983). The ferrous iron quadrupole-split doublet (Fig. 9b), however, is sometimes resolved into two doublets (Mysen et al., 1984), in particular for reduced samples (Mysen and Virgo, 1989). Optical absorption spectra of quenched melts in the system Na_2O-SiO_2-Fe-O (see Fig. 9c) have also been interpreted (Fox et al., 1982) to reflect two ferrous iron–oxygen polyhedra (absorption near 5000 and 10,000 cm^{-1}).

The relative proportion of two ferrous iron oxygen polyhedra (Fig. 10) appears sensitive to the degree of polymerization of silicate melts. With increasing *NBO/T*, the polyhedron with the highest isomer shift ($IS_{Fe^{2+}}$) and the lowest quadrupole-split ($QS_{Fe^{2+}}$) value [Fe^{3+}(I) in Fig. 10] becomes more dominant (Mysen et al., 1984, 1985a). As the iron becomes more reduced ($Fe^{3+}/\Sigma Fe$ decreases), the ferrous doublet with the largest isomer shift becomes more dominant, and the average value of the two doublets approaches that commonly associated with octahedrally-coordinated Fe^{2+}. It has been suggested (e.g., Dyar, 1985) that the doublet with the low value of isomer shift ($IS_{Fe^{2+}} = 0.85–0.90$ mm/s) may be the result of tetrahedrally coordinated ferrous iron. However, it is not clear whether these two fitted ferrous doublets have a structural interpretation or whether the effect merely reflects the limitations in the statistical fitting routine used to deconvolute the spectra (Mysen and Virgo, 1989). If the two ferrous iron doublets have a structural interpretation as suggested by Dyar (1985) and tetrahedral and octahedral Fe^{2+} coexist, octahedrally coordinated ferrous iron is prevalent in highly reduced samples, and is also increasingly important the more depolymerized is the melt.

On the basis of several hundred Mössbauer, absorption, luminescence, and Raman spectra of iron-bearing glasses and a comparison with Mössbauer spectra of ferrisilicate minerals with known ferric iron–oxygen polyhedra, it is generally concluded that with the isomer shift of ferric iron ($IS_{Fe^{3+}}$) less than about 0.3 mm/s (relative to metallic iron at 298 K), this ferric iron is in tetrahedral coordination. From spectra of crystalline material, $IS_{Fe^{3+}} > 0.4$ mm/s is diagnostic of octahedrally coordinated ferric iron (e.g., Hafner and Huckenholz, 1971; Annersten and Halenius, 1976; Amthauer et al., 1977; Waychunas and Rossman, 1983; Brown et al., 1978; Calas et al., 1980; Virgo and Mysen, 1985). From spectra of glassy materials, this latter value appears greater than about 0.5 mm/s. The quadrupole-split values are sensitive to polyhedral distortion (Seifert and Olesch, 1977; Calas and Petiau, 1983; Waychunas and Rossman, 1983). The higher the value, the greater the distortion of the tetrahedron.

The $QS_{Fe^{3+}(IV)}$ increases significantly in spectra of iron-bearing metal oxide—silica melts as the ionization potential (Z/r^2) of the metal cation increases (Fig. 11). Thus, it would seem that as the size decreases or the electric charge of the charge-balancing metal increases, the Fe^{3+}–O tetrahedra become increasingly distorted. For iron-bearing melts in the systems Na_2O–SiO_2, BaO–SiO_2, CaO–SiO_2, and MgO–SiO_2, the $IS_{Fe^{3+}}$ for tetrahedrally coordinated Fe^{3+} increases by about 10% as the ionization potential (Z/r^2) increases from that of Na^+ to that of Mg^{2+} with the same bulk melt NBO/T and amount of iron added

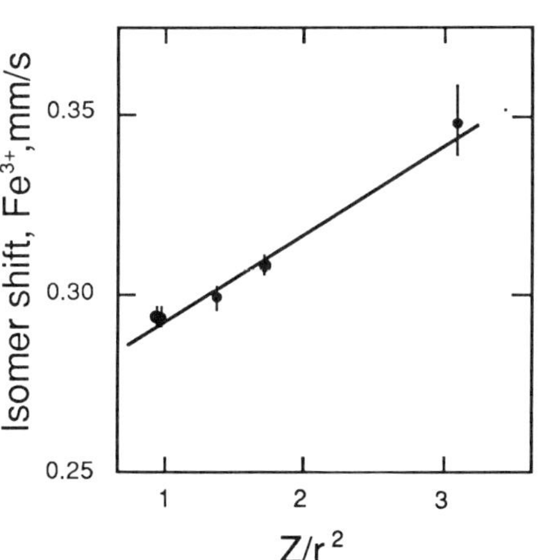

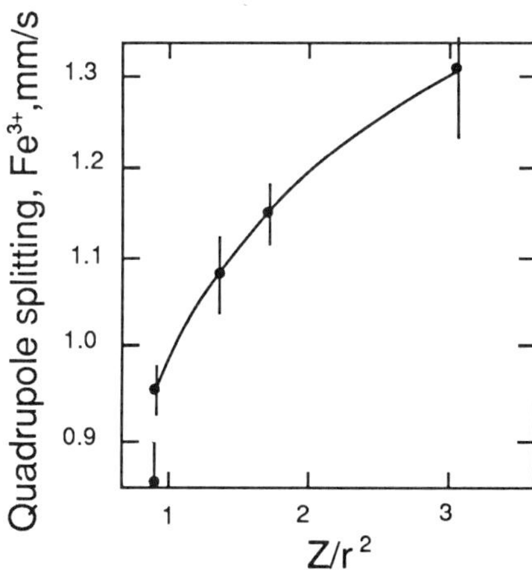

Fig. 11. Relationship between hyperfine parameters (IS and QS) for ferric iron in quenched binary metal oxide silicate melts as a function Of Z/r^2 of the metal cation with 5 wt% iron oxide added as Fe_2O_3 (data from Virgo et al., 1981; Mysen and Virgo, 1983; Mysen et al., 1984).

(Fig. 11). These inferred geometric changes of the Fe^{3+}–O tetrahedron probably reflect the increased difficulty with which the trivalent ferric iron can be charge-compensated in tetrahedral coordination as the electrical charge of the charge-balancing cation is increased or its ionic radius is reduced. As discussed in more detail below, increased $IS_{Fe^{3+}}$ and $QS_{Fe^{3+}}$ values are also associated with reduction of ferric to ferrous iron thus further attesting to the diminished stability of tetrahedrally coordinated Fe^{3+} in silicate melts with increasing Z/r^2 of the charge-balancing cation(s). It is notable that for crystalline materials tetrahedrally coordinated ferric iron is rare, but is found, for example, in some mica minerals (e.g., Annersten and Olesch, 1978).

The isomer shifts of ferric iron in silicate melts are systematic functions of $Fe^{3+}/\Sigma Fe$, controlled, for example, by the oxygen fugacity. On the basis of detailed Mössbauer studies of quenched melts in the systems Na_2O–Al_2O_3–SiO_2–Fe-O, Na_2O–SiO_2–Fe–O, CaO–Al_2O_3–SiO_2–Fe–O, and MgO–Al_2O_3–SiO_2–Fe–O (Mysen et al., 1984, 1985a; Dingwell and Virgo, 1987; Mysen and Virgo, 1989) it has been found that with up to about 10 wt% iron oxide added as Fe_2O_3, the $IS_{Fe^{3+}}$ is between 0.25 and 0.35 mm/s in the $Fe^{3+}/\Sigma Fe$ range 0.5–1.0 (Fig. 12). In the $Fe^{3+}/\Sigma Fe$ range 0.5–0.3, the $IS_{Fe^{3+}}$ increases systematically to above 0.50 mm/s and remains constant for $Fe^{3+}/\Sigma Fe$ values lower than about 0.3. This change in $IS_{Fe^{3+}}$ is also associated with a decrease in the intensities of Raman bands associated with Fe^{3+}(IV)–O bonds in the melts (Mysen et al., 1984;

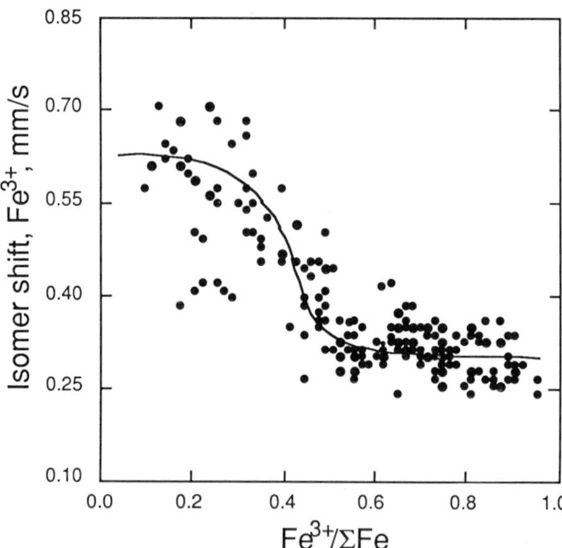

Fig. 12. Isomer shift for ferric iron (relative to Femetal) in silicate melts in the systems Na_2O–Al_2O_3–SiO_2, CaO–Al_2O_3–SiO_2, and MgO–Al_2O_3–SiO_2 as a function of $Fe^{3+}/\Sigma Fe$ with 5 and 10 wt% Fe_2O_3 added to the starting materials (data from Virgo and Mysen, 1985; Mysen et al., 1984, 1985a,c).

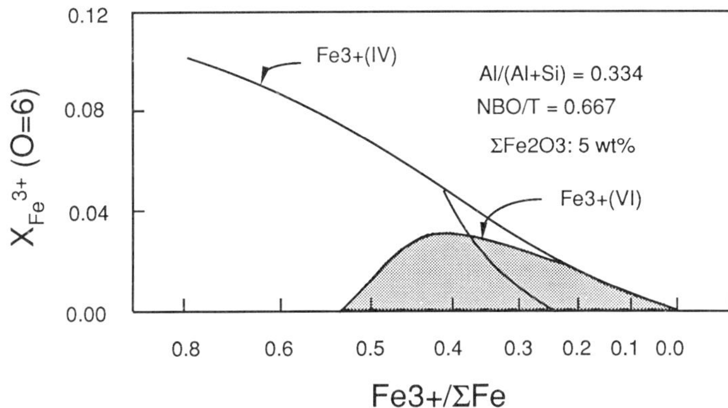

Fig. 13. Proportions of four- and six-coordinated ferric iron in Na_2O–Al_2O_3–SiO_2 melt with iron-free $NBO/T = 0.67$ and $Al/(Al + Si) = 0.33$ at 1550°C (from Virgo and Mysen, 1985).

Virgo and Mysen, 1985). These spectroscopic observations indicate that for oxidized, iron-bearing silicate melts ($Fe^{3+}/\Sigma Fe > 0.5$), ferric iron is in tetrahedral coordination, whereas for reduced ($Fe^{3+}/\Sigma Fe < 0.3$) melts, ferric iron is in octahedral coordination. In the intermediate $Fe^{3+}/\Sigma Fe$ range (0.5–0.3), the data are consistent with coexisting tetrahedrally and octahedrally coordinated ferric iron (Virgo and Mysen, 1985).

It has been suggested (Virgo and Mysen, 1985) on the basis of spectroscopic and magnetic data on iron-bearing silicate melts, that ferric iron occurs in clusters of Fe^{3+}–O tetrahedra. From the Raman spectroscopic data, these clusters either have invariant Fe/Si or there is no Si^{4+} present. From the systematic study of the variations in ${IS_{Fe}}^{3+}$ coupled with stoichiometric constraints in the system Na_2O–Al_2O_3–SiO_2–Fe–O, a model was suggested (Fig. 13) based on the existence of ${FeO_2}^-$ complexes, complexes (or units) stoichiometrically resembling Fe_3O_4 (with 50% ferric iron in tetrahedral and 50% ferric iron in octahedral coordination together with octahedrally coordinated Fe^{2+}), and Fe^{3+} as a network modifier (octahedral coordination). The proportions of these structural units vary as a function of $Fe^{3+}/\Sigma Fe$ (oxygen fugacity) as shown in Fig. 13 (Virgo and Mysen, 1985).

In the Mössbauer spectra the proportion of each quadrupole-split doublet relative to the total absorption envelope is a measure of the relative abundance of the particular iron-oxygen polyhedron giving rise to the doublet. For a simple spectrum with one ferric and one ferrous doublet (Fig. 9b) the proportions of each such doublet relative to the total absorption envelope are a measure of the relative abundance of ferric and ferrous iron. This method has long been known to yield accurate redox ratios of crystalline iron-bearing silicates (Bancroft et al., 1969). The accuracy with which this method can be employed to measure $Fe^{3+}/\Sigma Fe$ of quenched silicate glasses has been evaluated with duplicate analy-

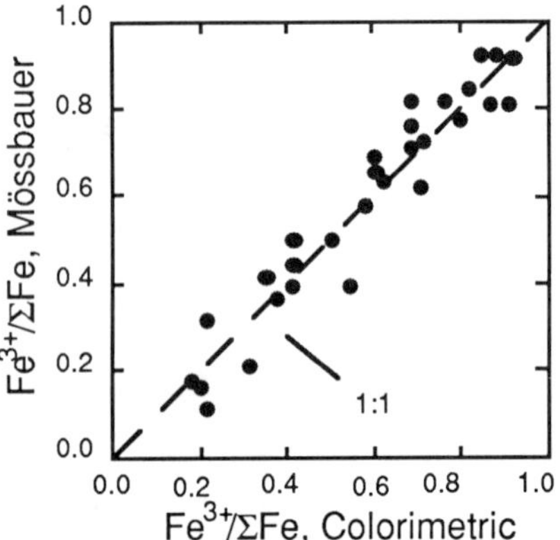

Fig. 14. Comparison of $Fe^{3+}/\Sigma Fe$ of iron-bearing glasses obtained by wet-chemical and ^{57}Fe Mössbauer resonant absorption methods. Iron oxide contents range from 2.2 to 34.7 wt% (calculated as Fe_2O_3). Bulk compositions range from binary metal oxide-silica systems (M = Na, Ca, and Mg) over ternary metal oxide-alumina-silica systems to laboratory-equilibrated natural rock compositions (data from Mysen et al., 1985d).

ses of numerous samples by both Mössbauer spectroscopy and wet-chemical methods (Fig. 14). As can be seen from the results (Mysen et al., 1985d; Fig. 14), the Mössbauer spectroscopic method is as accurate as wet-chemical methods. $Fe^{3+}/\Sigma Fe$ has also been determined from EPR and absorption spectra (Jones et al., 1981; Goldman, 1983). The results from the EPR spectra have not been compared with data obtained by other methods. The $Fe^{3+}/\Sigma Fe$ from near-infrared absorption spectroscopy accords with U.S. Geological Survey standard rock analyses to within 2%. Thus, $Fe^{3+}/\Sigma Fe$ obtained by wet-chemical methods (Wilson, 1960; Jones et al., 1981), Mössbauer spectroscopy (Mysen et al., 1985d), and optical absorption (Jones et al., 1981; Goldman, 1983) are in accord within analytical uncertainty.

Redox Equilibria and Extensive Variables

The structural information discussed above indicates that the types and proportions of iron-oxygen polyhedra are functions of bulk chemical composition and $Fe^{3+}/\Sigma Fe$ of the silicate liquids. Experimental studies with natural igneous rocks as starting materials (Kennedy, 1948; Fudali, 1965; Thornber et al., 1980; Sack et al., 1980; Kilinc et al., 1983) have revealed a systematic dependence of $Fe^{3+}/\Sigma Fe$ on silica, alumina, alkalies and alkaline earths, for example (Fig. 15). Thornber et al. (1980) found that in general there is a positive correlation between Fe^{3+}/Fe^{2+} and molar abundance of alkali metals, alkaline earths, alumina, and silica at 1200°C, although the dependence becomes less pronounced with increas-

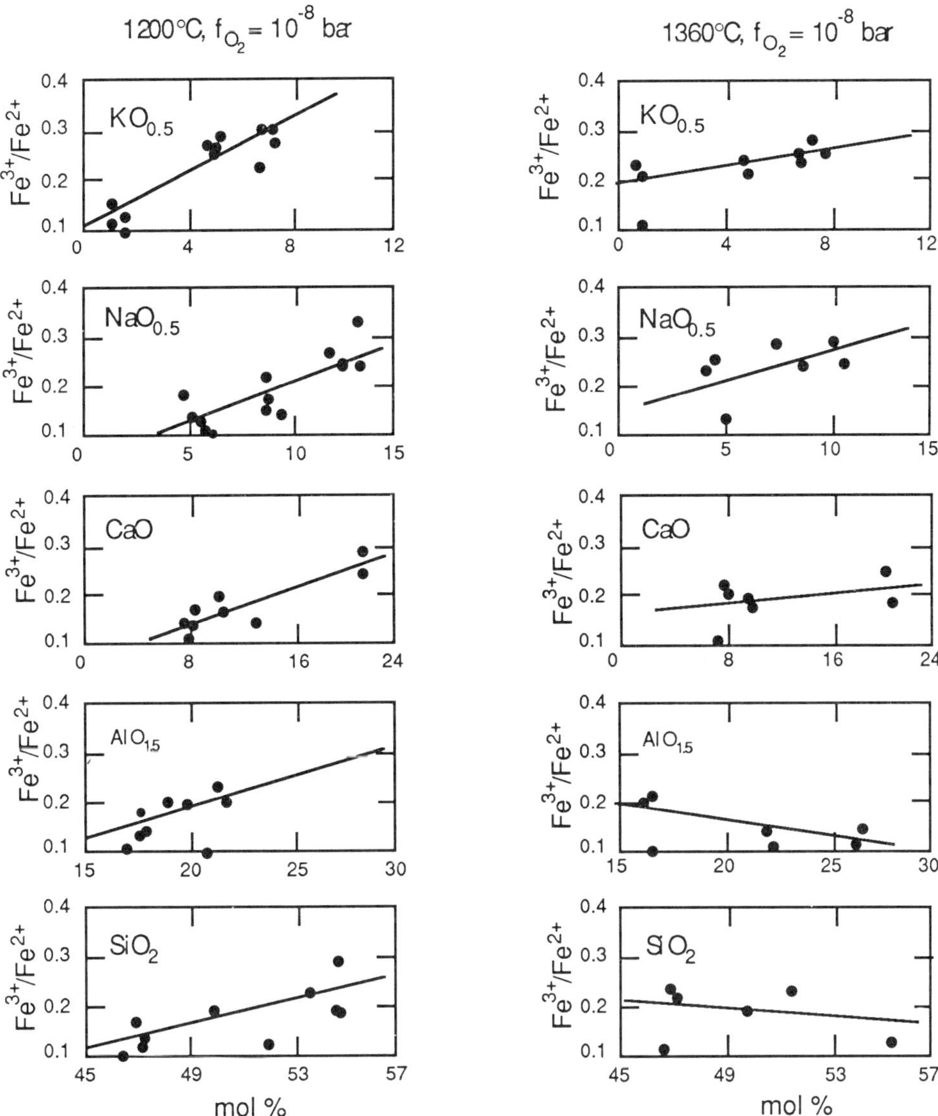

Fig. 15. Relationship between Fe^{2+}/Fe^{3+} and mole percent oxide component in natural rock compositions equilibrated at 1 bar pressure under temperature and oxygen fugacity conditions as indicated (data from Thornber et al., 1980).

ing temperature. In fact, the Fe^{3+}/Fe^{2+} decreases with increasing silica and alumina at 1360°C, whereas it increases at 1200°C (Fig. 15).

Sack et al. (1980) and Kilinc et al. (1983) used stepwise, multiple linear regression of ferric/ferrous to incorporate temperature, oxygen fugacity, and the abundance of various oxide components. These investigators employed an expression of the form;

$$\ln \frac{Fe_2O_3}{FeO} = a \ln f_{O_2} + \frac{b}{T} + c + \sum_{i=1}^{i} d_i X_i, \tag{1}$$

where a, b, c, and d_i are regression coefficients, T is absolute temperature, $\ln f_{O_2}$ is the natural logarithm of the oxygen fugacity, and X_i are concentrations of oxide components. By fitting 57 analyses from experimentally equilibrated liquids to this expression, Sack et al. (1980) found positive correlation of $\ln(Fe_2O_3/FeO)$ with Na_2O, K_2O, and CaO, whereas MgO, Al_2O_3, and FeO were negatively correlated. In a subsequent refinement of this treatment, Kilinc et al. (1983) concluded that the ferric/ferrous depended only on CaO, Na_2O, K_2O, FeO (all positively correlated), and Al_2O_3 remained negatively correlated (Table 2). Kilinc et al. (1983) concluded that Fe_2O_3/FeO was independent of MgO content of the liquid. Magnesium oxide was not identified as a variable in the experimental results reported by Thornber et al. (1980).

Significant bulk compositional differences were found between the samples used by Thornber et al. (1980), Sack et al. (1980), and Kilinc et al. (1983). Whereas Thornber et al. (1980) employed mostly basaltic liquids, with selective addition of specific oxides, Sack et al. (1980) and Kilinc et al. (1983) reported laboratory-calibrated $Fe^{3+}/\Sigma Fe$ with a wide range of bulk compositions from mafic to felsic. The discrepancies between these data sets (Fig. 15, Table 2) most probably arise from the fact that in neither treatment of the whole-rock analyses were the structural roles of the cations and the structural positions of ferric and ferrous

Table 2. Regression coefficients with standard errors ($\pm 1\sigma$) for equation (1).

Coefficient	Value	Standard error	Value	Standard error
a	0.218	0.007	0.219	0.004
b	13,184.7	959.0	12,670.0	900.0
b	−4.50	3.04	−7.54	0.55
$d_{Al_2O_3}$	−2.15	2.88	−2.24	1.03
d_{FeO}[a]	−4.50	3.69	1.55	1.03
d_{MgO}	−5.44	3.04	—	—
d_{CaO}	0.07	3.08	2.96	0.53
d_{Na_2O}	3.54	3.97	8.42	1.41
d_{K_2O}	4.19	4.12	9.59	1.45

[a] Total iron oxide FeO

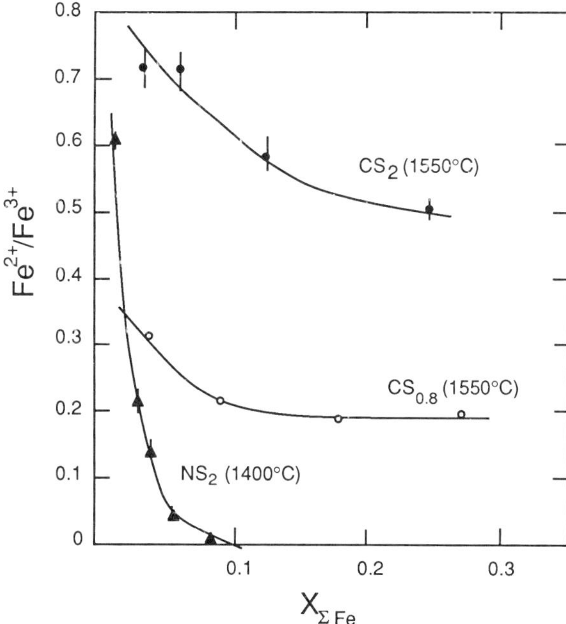

Fig. 16. Fe^{2+}/Fe^{3+} of quenched melts as a function of the amount of iron oxide (calculated as mole fraction on Fe^{3+}) for composition CS2 ($CaO-SiO_2$; NBO/Si = 1), CS0.8 ($CaO-SiO_2$; NBO/Si = 2.38), and NS2 (Na_2O-SiO_2; NBO/Si = 1.0) (data from Larson and Chipman, 1953; Virgo et al., 1981; Mysen et al., 1984, 1985a).

iron considered. The redox ratio of iron in silicate liquids may be related rigorously to the compositions after these compositional relations have been established in terms of their relationships to the structure of the melt.

The redox data for iron from natural melt compositions (Sack et al., 1980; Thornber et al., 1980; Kilinc et al., 1983) indicate that Fe^{3+}/Fe^{2+} depends on total iron content. Data from the simple systems $CaO-SiO_2-Fe-O$ and Na_2O-SiO_2-Fe-O (Fig. 16) show that $Fe^{3+}/\Sigma Fe$ increases rapidly and nonlinearly with increasing iron oxide content up to about 10 mole % Fe_2O_3. With higher iron contents, this dependence is less pronounced. Despite a suggestion to the contrary (Goldman, 1983), this observation indicates that dissolved iron oxide in simple silicate melts cannot be described with simple solution laws (e.g., Raoult's or Henry's law). Moreover, because the coordination state of Fe^{3+} (and possibly Fe^{2+}) depends on $Fe^{3+}/\Sigma Fe$, the coordination of iron probably also depends on the total iron content of the melt.

At constant oxygen fugacity, it has been suggested that the redox equilibrium is a simple function of the activity of nonbridging oxygens (O^-);

$$Fe^{3+} + O^- \Leftrightarrow Fe^{2+} + 1/2O_2, \tag{2}$$

with the equilibrium constant:

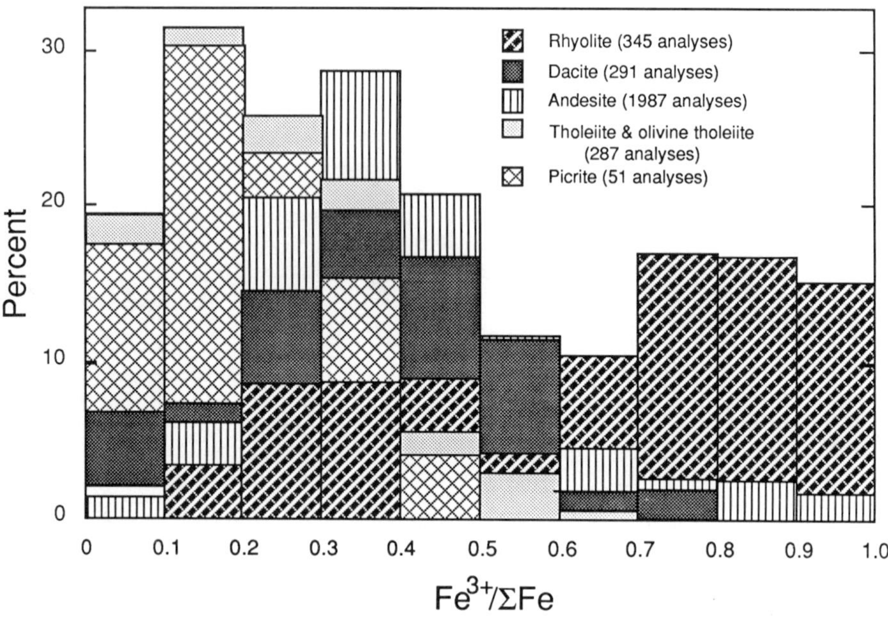

Fig. 17. $Fe^{3+}/\Sigma Fe$ of natural rocks (data from Chayes, 1975a,b, 1985).

$$K_2 = \frac{[Fe^{2+}]f_{O_2}{}^{1/2}}{[Fe^{3+}][O^-]}. \tag{3}$$

This simple relationship implies that the Fe^{3+}/Fe^{2+} decreases with increasing O^- activity (or increasing bulk melt NBO/T). This relationship [equation (2)] implies that at constant pressure, temperature, oxygen fugacity, and total iron content, silicate melts would become more reduced, the more depolymerized they are. From a survey of the $Fe^{3+}/\Sigma Fe$ of natural magmatic liquids (data from Chayes, 1975a,b, 1985) it appears that the more felsic the igneous rock, the more oxidized is the material (Fig. 17), in apparent discord with the prediction of equation (2). The redox ratio of igneous rocks, however, is not merely a function of degree of polymerization. Temperature, pressure, oxygen fugacity, volatile content and post-crystallization history also affect the $Fe^{3+}/\Sigma Fe$. In nature, these variables are not controlled.

Paul and Douglas (1965) observed that under controlled laboratory conditions there is a relationship opposite to that implied by equation (2). That is, at the same temperature, pressure, oxygen fugacity, and iron content, melts become more oxidized as they become more depolymerized (NBO/T increases). This latter conclusion subsequently was found to have general applicability to both alkali and alkaline earth silicate systems (Goldman, 1983; Virgo et al., 1981; Mysen et al., 1984, 1985a; Virgo and Mysen, 1985; Mysen and Virgo, 1989) and can also be observed in the early data on alkaline earth silicate melts of Larson

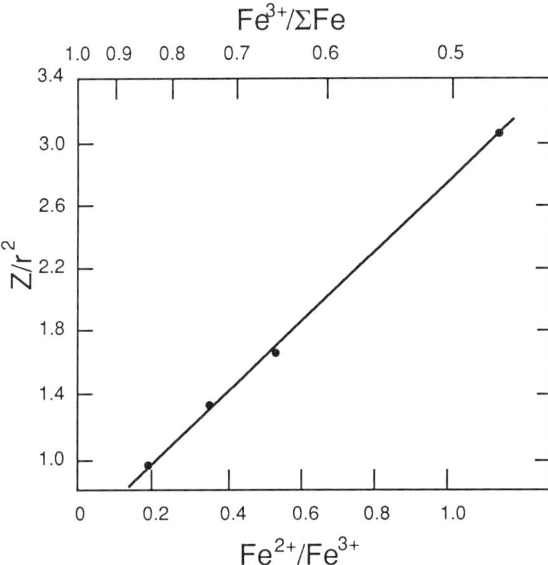

Fig. 18. Relationship between Fe^{2+}/Fe^{3+} and ionization potential (Z/r^2) of metal cation for binary metal oxide-silica melts ($NBO/T \sim 2.0$) with 5 wt% iron oxide added as Fe_2O_3 and equilibrated in air at 1550°C (data from Mysen et al., 1984; Virgo et al., 1981).

and Chipman (1953). It is also clear that at constant bulk melt NBO/T, temperature, and oxygen fugacity, the Fe^{2+}/Fe^{3+} is a systematic function of the type of metal cation in metal oxide silicate melts (Fig. 18). The smaller the cation or the more highly charged it is (increasing Z/r^2), the more reduced is the iron. This positive correlation between Fe^{2+}/Fe^{3+} and Z/r^2 of the network-modifying cation is also observed in ternary, aluminosilicate melts (Fig. 19), where, however, the available data seem to indicate that the more depolymerized the melt (higher NBO/T), the less pronounced is the dependence.

The negative correlation between Fe^{2+}/Fe^{3+} and bulk melt NBO/T also exists in aluminosilicate melts (Fig. 20). It can bc seen from the data in Fig. 20, however, that the Fe^{2+}/Fe^{3+} becomes less dependent on NBO/T, the lower the value of Z/r^2. [The slopes of the curves in Fig. 20 are less steep for Na_2O–Al_2O_3–SiO_2 (NAS) melts than for CaO–Al_2O_3–SiO_2 (CAS) melts and even more pronounced for MgO–Al_2O_3–SiO_2 (MAS) melts.].

It was noted above (Fig. 15) that in natural rock melts, a simple plot of mole % $AlO_{1.5}$ versus Fe^{3+}/Fe^{2+} showed a small positive correlation at 1200°C (Thornber et al., 1980), but all natural melt data taken together (Kilinc et al., 1983) indicated the opposite relationship (Table 2). In that treatment, the possible variations in bulk melt NBO/T and types of electrical charge-balance for Al^{3+} and Fe^{3+} in tetrahedral coordination were not considered. Instead, an empirically-based regression with the concentration of the oxide components, temperature

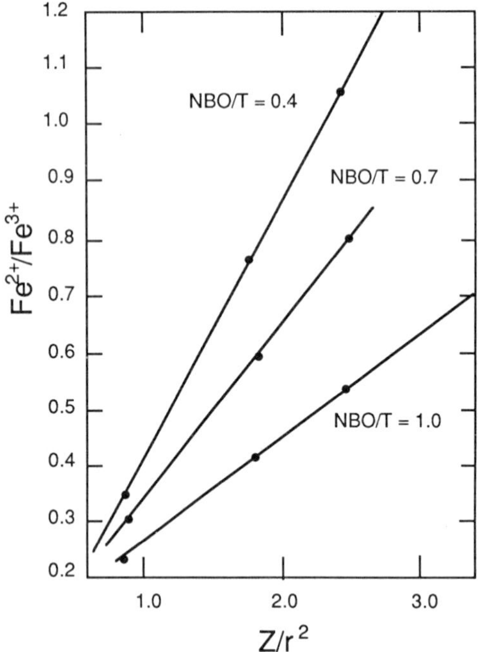

Fig. 19. Relationship between Fe^{2+}/Fe^{3+} and ionization potential of metal cation (Z/r^2) of melts in the systems Na_2O–Al_2O_3–SiO_2, CaO–Al_2O_3–SiO_2, and MgO–Al_2O_3–SiO_2 with 5 wt% iron oxide added as Fe_2O_3. All samples were equilibrated in air at 1550°C (data from Virgo and Mysen, 1985; Mysen et al., 1985a).

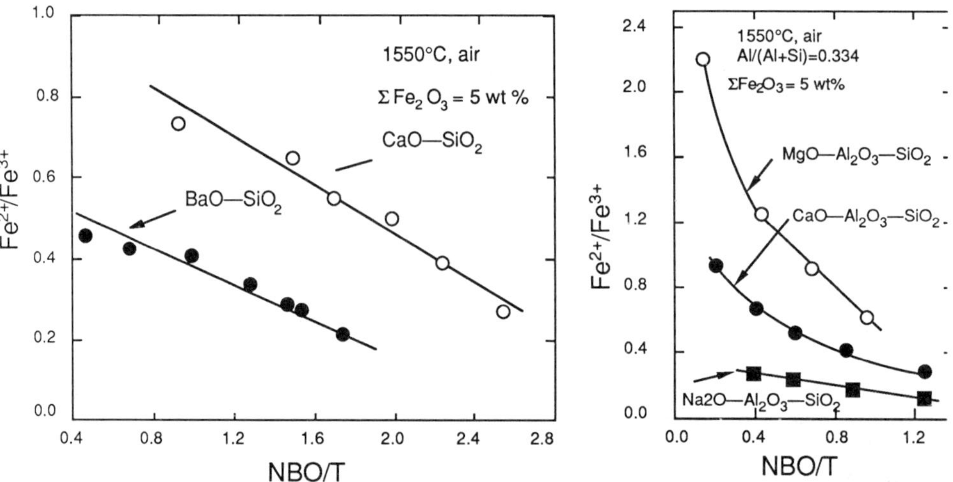

Fig. 20. Fe^{2+}/Fe^{3+} of binary metal oxide and ternary metal oxide-alumina-silica melts as a function of bulk melt *NBO/T* for melts equilibrated at 1 bar at 1550°C and oxygen fugacities indicated. In all samples 5 wt% iron oxide was added as Fe_2O_3 (data from Mysen et al., 1984, 1985a).

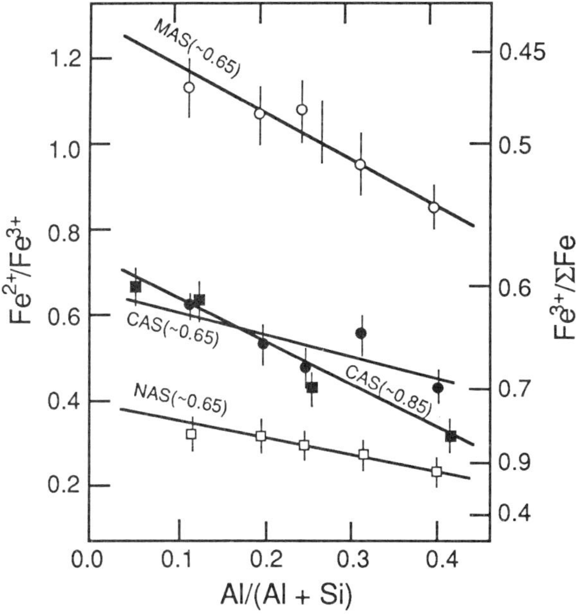

Fig. 21. Redox ratio of iron as a function of Al/(Al + Si) at 1550°C and $f_{O_2} = 10^{-0.68}$ (air). Numbers in parentheses indicate approximate NBO/T of the samples (this NBO/T is slightly dependent on $Fe^{3+}/\Sigma Fe$ of the samples). For samples with NBO/T variable, the Al/(Al + Si) of all compositions equals 0.334. Symbols NAS, CAS, and MAS denote systems Na_2O–Al_2O_3–SiO_2, CaO–Al_2O_3–SiO_2, and MgO–Al_2O_3–SiO_2, respectively (data from Virgo and Mysen, 1985; Mysen et al., 1985a).

and oxygen fugacity as independent variables [equation (1)] was used. Because of the complex roles of the oxides in controlling the melt structure, general conclusions and extrapolations from this regression analysis must be treated with caution. In simple ternary systems where only one network-modifying cation is present in addition to ferric and ferrous oxide with constant NBO/T, the $Fe^{3+}/\Sigma Fe$ increases with increasing Al/(Al+ Si) (Fig. 21). The relationship between Fe^{2+}/Fe^{3+} and Al/(Al + Si) in these simple ternary melt compositions is linear. At the same degree of polymerization (constant NBO/T), the slopes of the curves of Na^+ and Ca^+ aluminosilicate melts are similar, whereas in the analog magnesium system (denoted by MAS in Fig. 21), the slope of the curve is steeper. Furthermore, for a system with the same metal cation, the Fe^{2+}/Fe^{3+} is less dependent on Al/(Al + Si) the less polymerized (higher NBO/T) is the melt (Fig. 21). The implication of this type of analysis of redox relationships in natural magmatic liquids will be discussed further after treatment of relationships between redox equilibria and intensive variables below.

On the basis of the relationships between oxygen ion activity and redox ratio of iron, Holmquist (1966) proposed an expression somewhat different from that in equation (2) in order to account for the negative correlation between $Fe^{3+}/\Sigma Fe$ and degree of polymerization. In its simplest form, this expression is

$$4Fe(IV)O_2^- \Leftrightarrow 4Fe^{2+}(VI) + 6O^{2-} + O_2, \tag{4}$$

and

$$K_4 = \frac{[Fe^{2+}(VI)]^4}{[Fe(IV)O_{2^-}]^4}[O^{2-}]^6 f_{O_2}. \tag{5}$$

In this expression, the O^{2-} is related to the structure of the melt. The numbers in parentheses refer to the size of the oxygen coordination polyhedra. For silicate melts with anionic units having different *NBO/T* values, equation (4) may be expanded to relate Fe^{3+}/Fe^{2+} to the relative abundance of anionic units, an example of which is (Mysen et al., 1984)

$$4SiO_2 + Si_2O_5{}^{2-} + 4Fe(IV)O_{2^-} \Leftrightarrow 5SiO_3{}^{2-} + SiO_4{}^{4-} + 4Fe(VI)^{2+} + O_2, \tag{6}$$

with the equilibrium constant,

$$K_6 = \frac{[SiO_3{}^{2-}]^5[SiO_4{}^{4-}][Fe(VI)^{2+}]^4}{[SiO_2]^4[Si_2O_5{}^{2-}][Fe(IV)O_2{}^-]^4} f_{O_2}. \tag{7}$$

In equation (6), the SiO_2, $Si_2O_5{}^{2-}$ $SiO_3{}^{2-}$, and $SiO_4{}^{4-}$ are stoichiometric expressions (see also above and footnote *1*) of structural units with their individual *NBO/Si* equal to 0, 1, 2, and 4, respectively. The expressions $Fe(IV)_2{}^-$ and $Fe^{2+}(VI)$ represent Fe^{3+} in a tetrahedral oxygen complex and Fe^{2+} in octahedral coordination. The choice of anionic units in equation (6) is representative of that expected in CaO–SiO_2–Fe–O melts in the compositional range between di- and metasilicate (Mysen et al., 1985b).

Expression (6) will be shifted to the left with increasing activity of depolymerized structural units such as $SiO_4{}^{4-}$ and $SiO_3{}^{2-}$. By adding Al^{3+} to the system but keeping the bulk melt *NBO/T* constant, the net effect also is to shift this expression to the left because of the more rapid increase in concentration of the depolymerized units compared with that of the fully polymerized SiO_2 (or TO_2) unit (Dingwell, 1986; Mysen, 1990a; Mysen et al., 1985c). Thus, the increasing $Fe^{3+}/\Sigma Fe$ with increasing *NBO/T* and with increasing bulk melt *NBO/T* may be interpreted, qualitatively.

Redox Equilibria and Intensive Variables

It follows naturally from the expression;

$$\Delta G^\circ = -RT \ln K, \tag{8}$$

that the redox ratio of ferric and ferrous iron is a function of temperature and oxygen fugacity. Moreover, provided that significant volume change is associated with expressions such as those illustrated with equations (4) or (6), there will also be a pressure dependence of the redox ratio. Partial molar volume data on ferric and ferrous oxide (e.g., Lee and Gaskell, 1974; Shiraishi et al., 1978; Mo et al., 1982; Bottinga et al., 1983; Dingwell et al., 1988; see also direct experimental evidence below) in silicate melts indicate a significant volume change, suggesting that redox ratios of iron in silicate melts are pressure dependent.

The equilibrium constant for reaction (4) may be combined with equation (8) to yield the expression;

$$4\log\frac{Fe^{2+}}{Fe^{3+}} = \log K - [6\log a_O{}^{2-}(m) + \log f_{O_2}], \tag{9}$$

which can be reorganized to yield;

$$\log\frac{Fe^{2+}}{Fe^{3+}} = \frac{-0.0554 - \Delta G^\circ}{T} - \frac{1}{4}[6\log a_O{}^{2-}(m) + \log f_{O_2}], \tag{10}$$

where $a_O{}^{2-}$ (m) is the oxygen ion activity of the melt. It follows from this expression that a plot of log ($Fe2^{2+}/Fe^{3+}$) versus $1/T$ (absolute temperature) will result in a line where the slope expresses the standard free energy of reduction of ferric to ferrous iron.

It can be seen from the data summarized for simple binary metal oxide silicate and ternary metal oxide aluminosilicate melts (Fig. 22), that for melts equilibrated with air ($f_{O_2} = 10^{-0.68}$ bar), straight line relationships are generally obtained. Under the conditions of the experiments shown (Fig. 22) ferric iron is in tetrahedral coordination and ferrous iron is not. As also supported by Mössbauer and Raman spectroscopic data from the quenched melts (Virgo et al., 1983; Mysen et al., 1984, 1985a), this linear relationship indicates, therefore, that the solubility mechanisms of ferric and ferrous iron probably do not change as a function of temperature in the temperature range indicated. A change in solubility mechanism is likely to result in a temperature dependence of ΔG°. Such a dependence can be observed for iron in the system MgO–Al_2O_3–SiO_2 with 5 wt% iron oxide added as Fe_2O_3 (see also Mysen et al., 1985a).

Provided that the activity coefficient ratio of ferric/ferrous equals unity, and that the activity of oxygen ions does not change significantly as a result of the reduction of ferric to ferrous iron, the slopes of isothermal and isocompositional curves of log (Fe^{2+}/Fe^{3+}) versus log f_{O_2} curves should equal -0.25 [see equation (10)]. When this slope, x, differs from -0.25 with constant NBO/T, the activity coefficient ratio equals

$$\frac{\gamma Fe^{3+}}{\gamma Fe^{2+}} = 10^{(x+0.25)}. \tag{11}$$

The Fe^{2+}/Fe^{3+} also affects NBO/T. Consequently, the NBO/T is not, strictly speaking, independent of f_{O_2}. Variations in slope caused by this dependence result in about a 10% uncertainty in calculated $\gamma Fe^{3+}/\gamma Fe^{2+}$ values. Variations outside this uncertainty reflect real changes in the activity coefficient ratio.

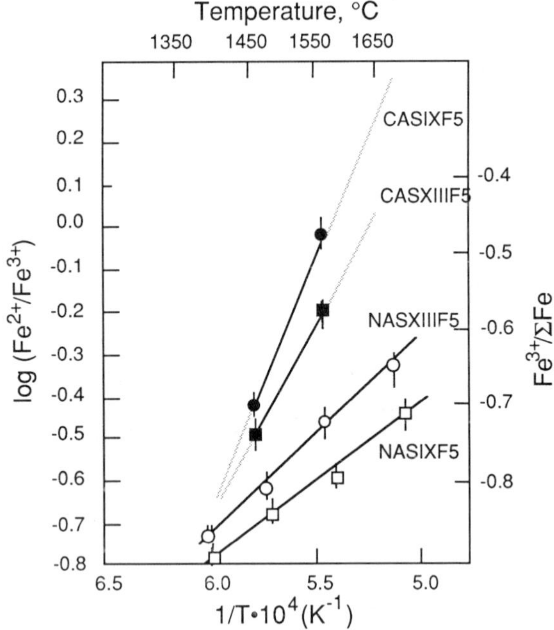

Fig. 22. Relationship between $1/T$ (absolute temperature) and log (Fe^{2+}/Fe^{3+}) for binary metal oxide silicate melts equilibrated with air and with 5 wt% iron oxide added as Fe_2O_3. Symbols: MAS, CAS, and NAS symbols as for Fig. 21. NASIX and CASIX; Al/(Al + Si) = 0.334 and $NBO/T = 0.15$, NASXIII and CASXIII, Al/(Al + SI) = 0.14 and $NBO/T \sim 0.65$. (data from Virgo et al., 1981; Mysen et al., 1985a,d; Virgo and Mysen, 1985).

Experimental data for iron-bearing melts in the systems Na_2O–Al_2O_3–SiO_2–Fe–O, CaO–Al_2O_3–SiO_2–Fe–O, and MgO–Al_2O_3–SiO_2–Fe–O as a function of oxygen fugacity exhibit straight line relationships between log (Fe^{2+}/Fe^{3+}) and log f_{O_2} (Fig. 23). It is clear, however, that the slopes of these lines are functions of bulk melt NBO/T at constant Al/(Al + Si) and also depend on Al/(Al + Si) at constant bulk melt NBO/T. The free energies of reduction and activity coefficient ratios of ferric to ferrous iron derived from the data in Figs. 22 and 23 are given in Table 3. With the same metal cation and constant Al/(Al + Si), the activity coefficient ratio, $\gamma Fe^{3+}/\gamma Fe^{3+}$, increases with decreasing degree of polymerization (increasing NBO/T, Table 3). Moreover, it appears that this ratio decreases with decreasing bulk melt Al/(Al + Si) and with increasing Z/r^2 of the metal cation (Table 3; see also Mysen et al., 1984, 1985a). It must be emphasized, though, that these are activity coefficient ratios. It cannot be determined from these data which of the two iron complexes is responsible for these activity coefficient changes.

The free energy of reduction of ferric to ferrous iron, ΔG°, also is significantly dependent on bulk composition. Its value decreases with increasing Al/(Al + Si)

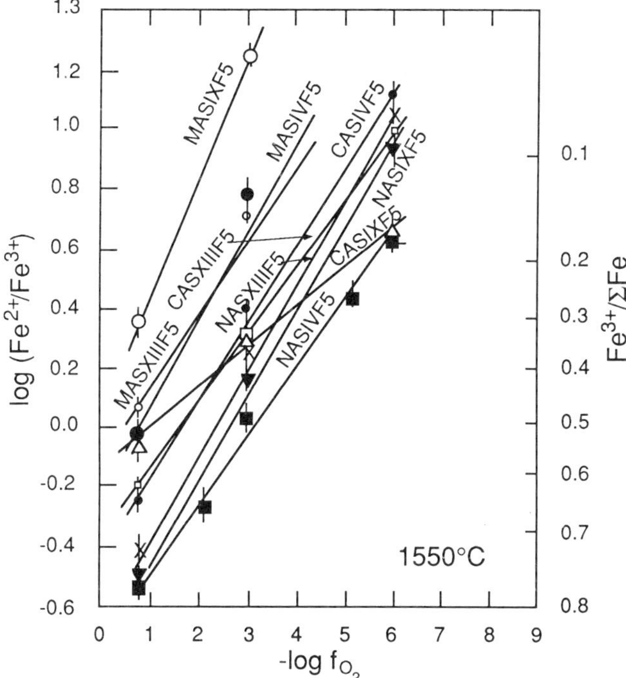

Fig. 23. Relationship between log (Fe^{2+}/Fe^{3+}) and log f_{O_2} at 1550°C for binary metal oxide-silica (MS and CS compositions) and ternary metal oxide-alumina-silica systems (NAS, CAS, and MAS compositions). The numerals after NAS, CAS, and MAS denote variations in NBO/T and Al/(Al + Si) as defined in Fig. 22. F5 and F10 denote 5 and 10 wt% iron oxide added as Fe_2O_3 (data from Mysen et al., 1984, 1985a; Virgo and Mysen, 1985).

and with increasing NBO/T of the melt (Table 3). The $\Delta G°$ increases with increasing Z/r^2 of the network-modifying cation (Table 3). Where ferric iron undergoes a gradual coordination transformation from largely octahedral to largely tetrahedral with decreasing temperature, the redox ratio of iron versus $1/T$ can be fitted to a quadratic equation:

$$\log \frac{Fe^{2+}}{Fe^{3+}} = a + \frac{b}{T} + \frac{c}{T^{2}}, \tag{12}$$

which for composition MASIVF5 (Mg-equivalent to CASIVF5) has $a = 21.5$, $b = 0.703 \cdot 10^5$,and $c = 0.578 \cdot 10^8$. In such an example, the free energy of reduction is a linear function of $1/T$ and decreases with increasing temperature. Thus, the $\Delta G°$ of reduction of octahedrally-coordinated Fe^{3+} to octahedrally-coordinated Fe^{2+} is smaller than when Fe^{3+} is in tetrahedral coordination.

Mo et al. (1982) suggested that because the partial molar volume of $FeO_{1.5}$ (with Fe^{3+} in tetrahedral coordination) is greater than that of FeO, at least in

Table 3. Thermodynamic data for reduction reactions in binary metal oxide-silica and ternary metal oxide-alumina-silica systems.

Metal cation	NBO/T	Al/(Al + Si)	ΔG° (kcal/mol)	$\gamma^{Fe^{2+}}/\gamma^{Fe^{3+}}$
Ca^{2+}	1.00	0.00	128.1	0.68
Ca^{2+}	1.40	0.00	79.5	0.83
Ca^{2+} [a]	1.40	0.00	—	0.98
Ca^{2+}	1.70	0.00	—	0.91
Ca^{2+}	2.00	0.00	—	1.00
Ca^{2+}	2.38	0.00	—	1.05
Mg^{2+}	1.40	0.00	—	0.79
Na^{+}	0.67	0.33	87.3	1.12
Na^{+}	0.17	0.33	66.0	0.89
Na^{+}	0.67	0.33	57.0	1.05
Na^{+}	0.67	0.13	85.2	1.00
Ca^{2+}	0.17	0.33	222.3	1.26
Ca^{2+}	0.67	0.33	150.0	0.98
Ca^{2+}	0.67	0.13	165.2	1.05
Mg^{2+}	0.17	0.33	—	0.74
Mg^{2+}	0.67	0.33	—	0.89
Mg^{2+}	0.67	0.13	—	1.00

[a] 10 wt% iron oxide added as Fe_2O_3. In all other samples 5 wt% iron oxide was added as Fe_2O_3.

Redox data from Mysen et al. (1984, 1985a) and Virgo and Mysen (1985). Note difference between corresponding data in this table and Table 3, Mysen et al. (1984) caused by arithmetic error.

silicate melts with bulk melt NBO/T less than about 1 (basaltic liquids typically have NBO/T between 0.9 and 0.7), it is likely that increasing pressure at constant oxygen fugacity and temperature will result in a lowering of $Fe^{3+}/\Sigma Fe$. Conversely, for constant $Fe^{3+}/\Sigma Fe$ at constant temperature, the f_{O_2} corresponding to a given ferric/ferrous increases with increasing pressure.

This predicted reduction in $Fe^{3+}/\Sigma Fe$ with increasing pressure has been observed experimentally for melts in the system $NaAlSi_2O_6$–$NaFeSi_2O_6$ (Mysen and Virgo, 1978). In that study, the reduction was ascribed to a possible pressure-induced aluminum coordination change with a concomitant increase in bulk melt NBO/T and an associated increase in redox ratio. Subsequent spectroscopic studies (Ohtani et al., 1985; Mysen et al., 1980b, 1983; Sharma et al., 1979a), as well as theoretical calculations (Angell et al., 1982), have revealed, however, that Al^{3+} coordination does not take place in this pressure range. Furthermore, it has been shown that $Fe^{3+}/\Sigma Fe$ also is lowered with increasing pressure for melts in the system Na_2O–SiO_2–Fe–O (Mysen and Virgo, 1985) where there is no Al^{3+} that could have undergone a coordination transformation (Fig. 24). In the Na_2O–SiO_2–Fe–O system, the $Fe^{3+}/\Sigma Fe$ decrease with pressure is also asso-

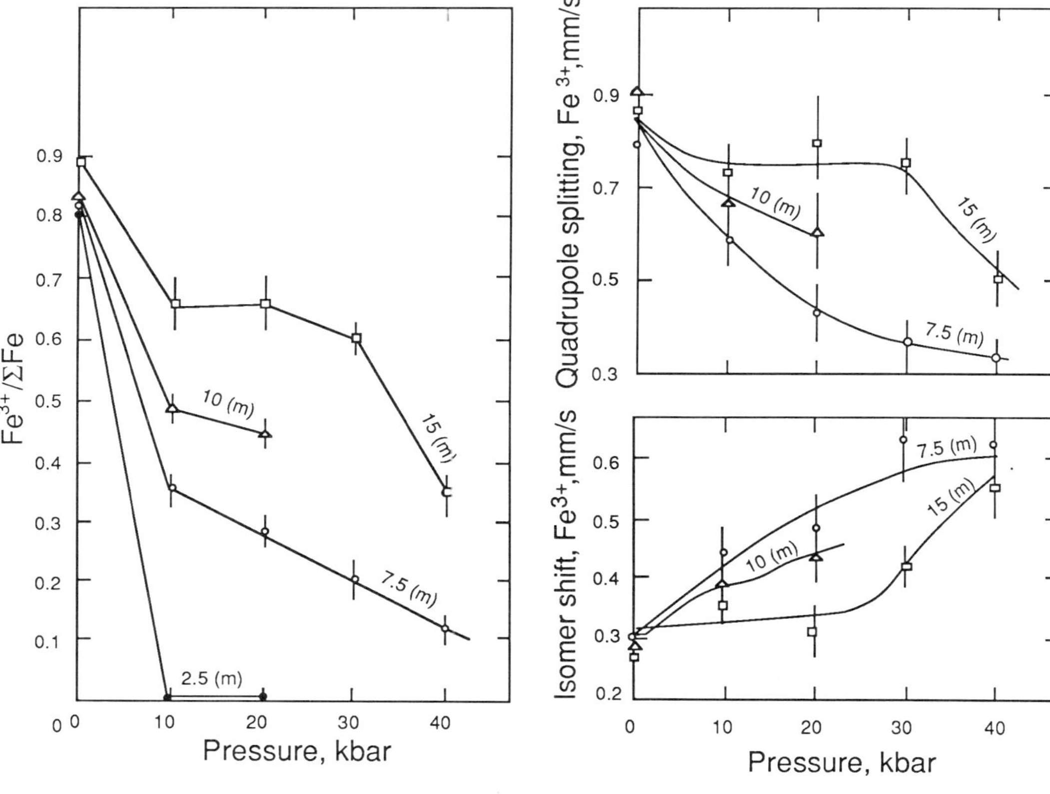

Fig. 24. Redox relationships of iron as a function of pressure for the composition $Na_2O \cdot 2SiO_2$ as a function of the amount of iron added [2.5(m) etc. refers to mole percent Fe_2O_3 added]. The $IS_{Fe_{3+}}$ and $QS_{Fe_{3+}}$ are isomer shifts and quadrupole-split values from Mössbauer spectra of quenched melts (data from Mysen and Virgo, 1983).

ciated with a significant increase in the isomer shifts of Fe^{3+} (Fig. 24), although the pressures at which the $IS_{Fe^{3+}}$ begins to increase range from less than 10 kbar for melts with 2.2 wt% iron oxide added as Fe_2O_3 to more than 20 kbar with iron contents of 13.4 wt% [curves 2.5(m) and 15(m), respectively in Fig. 24]. The form of the pressure dependence of $Fe^{3+}/\Sigma Fe$ is a function of the total amount of iron added. The proportion of ferric iron in tetrahedral and octahedral coordination depends on $Fe^{3+}/\Sigma Fe$ (Fig. 12), and the $Fe^{3+}/\Sigma Fe$ depends on the total amount of iron oxide in the system (Fig. 16). It is likely that the partial molar volumes of Fe^{3+} in octahedral and tetrahedral coordination differ with $\bar{V}_{FeO_{1.5}(VI)} < \bar{V}_{FeO_{1.5}(VI)}$ (Dingwell et al., 1988). Consequently, the difference in the pressure dependence of $Fe^{3+}/\Sigma Fe$ as a function of total iron content is understandable. Even for a fixed amount of iron oxide added, the $Fe^{3+}/\Sigma Fe$ is not a simple function of pressure. This observation probably relates to the finding (Mitra, 1982; Seifert et al., 1983; Mysen et al., 1983) that the compressibility of silicate melts is pressure dependent. In pure SiO_2 melt, for example, there is a more than 10% compression between 1 bar and 10 kbar (Mitra, 1982) associated with compression of Si–O–Si angles (Seifert et al., 1982, 1983; Mysen et al., 1983), whereas at pressures between 10 and 30 kbar the compressibility is of the order of 1%. Consequently, it is likely that the volume change upon reduction of ferric to ferrous iron is significantly greater in the pressure range between 1 bar and perhaps 10 kbar, so that at higher pressures, the effect is comparatively smaller. This suggestion is borne out by the data shown in Fig. 24.

Relationships Between Simple Synthetic and Complex Natural Systems

To apply structural information obtained from studies of simple binary, ternary, and quaternary systems to complex natural melts, it is necessary to combine the data from all the simple systems to the relationships in complex natural compositions. The redox equilibria of ferric and ferrous iron may be characterized in terms of structural components, pressure, temperature, and oxygen fugacity. A detailed description of a model that relates the composition to structure of natural magmatic liquids was presented by Mysen (1988), and only a brief summary of the model required to understand redox equilibria will be given here.

The first step in this process is to determine which cations are network-formers (in tetrahedral coordination) and the proportion of the various tetrahedrally-coordinated cations. In addition to Si^{4+}, these are Al^{3+} and Fe^{3+}, forming aluminate and ferrite complexes with suitable metal cations for electrical charge-balance. The proportion of aluminate complexes affects $Fe^{3+}/\Sigma Fe$. Moreover, the types of cations that electrically charge-balance this Al^{3+} in tetrahedral coordination are also important (Figs. 18 and 21) where the $Fe^{3+}/\Sigma Fe$ is more sensitive to $Al/(Al + Si)$, the larger the ionization potential of the metal cation. Furthermore, both ferric iron and aluminum in tetrahedral coordination form complexes requiring electrical charge-balance (Virgo et al., 1983). In natural rock compositions with alkalies, alkaline earths, and ferrous iron, a hierarchy of the preference of Al^{3+} and Fe^{3+} for these cations must be established. For aluminates, thermochemical (Navrotsky et al., 1982, 1985) and spectroscopic data (Mysen et al., 1980b, 1981a, 1982, 1985c; Seifert et al., 1982) indicate that the preference is $K^+ > Na^+ > Ca^{2+} > Mg^{2+}$. That is, the preference for Al^{3+} is greater the smaller the Z/r^2, as was also suggested by Bottinga and Weill (1972). The position of Fe^{2+} in this hierarchy is not known, but because the Z/r^2 of Fe^{2+} is less than that of Mg^{2+} and greater than that of Ca^{2+}, it is likely that the preference for Fe^{2+} is between that of Ca^{2+} and Mg^{2+}.

Similar data for Fe^{3+}(IV) complexes are less abundant. The observation that even alkali charge-balanced Fe^{3+} and Si^{4+} show great ordering in crystalline ferrisilicates, and the comparative scarcity of Fe^{3+}(IV)-bearing minerals, indicate that in competition for a specific cation between Al^{3+} and Fe^{3+}, the ferrite complex may be formed first. In the calculation scheme proposed here, for each metal cation, the proportion of a given metal cation required to charge-balance tetrahedrally coordinated Fe^{3+} is assigned before Al^{3+}. The proportion of tetrahedrally coordinated Al^{3+} and Fe^{3+} will be expressed as $Al^{3+}/(Al^{3+} + Si^{4+})$ and $Fe^{3+}/(Fe^{3+} + Si^{4+})$ with the understanding that these two trivalent cations are associated with the prerequisite cations for charge-balance.

The remaining metal cations are network modifiers, and the degree of polymerization of the melt is;

$$NBO/T = 1/T \sum_{i=1}^{i} M_i^{n+}, \tag{13}$$

where T is the total proportion of tetrahedrally coordinated cations, M_i is the proportion of network-modifying metal cation i after the proportion required for charge-balance of tetrahedrally-coordinated Al^{3+} and Fe^{3+} is subtracted, and n^+ is the electrical charge of this cation. The activities of the nonbridging oxygens associated with the various metal cations depend on the type of cation. This suggestion is supported, for example, by the fact that in binary and ternary metal oxide silicate and metal oxide aluminosilicate melt systems, the $Fe^{3+}/\Sigma Fe$ is not only a function of NBO/T (Fig. 18), but also depends on the cation itself (e.g., Figs. 19 and 20; see also Fig. 18). Thus, when $Fe^{3+}/\Sigma Fe$ is related to a natural melt composition, the proportion of nonbridging oxygen associated with each type of metal cation must be considered.

As summarized above Sack et al. (1980) suggested that stepwise, linear regression of oxide components in addition to ln f_{O_2} and 1/T from natural melt compositions equilibrated under laboratory conditions yielded an empirical expression [equation (1), with coefficients given in Table 2] that could be used to relate composition, temperature, and oxygen fugacity to ferric/ferrous of the melt. The limitation of this approach was evident in the disagreement between the functional relationships of the various oxide components depending on temperature, oxygen fugacity, and bulk composition itself (Sack et al., 1980; Thornber et al., 1980; Kilinc et al., 1983). By considering the individual structural components known to affect the redox ratio, the original empirical expression [equation (1)], with structural rather than oxide components, may be used to refine the relationships between redox ratio, oxygen fugacity, temperature and bulk composition. It is suggested that the expression of the form;

$$\ln\frac{Fe^{2+}}{Fe^{3+}} = a + \frac{b}{T} + c\ln f_{O_2} + d\frac{Al}{Al + Si} + e\frac{Fe^{3+}}{Fe^{3+} + Si^{4+}} + \sum_{j=1}^{j} f_j(NBO/T)_j, \tag{14}$$

adequately describes this relationship. The f_j and $(NBO/T)_j$ are the regression coefficients and NBO/T values for the individual network-modifying oxides, respectively. The coefficients a, b, c, and d together with f_j are obtained with stepwise, linear regression.[3]

The only three oxides for which extensive laboratory studies in simple systems have not been carried out are K_2O (limited data available in Dickenson and Hess, 1981; Virgo et al., 1981), TiO_2, and P_2O_5 [unpublished Raman and Mössbauer data by the author from the system $CaO-SiO_2-P_2O_5-Fe-O$ suggest a weak positive correlation between $Fe^{3+}/\Sigma Fe$ and $P^{5+}/(P^{5+} + Si^{4+})$]. In the present version of this calculation method, these oxides have been treated as follows. The potassium content has been recast into an equivalent proportion of sodium and titanium to an equivalent amount of silicon because Ti^{4+} apparently occupies a structural position in silicate melts similar to that of Si^{4+} in the TiO_2 concentration range of most natural magmatic systems. (Furukawa and White, 1979; Mysen et al., 1980c). Some disagreement exists as to the general applicability of this conclusion (e.g., Dickinson and Hess, 1985). Phosphorus also occurs in tetrahedral coordination, but available structural information indicates that it forms phosphate complexes with metal cations such as Na^+ or Ca^{2+} (Visser and Van Groos, 1979; Ryerson and Hess, 1980; Mysen et al., 1981b; Nelson and Tallant, 1984, 1986).

In analogy with the crystalline materials where the free energy of formation of $Ca_3(PO_4)_2$ is greater than that of $Na_3(PO_4)$, it is suggested that calcium phosphate complexing predominates in natural magmatic liquids. As a result, in calculating the proportion of structural units in natural melts, an amount of Ca^{2+} equivalent to that required to form a calcium phosphate complex is subtracted before Ca^{2+} is assigned to charge-balancing and network-modifying roles. Whether Fe^{3+}, or Fe^{2+}, or both, phosphate complexing takes place cannot be ascertained with available experimental data. In analogy with $AlPO_4$ complexing in certain P-bearing aluminosilicate melts (Dupree et al., 1987), it is possible that $Fe^{3+}PO_4$ may be present, for example. Ferrous iron complexes [e.g., $Fe^{2+}{}_3(PO_4)_2$] might also be stable. These and other complexes will affect the $Fe^{3+}/\Sigma Fe$. The phosphorus concentrations in most igneous rocks however, are low (< 1 wt%

[3] A FORTRAN program to calculate the structure of complex compositions such as natural magmatic liquids at 1 bar based on available experimentally obtained structural data for silicate melts is available from the author on request.

Table 4. Regression coefficients for equation (14).

	Simple system		Natural rocks		All analyses	
	Coefficient	Standard error	Coefficient	Standard error	Coefficient	Standard error
a(const.)	10.814	1.134	4.384	0.524	15.437	0.786
$b(1/T)$	−1.989	0.203	−0.9077	0.0915	−2.848	0.138
$c(\ln f_{O_2})$	−0.3210	0.0117	−0.1420	0.0081	−0.3484	0.0120
$d[Fe^{3+}/(Fe^{3+} + Si)$	−4.067	0.985	−9.875	0.952	−2.121	1.055
$e[Al^{3+}/Al^{3+} + Si)$	−1.535	0.467	1.621	0.418	−1.309	0.469
f_i $(NBO/T)^{Mg}$	0.494	0.134	0.8607	0.2093	0.6662	0.0966
f_i $(NBO/T)^{Ca}$	−0.5228	0.1095	−0.6560	−0.1617	−0.5255	0.1084
f_i $(NBO/T)^{Na}$	−1.584	0.238	−1.194	0.5112	−1.125	0.1790
f_i $(NBO/T)^{Fe^{2+}}$	−1.951	0.507	−2.310	0.422	−3.215	0.538

Number of analyses in regression: simple systems, 267; natural rocks, 190; all analyses, 460.
Experimental data for regression from Kennedy (1948), Fudali (1965), Sack et al. (1980), Thornber et al. (1980), Kilinc et al. (1983), Seifert et al. (1979), Virgo et al. (1981), Mysen and Virgo (1983), Virgo and Mysen (1985), Mysen et al. (1980a, 1984, 1985a,d).

P_2O_5), so that errors introduced by not considering these possible interactions are likely to be comparatively small.

Multiple, linear regression has been carried out with 267 analyses from simple systems (only binary metal oxide and ternary metal oxide-alumina-silica systems). The resulting coefficients are shown in Table 4 and are compared with those obtained by similar regression of available data for laboratory-calibrated, natural rock compositions (Kennedy, 1948; Fudali, 1965; Sack et al., 1980; Kilinc et al., 1983; Thornber et al., 1980) with a total of 193 such analyses. Finally, the redox ratios of both groups of compositions were regressed against melt structural parameters according to equation (14) (a total of 460 analyses). In each of these sets of coefficients, the standard errors are significantly smaller than those found from regression with simple oxide components (Sack et al., 1980; Kilinc et al., 1983). Thus, regression of ln (Fe^{2+}/Fe^{3+}) against independently established melt structural factors yields a more reliable formulation than one based on empirical relationships between redox ratio and oxide contents of the melts.

It is evident from this exercise that among the network-modifying cations the ln (Fe^{2+}/Fe^{3+}) is negatively correlated with the proportion of nonbridging oxygen associated with Ca^{2+}, Na^+, and Fe^{2+} (Table 4). A negative but less reliable correlation also exists in some of the data summarized in Table 2 (Kilinc et al., 1983). All three statistical analyses (Table 4) show that Fe^{2+}/Fe^{3+} increases with increasing NBO/T associated with Mg^{2+}. It is not clear why Mg^{2+} is an exception among the network-modifying cations in this respect. A possible explanation might be a structural role of Mg^{2+} different from the other network-modifying cations. Mg^{2+} is quite small for octahedral coordination and tetrahedral coordination (as observed in crystalline akermanite, for example) might

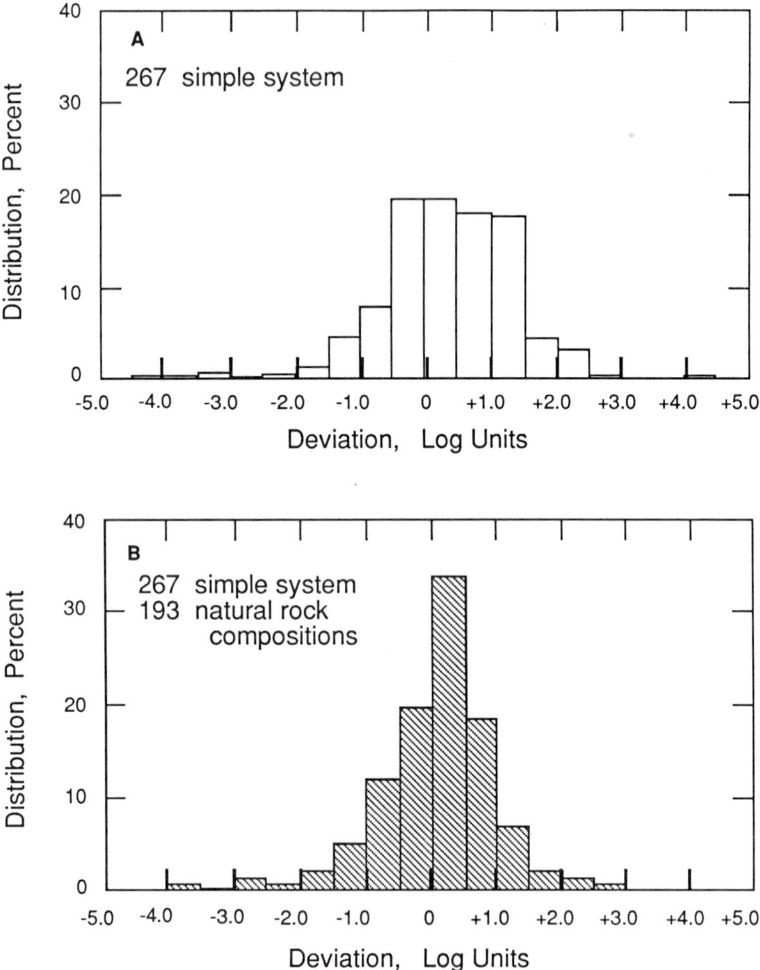

Fig. 25. Deviations of calculated oxygen fugacities relative to experimental values for samples used to obtain regression coefficients in Table 4 for equation (14). (a) Regression coefficients from calibration employing only experimental data from binary metal oxide-silica and ternary metal oxide-alumina-silica. (b) Regression coefficients obtained by employing laboratory calibration from all available data (data from Kennedy, 1948; Fudali, 1965; Larson and Chipman, 1953; Sack et al., 1980; Thornber et al., 1980; Seifert et al., 1979; Mysen et al., 1980a, 1984, 1985a,d; Kilinc et al., 1983; Virgo et al., 1981).

be possible (Scarfe, 1977) although Raman spectra of selected compositions (e.g., Sharma et al., 1979; Mysen et al., 1982) do not support such a hypothesis. There is also a rapid decrease in ln (Fe^{2+}/Fe^{3+}) with increasing $Fe^{3+}/(Fe^{3+} + Si^{4+})$. In the simple-system calibration, as well as that based on all analyses, the ln (Fe^{2+}/Fe^{3+}) will decrease with increasing Al/(Al + Si) (with all Al^{3+} charge-balanced in tetrahedral coordination), but the coefficient obtained by regression with only the rock compositions as input parameters has the opposite sign. Although there is no obvious explanation for this apparent difference, it should be remembered that the range in Al/(Al + Si) of the natural rock compositions is relatively small (0.15–0.25), whereas for the simple-system calibration, the range covered is between 0 and 0.43. More reliance is placed, therefore, on this latter analysis.

The coefficients in Table 4 may be inserted in equation (14), and this equation may be used, for example, as an oxygen fugacity barometer. The calculated f_{O_2} values for the samples in the data set are compared with the measured values in Fig. 25. From this comparison it is evident that as an oxygen fugacity barometer of natural igneous rocks based only on the simple-system calibration, 40% of the calculated values are within ± 0.5 log unit and 67% are within ± 1.0 log unit of f_{O_2}. Between 85 and 90% of the calculated values are within ± 1.5 log units of the measured value (Fig. 25a). When the whole data set of simple-system and natural melt compositions is employed (see Table 4), the deviation from measured values is smaller (Fig. 25b), and 54% of the analyses are within ± 0.5 log unit of oxygen fugacity and 85%, within ± 1.0 log unit. About 95% of the analyses fall within ± 1.5 log units. It is suggested that this model relating redox ratios of iron to temperature, oxygen fugacity, and melt structure is an adequate description and that equation (14), with the coefficients in Table 4, can be used with confidence to calculate oxygen fugacity conditions of natural magmatic liquids at 1 atm. Although some of the principles governing the pressure dependence of ferric/ferrous have been established (Fig. 24; see also Mo et al., 1982; Mysen and Virgo, 1983), the data base is at present insufficient to extend this treatment to high pressure.

Redox Ratios of Iron and Melt Polymerization

Numerous chemical and physical properties of silicate melts are affected by their degree of polymerization (NBO/T). For example, isothermal trace-element partition coefficients for incompatible elements may vary by more than an order of magnitude in the composition (polymerization) range of natural magmatic liquids (Watson, 1977; Irving, 1978; Hart and Davis, 1978; Mysen and Virgo, 1980). Other properties such as melt viscosity, expansion, and compressibility are also sensitive to the polymerization (e.g., Lacey, 1968; Bockris et al., 1955, 1956; Tomlinson et al., 1958; Bockris and Reddy, 1970; Bockris and Kojonen, 1960; Kushiro, 1976; Scarfe et al., 1987; Bottinga, 1985).

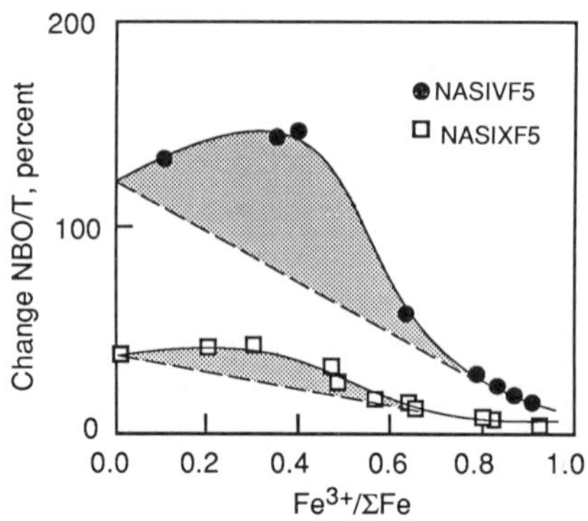

Fig. 26. NBO/T vs. $Fe^{3+}/\Sigma Fe$ (data from Mysen and Virgo, 1989). Symbols are as used in Figs. 22 and 23.

For iron-bearing silicate melts, the redox ratio of iron, as well as the oxygen coordination around ferric and ferrous iron, is related to the proportion of nonbridging oxygen (throughout this text referred to as nonbridging oxygen per tetrahedrally coordinated cation, *NBO/T*). With Fe^{3+} in tetrahedral and Fe^{2+} in octahedral coordination, relationships between polymerization and Fe^{3+}/Fe^{2+} may be illustrated by expressions such as those shown in equations (4) and (6). The straight lines in Fig. 26 show the relationships between *NBO/T* and $Fe^{3+}/\Sigma Fe$ with ferric iron in tetrahedral and ferrous iron in octahedral coordination in the entire $Fe^{3+}/\Sigma Fe$-range between 0 and 1. Both compositions in Fig. 26 have initial *NBO/T* near 0.6 and contain 5 wt% iron oxide calculated as Fe_2O_3 (data from Mysen and Virgo, 1989). The composition NASIVF5 is more aluminous [Al/(Al + Si) = 0.334] than NASIXF5 [Al/(Al + Si) = 0.13]. In either instance, the melts become systematically depolymerized as the $Fe^{3+}/\Sigma Fe$ is lowered. The transformation of Fe^{3+} from tetrahedral to octahedral coordination also results in an increase in *NBO/T*. For example, a coordination transformation of Fe^{3+} in SiO_2 melt may be illustrated with the equation;

$$4SiO_2 + Fe(IV)O_2^- \Leftrightarrow Fe(VI)^{3+} + 2Si_2O_5^{2-}, \tag{15}$$

with the equilibrium constant,

$$K_{15} = \frac{[Fe^{3+}(VI)][Si_2O_5^{2-}]^2}{[Fe(IV)O_2^-][SiO_2]^4}. \tag{16}$$

In this expression, the transformation of ferric iron from four- to six-fold coordination results in a completely polymerized melt (SiO_2 has $NBO/T = 0$) becoming one with $NBO/T = 1$ (disilicate stoichiometry). The solid lines in Fig. 26 are the *NBO/T* vs $Fe^{3+}/\Sigma Fe$ relationships taking into consideration both reduction of Fe^{3+} to Fe^{2+} and the coordination transformation of Fe^{3+} from tetrahedral to octahedral as a function of $Fe^{3+}/\Sigma Fe$. These combined effects lead to the maximum value in *NBO/T* near $Fe^{3+} > /\Sigma Fe = 0.3$, below which value reduction of $Fe^{3+}(VI)$ to $Fe^{2+}(VI)$ actually leads to polymerization of the melt [cf equation (2)].

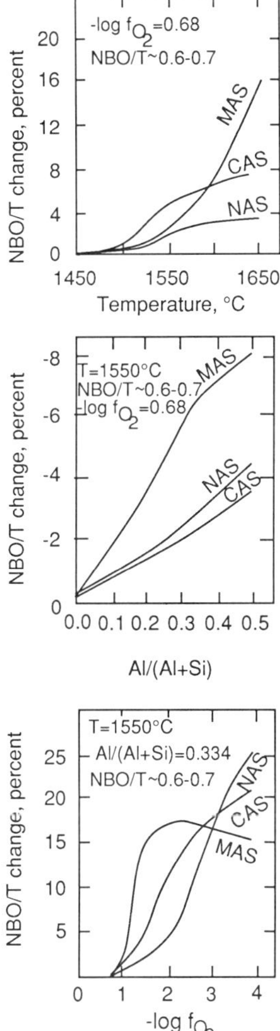

Fig. 27. Calculated variations in bulk melt NBO/T for compositions in metal oxide-alumina-silica systems with 5 wt% iron oxide added as Fe_2O_3, as a function of temperature (relative to value at 1450°C), oxygen fugacity (relative to value at $f_{O_2} = 10^{-0.68}$ bar) and Al/(Al + Si) [relative to value at Al/(Al + Si) = 0.0. $(NBO/T)_0$ refers to the reference points (data from Mysen et al., 1985a).

From the relationships between $Fe^{3+}/\Sigma Fe$, $Fe^{3+}(IV)/Fe^{3+}(VI)$, and the intensive and extensive variables summarized above (Figs. 12, 13, 16–23), changes in polymerization of aluminosilicate melts can be calculated as a function of the redox ratio of iron with $Fe^{3+}/\Sigma Fe$ controlled by temperature, pressure, oxygen fugacity and bulk composition (Figs. 27, 28). At constant oxygen fugacity and bulk melt Al/(Al + Si), aluminosilicate melts become more depolymerized (NBO/T increases) with increasing temperature, because $Fe^{3+}/\Sigma Fe$ decreases systematically with increasing temperature (Fig. 22). As a result of the greater temperature dependence of $Fe^{3+}/\Sigma Fe$ with increasing Z/r^2 of the metal cation (Fig. 22), polymerization of magnesium aluminosilicate melt is more sensitive to temperature than that of calcium aluminosilicate, and polymerization of calcium alumi-

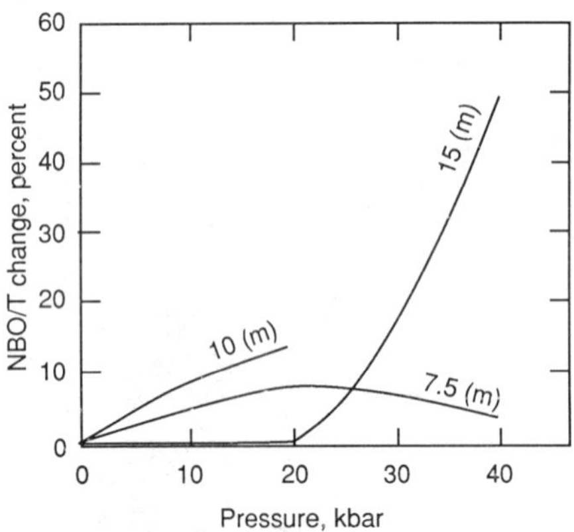

Fig. 28. Effect of pressure on degree of polymerization of silicate melts in the system Na_2O–SiO_2–Fe–O as a function of amount of iron added [7.5(m) etc. refer to mole percent Fe_2O_3 added]. The symbol (NBO/T)(VI) is reference NBO/T at 1 bar pressure (data from Mysen and Virgo, 1983).

nosilicate is more sensitive than that of sodium aluminosilicate composition. Among natural magmatic liquids with similar NBO/T, alkali basalt has approximately 30–40% of its Al^{3+} charge-balanced with alkali metals, whereas in those of abyssal tholeiite composition, typically less than 20% of the Al^{3+} is charge-balanced with alkalies. The remaining Al^{3+} is charge-balanced with alkaline earths. One may infer from these observations that the temperature dependence of their polymerization, controlled by the relationship between temperature and ferric/ferrous, is greater for abyssal tholeiite than for alkali basalt. Consequently, NBO/T-dependent properties of olivine tholeiite are probably more sensitive to temperature than those of alkali basalt.

The redox ratio of iron is also strongly dependent on oxygen fugacity (Fig. 23) as is, therefore, NBO/T of the melt. Inasmuch as decreasing oxygen fugacity typically results in a lowering of $Fe^{3+}/\Sigma Fe$ into the range where coordination transformation of ferric iron will also occur (Fig. 12), decreasing f_{O_2} from that of air to lower values results in depolymerization from both $Fe^{3+}/\Sigma Fe$ reduction and coordination transformation (compare also Fig. 26). These considerations are built into the results shown in Fig. 27. When Fe^{3+} is in tetrahedral coordination, the relative order of the depolymerization relations as a function of the type of metal cation is similar to that controlled by temperature with melt equilibrated in air ($Mg^{2+} > Ca^{2+} > Na^+$). As Fe^{3+} enters octahedral coordination, further reduction of Fe^{3+} to Fe^{2+} actually results in *polymerization* as illustrated by the expression:

$$4Fe^{3+}(VI) + 2Si_2O_5{}^{2-} \Leftrightarrow 4Fe^{2+} + 4SiO_2 + O_2, \tag{17}$$

with the equilibrium constant,

$$K_{17} = \left(\frac{[Fe^{2+}]}{[Fe^{3+}(VI)]}\right)^4 \frac{[SiO_2]^4}{[Si_2O_5{}^{2-}]^2} f_{O_2}. \tag{18}$$

In the curves shown in Fig. 27, the decreasing slope (for magnesium aluminosilicate the slope reverses in the lower f_{O_2} range) with decreasing f_{O_2} occurs because reactions such as that illustrated in equation (17) become important.

Finally, the $Fe^{3+}/\Sigma Fe$ is strongly dependent on Al/(Al + Si) (Fig. 21), where this dependence is shown to become increasingly pronounced the higher the Z/r^2 of the charge-balancing cation. Consequently, the NBO/T of iron-bearing melts is more sensitive to Al/(Al + Si) the greater the Mg/Ca and the Ca/Na. This conclusion leads to the inference that the relative dependence of polymerization of natural magmatic liquids on their Al/(Al + Si) for the same total iron content is more important for mafic and ultramafic melts than it is for felsic melts.

Pressure also affects melt polymerization because increasing pressure results in decreasing $Fe^{3+}/\Sigma Fe$ (Figs. 24, 28). For melts with relatively low iron contents [e.g., 7.5(m) in Figs. 24 and 28] all ferric iron is transformed to octahedral coordination at relatively low pressures [<20 kbar for NS2F7.5(m), for example]. Consequently, a further pressure increase (and decrease in $Fe^{3+}/\Sigma Fe$) results in a decrease in NBO/T. In the pressure range shown in Fig. 28, for the more iron-rich melts [e.g., 15(m)], there is a continuous increase in NBO/T with increasing pressure. These relationships imply that not only will a melt property such as viscosity decrease with increasing pressure because of the weakening of bridging Si–O–Si bonds as their bond angles are decreased (Mysen et al., 1983), but the pressure-induced melt depolymerization from reduction of Fe^{3+} to Fe^{2+} will enhance this effect (Fig. 29). In calculating the viscosity versus pressure trajectory in Fig. 29 it was assumed that the NBO/T is a principal factor in controlling melt viscosity. Although this assumption clearly is an over-simplification for complex silicate melts (e.g., Dingwell, 1989), simple positive correlations between NBO/Si and melt viscosity have been found for binary alkali and alkaline earth silicate melts (Bockris et al., 1955, 1956). The difference between the calculated pressure versus viscosity trajectories of NS2 (sodium disilicate melt with $NBO/Si = 1$) and NS2F15(m) (sodium disilicate with 15 mol% iron oxide added as Fe_2O_3) is based on the known pressure dependence of $Fe^{3+}/\Sigma Fe$ for NS2F15(m) melt and the resulting effect of pressure on the NBO/T of this melt (Mysen and Virgo, 1985).

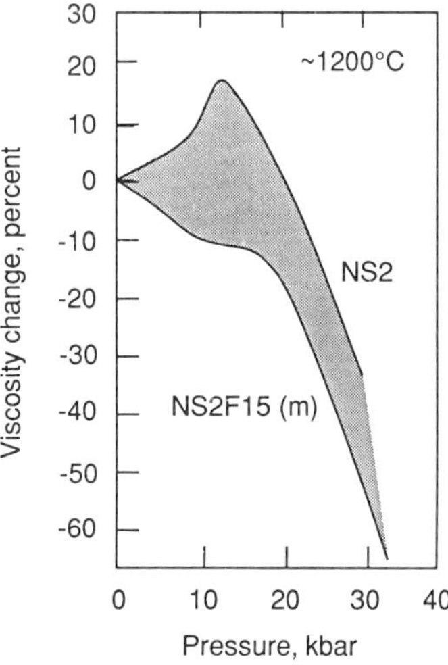

Fig. 29. Viscosity vs. pressure form $Na_2Si_2O_5$ melt with 13.4 wt% Fe_2O_3 (15 mol %) based on redox data from Mysen and Virgo (1985).

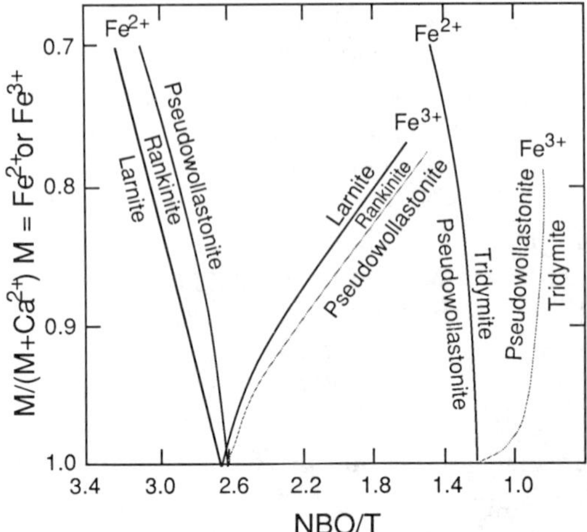

Fig. 30. Shifts in liquidus boundaries of calcium silicate minerals with different degrees of polymerization as a function of the proportion of iron added as Fe^{3+} or Fe^{2+} (phase-equilibrium base data from Osborn and Muan, 1960a, b).

Liquidus Phase Relations in Iron-Bearing Systems

Liquidus boundaries between phases of different degrees of polymerization may be employed to define activities of these components in the melt (Carmichael et al., 1974). The composition of the melts corresponding to a particular liquidus boundary varies as a function of an added third component (e.g., Fe_2O_3). Such changes in liquidus boundaries as a function of added components have been used to deduce solubility mechanisms of this component in the melt (Kushiro, 1975; Ryerson, 1985). The relationships between iron content and calcium silicate phase boundaries in the system $CaO-FeO-Fe_2O_3-SiO_2$ (Osborn and Muan, 1960a,b) serve to illustrate this effect (Fig. 30). The liquidus boundaries of interest are those between larnite (Ca_2SiO_4; $NBO/Si = 4$) and rankinite ($Ca_3Si_2O_7$; NBO/Si = 3), between rankinite and pseudowollastonite ($CaSiO_3$; NBO/Si = 2), and between pseudowollastonite and tridymite (SiO_2; NBO/Si = 0). The activities of nonbridging oxygen along each of these liquidus boundaries may be defined as unity, and any shift in compositional space of one of these boundaries as a function of added iron oxide reflects activity changes of the nonbridging oxygen in the melt as a function of the iron oxide addition. From the structural information in this system, Fe^{3+} is in tetrahedral coordination whenever $Fe^{3+}/\Sigma Fe > 0.5$ at the liquidus temperatures and Fe^{2+} is in octahedral coordination. Therefore, the NBO/T of any liquidus boundary may be calculated from bulk composition.

It can be seen from the liquidus boundary relations recast as NBO/T of the melt versus the proportion of calcium replaced by iron that with the addition of Fe^{2+}, the liquidus boundary shifts to less polymerized compositions. Replacement of 10% of the Ca^{2+} by Fe^{2+} results in an increase in NBO/T of about 8%, for example, for the larnite-rankinite boundary and about 5% for the rankinite-pseudowollastonite boundary. The addition of FeO to a pyrosilicate melt ($Si_2O_7^{6-}$) may be illustrated with the expression

$$Si_2O_7^{6-} + FeO \Leftrightarrow Fe^{2+} + 2SiO_4^{4-}, \quad (19)$$

The more FeO that is added to the system, the larger will be the concentration of SiO_4^{4-} in the melt relative to $Si_2O_7^{6-}$. This observation is consistent with the structural information that the solution of Fe^{2+} in silicate melts results in depolymerization of the melts. As a result, the activity of nonbridging oxygen increases, and the liquidus boundary shifts to lower silica content (*NBO/T* of the melt has decreased). Furthermore, it is likely that with a simple replacement of Ca^{2+} with Fe^{2+} *without* changes in the overall *NBO/T*, the pyrosilicate–orthosilicate liquidus boundary will be affected. This possibility is related to the observation (Mysen et al., 1982) that various possible network-modifying cations are not randomly distributed among the depolymerized anionic units in the melt. The larger the ionization potential of the metal cation, the greater is its tendency to form polyhedra with nonbridging oxygen from the least polymerized structural unit. For example, an anionic equilibrium of the form;

$$Si_2O_7^{6-} \Leftrightarrow SiO_3^{2-} + SiO_4^{4-}, \quad (20)$$

will shift to the right with increasing $Fe^{2+}/(Fe^{2+} + Ca^{2+})$ in the system CaO–FeO–SiO_2, thus resulting in an increase in the $X_{SiO_4^{4-}}/X_{Si_2O_7^{6-}}$. The result will be shrinkage of the pyrosilicate liquidus volume in accordance with experimental observation (Osborn and Muan, 1960a).

For ferric iron, on the other hand, the same liquidus boundaries shift to higher silica contents with increasing $Fe^{3+}/(Fe^{3+} + Ca^{2+})$ (Osborn and Muan, 1960b). Thus, the activity of nonbridging oxygen in the melt, expressed as *NBO/T*, represented by the position of the liquidus boundary, decreases with increasing ferric iron content. This conclusion is consistent with Fe_2O_3 dissolving to form ferrite complexes with Ca^{2+} for electrical charge-balance in the melt as suggested by the above structural information. The following expression illustrates the relationships for the pyro-orthosilicate boundary:

$$2SiO_4^{4-} + 4Ca^{2+} + Fe_2O_3 \Leftrightarrow CaFe_2O_4 + Si_2O_7^{6-} + 3Ca^{2+}. \quad (21)$$

Expressions similar to (19) and (21) can be written for other boundaries of minerals with different NBO/Si values.

Because the melt structure depends on $Fe^{3+}/\Sigma Fe$, liquidus phase equilibria of iron-free silicate minerals in iron-bearing systems may vary with $Fe^{3+}/\Sigma Fe$. Some consequences of the relationships may be pursued with the phase equilibria in the systems CaO–FeO–Fe_2O_3–SiO_2 and MgO–FeO–Fe_2O_3–SiO_2 (Osborn and Muan, 1960a,b; Muan and Osborn, 1956). In Fig. 31 the liquidus equilibria for two compositions, CS1.4F10 and MS1.4F10 (Ca/Si = 0.7 and Mg/Si = 0.7 with 10 wt% Fe_2O_3 added), have been estimated as a function of $Fe^{3+}/\Sigma Fe$ of the liquid. The abscissa could be replaced with oxygen fugacity because $Fe^{3+}/\Sigma Fe$, the redox ratio, is a simple function of oxygen fugacity (Fig. 23). The calculated liquidus surface topologies depend solely on the chosen $Fe^{3+}/\Sigma Fe$ and

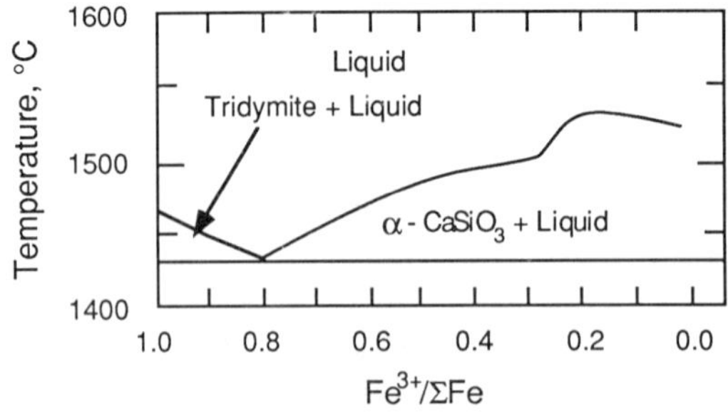

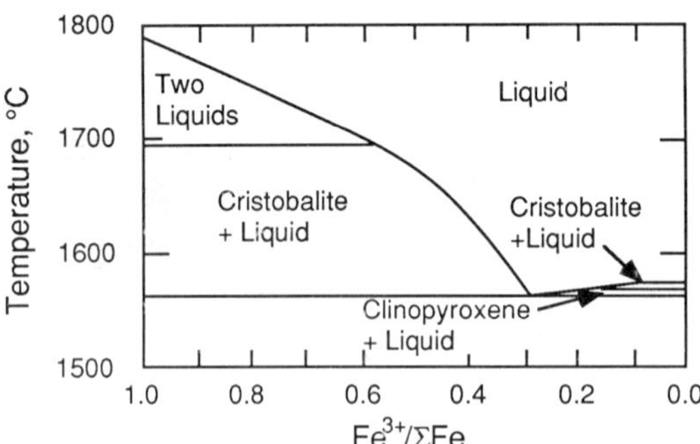

Fig. 31. Calculated liquidus equilibria for compositions CS1.4F10 ($CaO \cdot 1.4SiO_2$ with 10 wt% Fe_2O_3 added) and MS1.4F10 ($MgO \cdot 1.4SiO_2$ with 10 wt% Fe_2O_3 added).

the coordination polyhedra of Fe^{2+} and Fe^{3+} in the melt. The phase equilibria (Mysen et al., 1984) have been obtained by calculating Ca/Si and Mg/Si of iron-free melts with NBO/Si on the binary MgO–SiO_2 and CaO–SiO_2 joins corresponding to the *NBO/T* values obtained for the various values of $Fe^{3+}/\Sigma Fe$. In these calculations it was assumed that Fe^{3+}(IV) occurs in separate complexes in the melt. The compositions of these complexes were assumed not to be affected by the intensive and extensive variables in this system.

In the system CaO–SiO_2 (Phillips and Muan, 1959) the melt polymerization (decreasing *NBO/Si*) resulting from the addition of 10 wt% Fe_2O_3 to the CS1.4 base composition results in a change from pseudowollastonite at 1500°C to tridymite at 1470°C on the liquidus. The tridymite liquidus field extends to $Fe^{3+}/\Sigma Fe < 0.8$ at the minimum liquidus temperature in this system (<1435°C; see Fig. 31). A further reduction in $Fe^{3+}/\Sigma Fe$ results in an increase in the liquidus temperature and pseudowollastonite as the liquidus phase. The change in slope of the liquidus curve is due to the transformation of Fe^{3+} from tetrahedral to octahedral coordination in this range of $Fe^{3+}/\Sigma Fe$ (see also Fig. 31).

The composition MS1.4 in the system MgO–SiO_2 has the same NBO/Si as composition CS1.4 in the system CaO–SiO_2. The liquidus phase equilibria of the magnesium system (Greig, 1927) are, however, considerably different from those of the calcium oxide-silica system. First, the composition MS1.4 has the fully polymerized (NBO/Si = 0) cristobalite on the liquidus, whereas for the system CaO–SiO_2 the liquidus phase is the metasilicate (NBO/Si = 2) pseudowollastonite. This difference extends into the system $MgO-FeO-Fe_2O_3-SiO_2$ (Muan and Osborn, 1956). In equilibrium with air, composition MS1.4 + 10 wt% Fe_2O_3 has two immiscible liquids coexisting at 1780°C, whereas for CS1.4F10, tridymite is the liquidus phase at 1470°C. As for the system CS1.4F10, decreasing $Fe^{3+}/\Sigma Fe$ (resulting, for example, from a decrease in f_{O_2}), is associated with a rapid decrease in liquidus temperature (Fig. 31) to 1570°C at $Fe^{3+}/\Sigma Fe < 0.3$ where cristobalite coexists with clinoenstatite. A tectosilicate-metasilicate liquidus boundary also exists in the system CS1.4F10, but at a much higher value of $Fe^{3+}/\Sigma Fe$ (near 0.8). Furthermore, decreasing $Fe^{3+}/\Sigma Fe$ in the system MS1.4F10 has a more pronounced effect on the liquidus temperature (<200°C decrease) than in the system CS1.4F10 (<35°C). One might infer from these observations that in natural magmatic liquids, liquidus phase relations are significantly dependent on $Fe^{3+}/\Sigma Fe$. The liquidus phase relations are, however, dependent on the Ca/Mg of the liquid. Mysen et al. (1985a) performed analogous calculations of liquidus equilibria in the systems $CaO-FeO-Fe_2O_3-Al_2O_3-SiO_2$ and $MgO-FeO-Fe_2O_3-Al_2O_3-SiO_2$ and applied the principal conclusions to the relationships between $Fe^{3+}/\Sigma Fe$ and liquidus equilibria in natural magmatic systems. For tholeiitic liquid, for example, it was suggested that under highly oxidizing conditions, tridymite (or quartz) would be a liquidus phase. Reduction of ferric iron to ferrous iron would result in the appearance of a metasilicate phase (pyroxene) and complete reduction in the stabilization of an orthosilicate (olivine) liquidus phase. Fractionation trends controlled by silicate minerals therefore, are significantly dependent on the iron content and oxygen fugacity whether or not iron-bearing phases crystallize during such processes.

Physical Properties of Iron-Bearing Silicate Melts

Densities and molar volumes of components in silicate melts have numerous thermodynamic and physical applications. For example, Mo et al. (1982) pointed out that the relationship between oxygen fugacity and redox ratio in silicate melts is pressure dependent because the $\bar{V}_{Fe_2O_3}-\bar{V}_{FeO}$ differs from 0 (Shiraishi et al., 1978; Bottinga et al., 1983; Mo et al., 1982). Under the assumption that the partial molar volumes of ferric and ferrous oxide in silicate melts are pressure independent (not necessarily correct), the following relationships between 1-bar and high-pressure oxygen fugacity holds:

$$\ln f_{O_2}{}^{P} = \ln f_{O_2}{}^{1\ \mathrm{bar}} + \frac{1}{RT}(2\bar{V}_{Fe_2O_3} - 4\bar{V}_{FeO})(P - 1). \quad (22)$$

From this expression, Mo et al. (1982) concluded that the $Fe^{3+}/\Sigma Fe$ determined by the f_{O_2} of the nickel-nickel oxide oxygen buffer at 1 bar (Wones and Eugster, 1962) corresponds to an oxygen fugacity approximately one order of magnitude higher at a total pressure of 10 kbar.

Not only are partial molar volume data important for calculations such as those above, but together with data for all the components in magmatic liquids, they provide a basis on which to calculate melt densities and compressibilities (see also Bottinga et al., 1982, 1983; Bottinga, 1985). These data are of central importance when deducing the physics and chemistry associated with partial melting, magma aggregation and magma ascent from depth in the upper mantle (e.g., Stolper et al., 1981; Herzberg, 1987; Rigden et al., 1984, 1988; Agee and Walker, 1988).

Experimental determinations of partial molar volume of oxide components in silicate melts are often conducted with stepwise, linear least-squares regression, so that the partial molar volume of a component, i, is

$$\bar{V}_i = \frac{\text{melt volume/gfw} - \sum_{j=1}^{j} X_j \bar{V}_j}{M_i}, \quad (23)$$

where $\bar{V}^i$ is partial molar volume, gfw is gram formula weight, X_j and $\bar{V}_j$ are mole fraction and partial molar volume, respectively, of component j, and M_i is the molecular weight of component i. On this basis, both Mo et al. (1982) and Bottinga et al. (1983) derived partial molar volumes of ferric and ferrous oxide in silicate melts together with the volumes of the other oxide components. Both publications suggested that the partial molar volumes (Table 5) were independent of bulk composition over a very broad range of bulk compositions, including that of natural magmatic liquids. This conclusion however, has been challenged by Dingwell et al. (1988) in a recent study of partial molar volume of Fe_2O_3 in the simple system Na_2O–FeO–Fe_2O_3–SiO_2. A comparison of the data on partial molar volume for individual oxide components in Al-free and Al-bearing melts (Table 5) also indicates that there may be a compositional dependence of the partial molar volumes on the bulk composition of the melt. This dependence may exist in aluminosilicate melts because a portion of the metal cations is

Table 5. Partial molar volumes of oxide components (cm^3/mole).

	1250°C[a]	1300°C[a]	1400°C[a]	1400°C[b]	1500°C[a]	1600°C[a]
SiO_2	27.07	27.04	27.01	26.75	26.99	27.01
TiO_2	21.67	22.00	22.64	22.45	23.33	23.79
Al_2O_3	35.81	36.09	36.64	—	37.18	37.70
Fe_2O_3	42.86	43.16	43.89	44.40	44.37	44.68
FeO	13.28	13.44	13.73	13.94	14.27	14.82
MgO	11.25	11.32	11.46	12.32	11.57	11.62
CaO	15.38	15.70	16.32	12.32	16.96	17.57
Na_2O	27.74	28.10	28.80	29.03	29.48	30.10
K_2O	44.18	44.79	45.97	46.30	47.12	48.18

[a] Data from Mo et al. (1982).
[b] Data from Bottinga et al. (1983) for Al-free systems with thermal expansion coefficient given in their text.

associated with Al^{3+} (and Fe^{3+}) for charge-compensation, and another portion of the same oxide, with nonbridging oxygen. Such dual structural roles for individual melt components (the proportion of each is dependent on bulk composition) are likely to affect the measured partial molar volume of any such oxide. Without consideration of relevant structural components in the various simple and complex melt compositions studied, apparent discrepancies between densities and molar volume data from chemically complex systems such as natural magmatic liquids are not likely to be resolved.

The treatment of partial molar volumes in these recent papers is similar to the treatment of redox ratios of iron, as summarized in equation (1). The use of oxide components rather than melt structural components results in statistically reasonable data, but, from a melt structural point of view, it should be remembered that the approach is empirical rather than fundamental. The success with which one may treat volume relationships on the basis on melt structural considerations has been demonstrated by Gaskell (1982) for the system $FeO-SiO_2$ (Fig. 32). For highly basic melts in this system ($X_{SiO_2} < 0.4$), Gaskell made two observations. First, if the densities of the melts were calculated as simple mechanical mixtures of the end-member components (FeO and SiO_2), the resulting melt densities were significantly lower than the observed values (see dashed

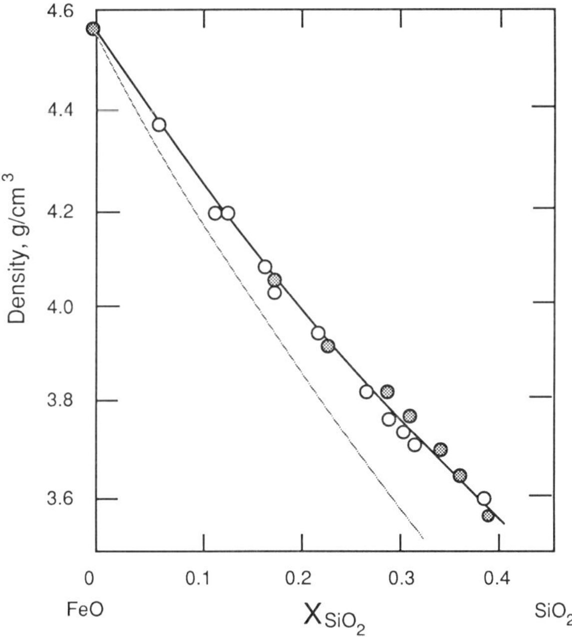

Fig. 32. Density of melts on the join $FeO-SiO_2$. Open circles, from Lee and Gaskell (1974); closed circles, from Shiraishi et al. (1978); dashed line, calculated density assuming mechanical mixing of SiO_2 and FeO; solid line, calculated density assuming coexisting anionic units in melts (see text and Gaskell, 1982).

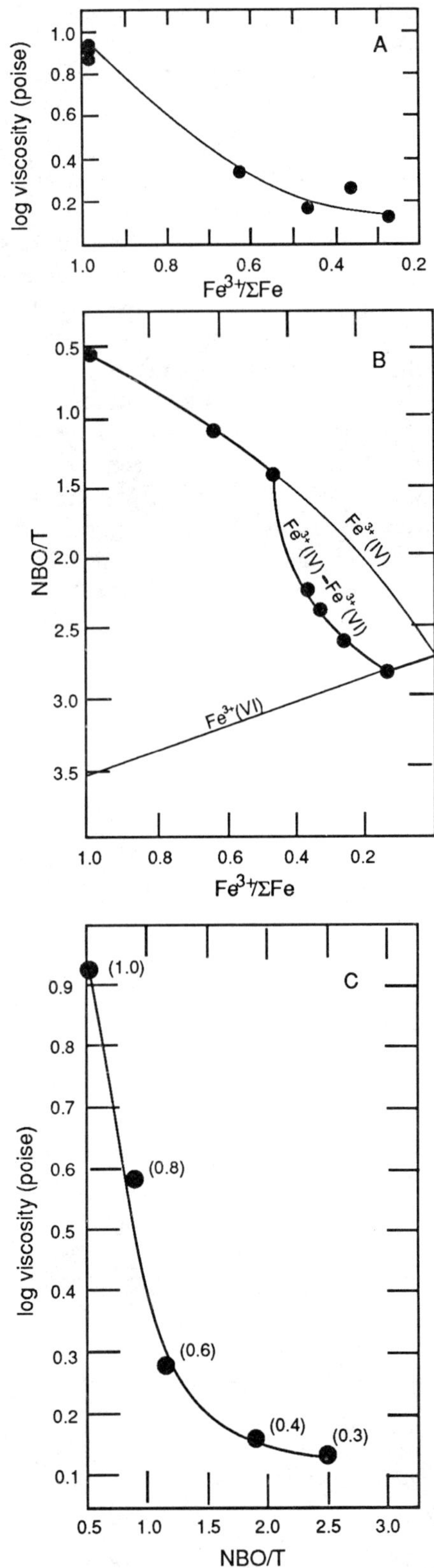

Fig. 33. Relationships between ferric/ferrous, *NBO/T* and viscosity of $Na_2Si_4O_9$ composition melt with 40 mol% Fe^{3+} (nominal) in replacement for Si^{4+} (data from Dingwell and Virgo, 1987). Numbers in parentheses in Fig. 33c represent $Fe^{3+}/\Sigma Fe$.

and solid lines in Fig. 32). As an alternative, Gaskell (1982) considered the melts as a mixture of free oxygen (O^{2-}), SiO_4^{4-}, and SiO_3^{2-} units these are most likely the structural units coexisting in melts as depolymerized as those shown in Fig. 32 (Mysen et al., 1982). The densities of the melts along the join $FeO-SiO_2$ could be modeled accurately. The solid line in Fig. 32 is calculated with this model (Gaskell, 1982). Although the relevant data base is not yet available for this kind of exercise for ferric iron, it is likely that similar conclusions might be reached for this (and other silicate melt) components.

Viscosity of Iron-Bearing Silicate Melts

The viscosities of natural silicate melts vary by several orders of magnitude as a function of temperature and bulk composition (see Shaw, 1972; Bottinga and Weill, 1972; for review of data). Available data also show that among the most important structural parameters affecting the melt viscosity are NBO/T (e.g., Bockris et al., 1955; Lacey, 1968), the bulk melt Al/(Al + Si) (e.g., Riebling, 1966; Rossin et al., 1964; Urbain et al., 1982), and the type of network-modifying cation (Bockris et al., 1955). The iron content also affects melt viscosity whether it occurs as FeO (Rontgen et al., 1960), as Fe_2O_3 (Klein et al., 1981), or as mixtures of the two (Cukierman and Uhlmann, 1974; Mysen et al., 1985d). Because melt polymerization (NBO/T) depends on $Fe^{3+}/\Sigma Fe$, it is likely, as shown by Cukierman and Uhlmann (1974) and Dingwell and Virgo (1987), that melt viscosity depends on the redox ratio of iron. In particular Dingwell and Virgo (1987) found that their viscosity data for melts in the system $Na_2O-FeO-Fe_2O_3-SiO_2$ could be rationalized by considering the observed changes in $Fe^{3+}/\Sigma Fe$ and the relationships between $Fe^{3+}/\Sigma Fe$ and the coordination state of ferric iron on the degree of polymerization of the melts (Fig. 33; see also Fig. 26). A rapid lowering of melt viscosity with lowering of $Fe^{3+}/\Sigma Fe$ until $Fe^{3+}/\Sigma Fe \sim 0.5$ (Fig. 33a) was ascribed to increasing NBO/T as illustrated in equation (4). With $Fe^{3+}/\Sigma Fe$ near 0.5 the Fe^{3+} begins to transform into octahedral coordination, and a further reduction in $Fe^{3+}/\Sigma Fe$ causes a small NBO/T increase [Fig. 33b; see also equation (17)], and a near independence of the melt viscosity on the $Fe^{3+}/\Sigma Fe$ (Fig. 33c). In Fig. 34, the effect on melt viscosity of Fe^{3+}(VI)/Fe^{2+}(VI) [labelled (VI) in Fig. 34] and Fe^{3+}(IV)/Fe^{2+}(VI) [labelled (IV) in Fig. 34)] are compared with the more probable situation where ferric iron undergoes a gradual coordination transformation as a function of the $Fe^{3+}/\Sigma Fe$ (labelled M in Fig. 34). Without coordination transformation of ferric iron, the melt viscosity will decrease continuously throughout the entire range of $Fe^{3+}/\Sigma Fe$ between 1 and 0. With Fe^{3+} in octahedral coordination over the entire range of $Fe^{3+}/\Sigma Fe$, the melt becomes polymerized as ferric iron is reduced and the viscosity increases. In the more realistic situation, as soon as the ferric iron begins to undergo a coordination transformation, the rate of increase of NBO/T initially increases and the curve labelled M shifts below the IV-curve in Fig. 34, but then slows

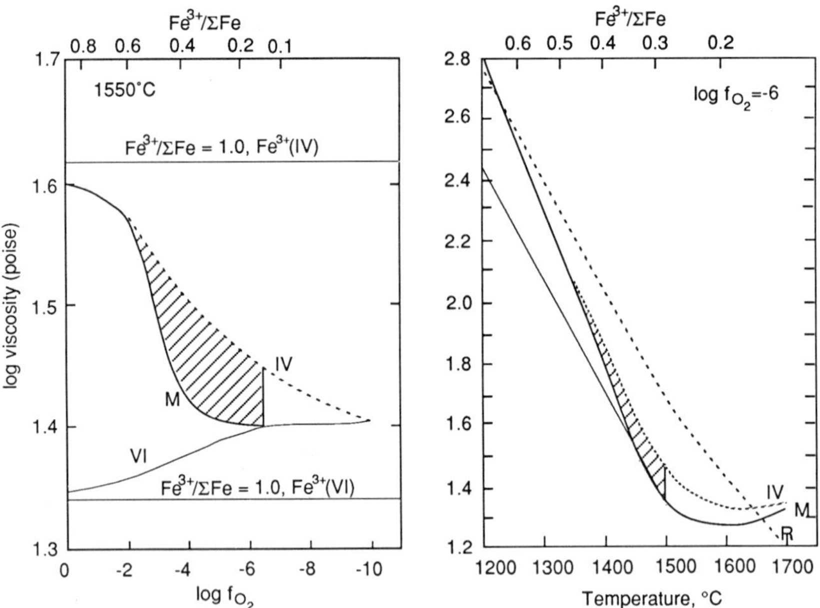

Fig. 34. Relationships between melt viscosity and ferric/ferrous for melt composition NASIVF5 [$NBO/T = 0.65$ and Al/(Al + Si) = 0.334 with 5 wt% iron oxide as Fe_2O_3]. Data from Mysen and Virgo (1989).

down as a larger and larger fraction of the remaining Fe^{3+} is in octahedral coordination. When all remaining ferric iron is in octahedral coordination, the curves M and VI will coincide (Fig. 34).

At times Fe^{3+} and Al^{3+} have been considered to behave similarly in silicate melt structures. Numerous exceptions to this suggestion have been discussed above. Differences between ferric iron and aluminum can also be seen in the viscous behavior of equivalent iron- and aluminum-bearing silicate melts. For example, from a comparative study of melt viscosity in the systems CaO–Al_2O_3–SiO_2 and CaO–Fe_2O_3–SiO_2 at 1 bar pressure, Mysen et al. (1985b) made two observations (Fig. 35). First, whereas increasing alumina content at constant temperature resulted in an increase in viscosity, increasing ferric iron of the same base composition on the join CaO–SiO_2 content resulted in a decrease (compare analogous F5, F10 and A5, A10 curves in Fig. 35). Second, in the aluminous system, the activation energies were independent of temperature (straight lines in Fig. 35). In the iron-bearing analogous system, on the other hand, the viscosity versus $1/T$ lines were distinctly curved. The viscosity curves can be fitted to a quadratic equation;

$$\log \eta = a + b\frac{1}{T} + c\frac{1}{T^2}, \tag{24}$$

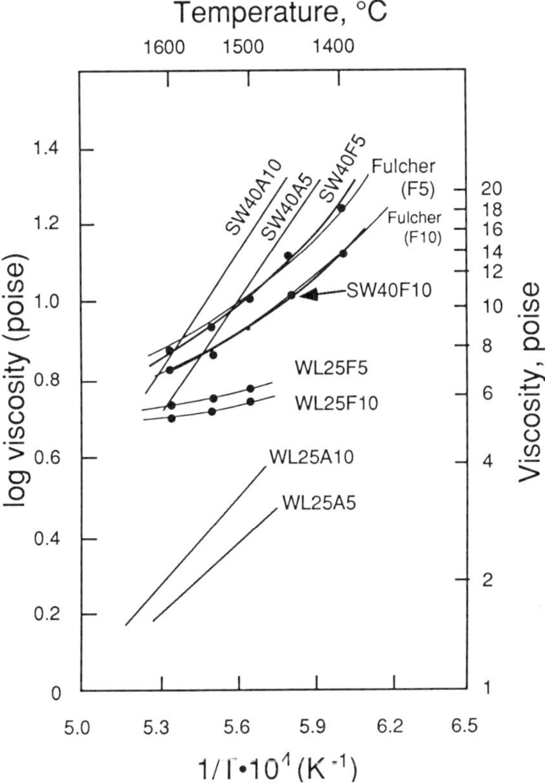

Fig. 35. Viscosity of melts in the system $CaO–SiO_2–Fe–O$ (Mysen et al., 1985c) and $CaO–Al_2O_3–SiO_2$ (Rossin et al., 1964) as a function of temperature. Symbols F5, F10, A5, and A10 denote wt percent Fe_2O_3 or Al_2O_3 added to base compositions (SW40: NBO/Si = 1.4; WL25: NBO/Si = 2.38).

which shows that the activation energy of viscous flow is a linear function of $1/T$:

$$E_\eta = 2.303\left(b + 2c\frac{1}{T}\right). \tag{25}$$

From this treatment, it is evident that activation energies of viscous flow of the iron-bearing melts decreased with increasing temperature. Raman and Mössbauer spectral data of the melts reveal that both Fe^{3+} and Al^{3+} were in tetrahedral coordination throughout the composition and temperature ranges of the study (Mysen et al., 1985b). From the measured $Fe^{3+}/\Sigma Fe$ of the liquids as a function of temperature, Mysen et al. (1985b) concluded that the decrease in bulk melt NBO/T that resulted from decreasing $Fe^{3+}/\Sigma Fe$ with increasing temperature could not account for the decrease in activation energy.

In contrast, it was suggested (Mysen et al., 1985b) that the activation energy changes might be explained with the aid of the melt structural data in conjunction with the configurational entropy model for viscous flow (see also Adam and Gibbs, 1965; Richet, 1984).

In the superliquidus temperature region many silicate liquids exhibit viscous behavior that may be described with the Arrhenius equation:

$$\log \eta = \log \eta_0 + E_\eta/\mathrm{R}T, \tag{26}$$

where η is viscosity, η_0, is the pre-exponential constant, E_η is the activation energy of viscous flow, R is the gas constant, and T is the absolute temperature. The activation energy of viscous flow may be a function of temperature, however, if the melt structure is temperature dependent. One such structural variable is nonbridging oxygens per tetrahedral cations in iron-bearing melts. Another may be variations in proportions of structural units in the melts as a function of temperature, leading to changes in configurational entropy of the melt. The configurational entropy may be related to activation energy from the expression (Richet, 1984):

$$E_\eta = \frac{\mathrm{R}B_e}{S_{conf}}, \tag{27}$$

where B_e includes the molar free hindrance energy (Adam and Gibbs, 1965) and S_{conf} is the configurational entropy. The value of the configuration entropy term in equation (27) may be assessed with knowledge of the temperature dependence of the proportion of structural units in the iron- and aluminum-bearing silicate melts. From this information, the mixing term:

$$S^{mix} = -\mathrm{R} \sum_{i=1}^{i} x_i \ln x_i, \tag{28}$$

where x_i is the mole fraction of structural unit i in the melt and may be calculated from Raman spectra (Seifert et al., 1981). Then changes in activation energy from condition a $[E_\eta(\mathrm{a})]$ to condition b $[E_\eta(\mathrm{b})]$ can be calculated:

$$\frac{E_\eta(\mathrm{a})}{E_\eta(\mathrm{a}) + E_\eta(\mathrm{b})} = \frac{\dfrac{1}{\sum_{i=1}^{i} x_i \ln x_i(\mathrm{a})}}{\dfrac{1}{\sum_{i=1}^{i} x_i \ln x_i(\mathrm{a})} + \dfrac{1}{\sum_{i=1}^{i} x_i \ln x_i(\mathrm{b})}}. \tag{29}$$

It was found (Mysen et al., 1985b) that changing iron oxide contents of SW40 base melts (NBO/Si = 1.40) from 5 to 10 wt% Fe_2O_3 at 1550°C, for example, resulted in agreement between observed and calculated changes in activation energies within 10%. For the aluminous system, calculated and observed changes in activation energies agreed within 15%. Although relevant detailed data on melt structure are relatively limited, both the data summarized by Richet (1984) and those of Mysen et al. (1985b) indicate that this treatment of viscous flow is promising.

Summary

The structure and properties of iron-bearing silicate melts may be summarized as follows:

1. A wide range of spectroscopic investigations has shown that ferrous iron is a network modifier in silicate melts. Ferric iron, on the other hand, occurs both as a network former (tetrahedral coordination) and as a modifier. As a network former, in oxidized melts, Fe^{3+} is charge-balanced with alkali metals and alkaline earths. There is an intermediate $Fe^{3+}/\Sigma Fe$ range (0.5–0.3) where ferric iron occurs predominantly in clusters compositionally resembling Fe_3O_4 with both tetrahedral and octahedral Fe^{3+}.
2. From data in simple binary metal oxide-silica and ternary metal oxide-alumina-silica systems, Fe^{2+}/Fe^{3+} is positively and generally linearly correlated with bulk melt Al/(Al + Si) and NBO/T, where, however, the free energy of reduction of ferric to ferrous iron decreases with increasing Al/(Al + Si) and NBO/T. The redox ratio of iron, Fe^{2+}/Fe^{3+}, is also positively correlated with Z/r^2 of the metal cation.
3. The redox ratio of iron [expressed as ln (Fe^{2+}/Fe^{3+})] can be related to the expression:

$$\ln\frac{Fe^{2+}}{Fe^{3+}} = a + \frac{b}{T} + c\ln f_{O_2} + d\frac{Al}{Al + Si} + e\frac{Fe^{3+}}{Fe^{3+} + Si^{4+}} + \sum_{j=1}^{j} f_j(NBO/T)_j$$

   This expression, calibrated against available laboratory-determined relationships from simple and complex natural systems (total of 460 analyses) yields calculated f_{O_2} values that accord with those determined in the laboratory for the same samples to within ± 1 log unit for 85% of the analyses.
4. Chemical and physical properties of silicate melts depend on melt polymerization (NBO/T). The NBO/T depends on $Fe^{3+}/\Sigma Fe$. It has been shown that properties such as crystal-liquid trace-element partitioning and liquidus phase equilibria can be related to $Fe^{3+}/\Sigma Fe$ and the structural positions of ferric and ferrous iron in the melts.
5. Physical properties such as melt density and viscosity are simple functions of the proportions of structural units in the silicate melts. These proportions can be determined for iron-bearing (and other) silicate melts from spectroscopic measurements. On this structural basis, the values of these properties can be predicted within 10–15% relative uncertainty.

References

Adam, G., and Gibbs, J.H. (1965) On the temperature dependence of cooperative relaxation properties in glass-forming liquids. *J. Chem. Phys.*, **43**, 139–146.

Agee, C.B., and Walker, D. (1988) Static compression and olivine flotation in ultrabasic silicate liquid. *J. Geophys. Res.*, **93**, 3437–3449.

Amthauer, G., Annersten, H., and Hafner, S.S. (1977) The Mössbauer spectrum of ^{57}Fe in titanium-bearing andradites. *Phys. Chem. Miner.*, **1**, 399–413.

Angell, C.A., Cheeseman. P.A., and Tamaddon, S. (1982) Pressure enhancement of ion mobilities in liquid silicates from computer simulation studies to 800 kbar. *Science*, **218**, 885–888.

Annersten, H., and Halenius, U. (1976) Ion distribution in pink muscovite: a discussion. *Amer. Mineral.*, **61**, 1045–1050.

Annersten, H., and Olesch, M. (1978) Distribution of ferrous and ferric iron in clintonite and the Mössbauer characteristics of ferric iron in tetrahedral coordination. *Can. Mineral.*, **16**, 199–204.

Bancroft, G.M., Williams, P.G.L., and Essene, E.J. (1969) Mössbauer spectra of omphacites. *Mineral. Soc. Amer. Spec. Pap.*, **2**, 59–65.

Bockris, J.O'M., and Kojonen, F. (1960) The compressibility of certain alkali silicates and borates. *J. Amer. Chem. Soc.*, **82**, 4493–4497.

Bockris, J.O'M., and Reddy, A.K.N. (1970) *Modern Electrochemistry, Vol. 1.* Plenum Press, New York.

Bockris, J.O'M., MacKenzie, J.D., and Kitchner, J.A. (1955) Viscous flow of silica and binary silicate liquids. *Trans. Faraday Soc.*, **51**, 1734–1748.

Bockris, J.O'M., Tomlinson, J.W., and White, J.L. (1956) The structure of liquid silicates. *Trans. Faraday Soc.*, **52**, 299–311.

Bottinga, Y. (1985) On isothermal compressibility of silicate liquids at high pressure. *Earth Planet. Sci. Lett.*, **74**, 350–360.

Bottinga, Y., and Weill, D.F. (1972) The viscosity of magmatic silicate liquids: a model for calculation. *Amer. J. Sci.*, **272**, 438–475.

Bottinga, Y., Weill, D.F., and Richet, P. (1981) Thermodynamic modeling of silicate melts. In: *Thermodynamics of Minerals and Melts*, pp. 207–247. Edited by R.C. Newton, A. Navrotsky and B.J. Wood. Springer Verlag, New York.

Bottinga, Y., Weill, D.F., and Richet, P. (1982) Density calculations for silicate liquids. I. Revised method for aluminosilicate components. *Geochim. Cosmochim. Acta*, **46**, 909–919.

Bottinga, Y., Richet, P., and Weill, D.F. (1983) Calculation of the density and thermal expansion coefficients of silicate liquids. *Bull. Mineral.*, **106**, 129–138.

Brandriss, M.E., and Stebbins, J.F. (1988) Effects of temperature on the structures of silicate liquids: ^{29}Si NMR results. *Geochim. Cosmochim. Acta*, **52**, 2659–2669.

Brawer, S.A. (1975) Theory of vibrational spectra of some network and molecular glasses. *Phys. Rev.*, **B11**, 3173–3194.

Brawer, S.A., and White, W.B. (1975) Raman spectroscopic investigation of the structure of silicate glasses. I. The binary silicate glasses. *J. Chem. Phys.*, **63**, 2421–2432.

Brawer, S.A., and White, W.B. (1977) Raman spectroscopic investigation of the structure silicate glasses. II. soda-alkaline earth-alumina ternary and quaternary glasses. *J. Noncrystall. Solids*, **23**, 261–278.

Brown, G.E., Keefer, K.D., and Fenn, P.M. (1978) Extended X-ray fine structure (EXAFS) study of iron-bearing silicate glass (Abstr.). *Progr. Geol. Soc. Amer.*, **10**, 373.

Calas, G., and Petiau, J. (1983) Structure of oxide glasses: spectroscopic studies of local order and crystallochemistry; geochemical implications. *Bull. Mineral.*, **106**, 33–55.

Calas, G., Levitz, P., Petiau, J., Bondot, P., and Loupias, G. (1980) Etude de l'order local autor du fer sans les verres silicate's naturels and synthetiques a l'aide de la spectrometrie d'absorption. *X. Rev. Phys. Appl.*, **15**, 1161–1167.

Carmichael, I.S.E., and Ghiorso, M.S. (1986) Oxidation-reduction relations in basic magma: a case for homogeneous equilibria. *Earth Planet. Sci. Lett.*, **78**, 200–210.

Carmichael, I.S.E., Turner, F.J., and Verhoogen, J. (1974) *Igneous Petrology*. McGraw-Hill, New York.

Chayes, F. (1975a) Average composition of the commoner cenozoic volcanic rocks. *Carnegie Instn. Wash. Year Book.*, **74**, 547–549.

Chayes, F. (1975b) A world data base for igneous petrology. *Carnegie Instn. Washington Year Book*, **74**, 549–550.

Chayes, F. (1985) *Version NTRM2 of System RKNFSYS*. Unpublished document, Geophysical Laboratory.

Cooney, T., Sharma S.K., and Urmos, J.P. (1987) Structure of glasses along the join forsterite–fayalite and tephroite–fayalite. EOS, **68**, 436.

Cukierman, M., and Uhlmann, D. R. (1974) Effect of iron oxidation state on viscosity, lunar composition 1555. *J. Geophys. Res.*, **79**, 1594–1598.

Dickenson, M.P., and Hess, P.C. (1981) Redox equilibria and the structural role of iron in aluminosilicate melts. *Contrib. Mineral. Petrol.*, **78**, 352–358.

Dickinson, J.E., and Hess, P.C. (1985) Rutile solubility and titanium coordination in silicate melts. *Geochim. Cosmochim. Acta*, **49**, 2289–2296.

Dingwell, D.B. (1986) Viscosity-temperature relationships in the system $Na_2Si_2O_5$–$Na_4Al_2O_5$. *Geochim. Cosmochim. Acta*, **50**, 1261–1265.

Dingwell, D.B. (1989) Effect of fluorine on the viscosity of diopside liquid. *Amer. Mineral.*, **74**, 333–338.

Dingwell, D.B., and Brearley, M. (1988) Melt densities in the CaO–FeO–Fe_2O_3–SiO_2 system and the compositional dependence of the partial molar volume of ferric iron in silicate melts. *Geochim. Cosmochim. Acta*, **52**, 2815–2825.

Dingwell, D.B., and Virgo, D. (1987) The effect of oxidation state on the viscosity of melts in the system Na_2O–FeO–Fe_2O_3–SiO_2. *Geochim. Cosmochim. Acta*, **51**, 195–205.

Domine, F., and Piriou, B. (1986) Raman spectroscopic study of the SiO_2–Al_2O_3–K_2O vitreous system: Distribution of silicon and second neighbors. *Amer. Mineral.*, **71**, 38–50.

Dupree, R., Holland D., and Mortuza, M.G. (1987) Six-coordinated silicon in glasses. *Nature*, **328**, 416–417.

Dyar, D.M. (1983) Effect of quench media on iron-bearing glasses quenched from melts. *Geol. Soc. Amer. Ann. Meeting*, **15**(6), 564.

Dyar, M.D. (1985) A review of Mossbauer data on inorganic glasses: The effect of composition on iron valency and coordination. *Am. Mineral.*, **70**, 304–317.

Dyar, M.D., and Bums, R.G. (1981) Chemistry of iron in glass contributing to remote-sensed spectra of the moon. *Proc. Lunar Planer. Sci. Conf.*, **12B**, 965–702.

Fox, K.E., Furukawa, T., and White, W.B. (1982) Transition metal ions in silicate melts. 2. Iron in sodium silicate glasses. *Phys. Chem. Glasses*, **23**, 169–178.

Fudali, F. (1965) Oxygen fugacity of basaltic and andesitic magmas. *Geochim. Cosmochim. Acta*, **29**, 1063–1075.

Fulcher, G.S. (1925) Analysis of recent measurements of the viscosity of glasses. *Amer. Ceram. Soc. Bull.*, **8**, 339–355.

Furukawa, T., and White, W.B. (1979) Structure and crystallization of glasses in the $Li_2Si_2O_5$–TiO_2 system determined by Raman spectroscopy. *Phys. Chem. Glasses*, **20**, 69–80.

Furukawa, T., Fox, K.E., and White, W.B. (1981) Raman spectroscopic investigation of the structure of silicate glasses. III. Raman intensities and structural units in sodium silicate glasses. *J. Chem. Phys.*, **75**, 3226–3237.

Gaskell, D.M. (1982) The densities and structures of silicate melts. In *Advances in Physical Geochemistry, Vol. 2*, edited by S.K. Saxena, pp. 153–171. Springer-Verlag, New York.

Goldman, D.S. (1983) Oxidation equilibrium of iron in borosilicate glass. *J. Amer. Ceram. Soc.*, **66**, 205–209.

Greig, J.W. (1927) Immiscibility in silicate melts, pt. II. *Amer. J. Sci.*, **13**, 133–155.

Hafner, S.S., and Huckenholz, H.G. (1971) Mössbauer spectrum of synthetic ferridiopside. *Nature Phys. Sci.*, **233**, 9–11.

Hart, S.R., and Davis, K.E. (1978) Nickel partitioning between olivine and silicate melt. *Earth Planet. Sci. Lett.*, **40**, 203–220.

Herzberg, C.T. (1987) Magma density at high pressure. Part 1. The effect of composition on the elastic properties of silicate liquids. In: *Magmatic Processes: Physicochemical Principles*, edited by B.O. Mysen, pp. 25–46. Geochemical Society of America. Spec. Publ. No. 1. 500 pp.

Hess, P.C. (1980) Polymerization model for silicate melts. Ch. 1. In: *Physics of Magmatic Processes*, edited by R.B. Hargraves, Princeton University Press, Princeton, 587 pp.

Holmquist, S. (1966) Ionic formulation of redox equilibria in glass melts. *J. Amer. Ceram. Soc.*, **49**, 228–229.

Irving, A.J. (1978) A review of experimental studies of crystal/liquid trace element partitioning. *Geochim. Cosmochim. Acta*, **42**, 743–771.

Jones, D.R., IV, Jansheski, W.C., and Goldman, D.S. (1981) Spectrometric determination of reduced and total iron in glass with 1,10-phenantroline. *Anal. Chem.*, **53**, 923–924.

Kennedy, G.C. (1948) Equilibrium between volatiles and iron oxides in rocks. *Amer. J. Sci.*, **246**, 529–549.

Kilinc, A., Carmichael, I.S.E., Rivers, M.L., and Sack, R.O. (1983) The ferric-ferrous ratio of natural silicate liquids equilibrated in air. *Contrib. Mineral. Petrol.*, **83**, 136–141.

Kirkpatrick, R.J., Smith, K.A., Kinsley, R.A., and Oldfield, E. (1982) High-resolution ^{29}Si NMR of silicate glasses and crystals in the system $CaO-MgO-SiO_2$ (abstr.). *EOS*, **63**, 1140.

Kirkpatrick, R.J., Oestrike, R., Weiss, C.A., Smith, K.A., and Oldfield, E. (1986) High-resolution ^{27}Al and ^{29}Si NMR spectroscopy of glasses and crystals along the join $CaMgSi_2O_6-CaAl_2SiO_6$. *Amer. Mineral.*, **71**, 705–711.

Klein, L.C., Fassano, B.V., and Wu, J.M. (1981) Flow behavior of ten iron-containing silicate compositions. *Proc. Lunar Sci. Conf.*, **12B**, 1759–1767.

Konnert, J.H., and Karle, L. (1974) The computation of radial distribution functions for glassy materials. *Acta Crystallogr.*, **A29**, 702–710.

Kushiro, I. (1975) On the nature of silicate melt and its significance in magma genesis: regularities in the shift of liquidus boundaries involving olivine, pyroxene and silica minerals. *Amer. J. Sci.*, **275**, 411–431.

Kushiro, I. (1976) Changes in viscosity and structure of melt of $NaAlSi_2O_6$ composition at high pressures. *J. Geophys. Res.*, **81**, 6347–6350.

Kushiro, I. (1980) Viscosity, density and structure of silicate melts at high pressures, and their petrological applications. In: *Physics of Magmatic Processes*, edited R.B. Hargraves, pp. 93–121. Princeton University Press, Princeton, NJ.

Kushiro, I. (1981) Viscosity change with pressure of the melts in the system $CaO-Al_2O_3-SiO_2$, *Carnegie Inst. Washington Year Book*, **80**, 339–311.

Lacey, E.D. (1968) Structure transition in alkali silicate glasses. *J. Amer. Ceram. Soc.*, **51**, 150–157.

Larson, H., and Chipman, J. (1953) Oxygen activity in iron oxide slags. *Trans. AIME*, **196**, 1089–1096.

Lee, Y.E., and Gaskell, D.M. (1974) The densities and structures of melts in the system $CaO-FeO-SiO_2$, *Metall. Trans.*, **5**, 853.

Levy, R.A., Lupis, C.H.P., and Flinn, P.A. (1976) Mössbauer analysis of the valence and coordination of iron cations in SiO_2–Na_2O–CaO glasses. *Phys. Chem. Glasses*, **17**, 94–103.

Mao, H.-K., Virgo, D., and Bell, P.M. (1973) Analytical study of the orange soil returned by the Apollo 17 astronauts. *Carnegie Inst. Washington Year Book*, **72**, 631–638.

Matson, D.W., Sharma, S.K., and Philpotts, J.A. (1983) The structure of high-silica alkali-silicate glasses—a Raman spectroscopic investigation. *J. Noncrystall. Solids*, **58**, 323–352.

McKeown, D. A., Galeener, F.L., and Brown, G.E. (1984) Raman studies of the Al-coordination in silica-rich sodium aluminosilicate glasses and some related minerals. *J. Noncrystall. Solids*, **68**, 361–378.

McMillan, P. (1984) A Raman spectroscopic study of glasses in the system CaO–MgO–SiO_2, *Amer. Mineral.*, **69**, 645–659.

McMillan, P., and Piriou, B. (1983) Raman spectroscopic studies of silicate and related glass structure. *Bull. Mineral.*, **106**, 57–77.

McMillan, P., Piriou, B., and Navrotsky, A. (1982) A Raman spectroscopic study of glasses along the joins silica-calcium aluminate, silica-sodium aluminate and silica-potassium aluminate. *Geochim. Cosmochim. Acta*, **46**, 2021–2037.

Mitra, S.K. (1982) Molecular dynamics simulation of silicon dioxide glass. *Phil. Mag.*, **B45**, 529–548.

Mo, X., Carmichael, I.S.E., Rivers, M., and Stebbins, J. (1982) The partial molar volume of Fe_2O_3 in multicomponent silicate liquids and the pressure dependence of oxygen fugacity in magmas. *Mineral. Mag.*, **45**, 237–245.

Muan, A., and Osborn, E.F. (1956) Phase equilibria and liquidus temperatures in the system MgO–FeO–Fe_2O_3–SiO_2, *J. Amer. Ceram. Soc.*, **39**, 121–140.

Murdoch, J.B., Stebbins, J.F., and Carmichael, I.S.E. (1985) High-resolution ^{29}Si NMR study of silicate and aluminosilicate glasses: The effect of network-modifying cations *Amer. Mineral.*, **70**, 332–343.

Mysen, B.O. (1988) *Structure and Properties of Silicate Melts*. Amsterdam, Elsevier, 354 pp.

Mysen, B.O. (1990a) The role of aluminum in depolymerized, peralkaline aluminosilicate melts in the systems Li_2O–Al_2O_3–SiO_2, Na_2O–Al_2O_3–SiO_2 and K_2O–Al_2O_3–SiO_2. *Amer. Mineral.*, **57**, 120–134.

Mysen, B.O. (1989b) Effect of pressure, temperature, and bulk composition on the structure and species distribution in depolymerized alkali aluminosilicate melts and quenched melts. *J. Geophys. Res.*, **95**, 15733–15744.

Mysen, B.O., and Virgo, D. (1978) Influence of pressure, temperature and bulk composition on melt structures in the system $NaAlSi_2O_6$–$NaFe^{3+}Si_2O_6$. *Am. J. Sci.*, **278**, 1307–1322.

Mysen, B.O., and Virgo, D. (1980) Trace element partitioning and melt structure: an experimental study at 1 atm pressure. *Geochim. Cosmochim. Acta*, **44**, 1917–1930.

Mysen, B.O., and Virgo, D. (1983) Effect of pressure on the structure of iron-bearing silicate melts. *Carnegie Inst. Washington Year Book*, **82**, 321–325.

Mysen, B.O., and Virgo, D. (1985) Iron-bearing silicate melts: relations between pressure and redox equilibria. *Phys. Chem. Minerals.*, **12**, 191–200.

Mysen, B.O., and Virgo, D. (1989) Redox equilibria, structure, and properties of Fe-bearing aluminosilicate melts: Relationships between temperature, composition, and oxygen fugacity in the system Na_2O–Al_2O_3–SiO_2–Fe–O. *Amer. Mineral.*, **74**, 58–76.

Mysen, B.O., Seifert, F.A., and Virgo, D. (1980a) Structure and redox equilibria of iron-bearing silicate melts. *Amer. Mineral.*, **65**, 867–884.

Mysen, B.O., Virgo, D., and Scarfe, C.M. (1980b) Relations between the anionic structure and viscosity of silicate melts—a Raman spectroscopic study. *Amer. Mineral.*, **65**, 690–710.

Mysen, B.O., Ryerson, F.J., and Virgo, D. (1980c) The influence of TiO_2 on the structure and derivative properties of silicate melts. *Amer. Mineral.*, **65**, 1150–1165.

Mysen, B.O., Virgo, D., and Kushiro, I. (1981a) The structural role of aluminum in silicate melts—a Raman spectroscopic study at 1 atmosphere. *Amer. Mineral.*, **66**, 678–701.

Mysen, B.O., Ryerson, F.J., and Virgo, D. (1981b) The structural role of phosphorus in silicate melts. *Amer. Mineral.*, **66**, 106–117.

Mysen, B.O., Virgo, D., and Seifert, F.A. (1982) The structure of silicate melts: implications for chemical and physical properties of natural magma. *Rev. Geophys.*, **20**, 353–383.

Mysen, B.O., Virgo, D., Danckwerth, P., Seifert, F.A., and Kushiro, I. (1983) Influence of pressure on the structure of melts on the joins $NaAlO_2$–SiO_2, $CaAl_2O_4$–SiO_2 and $MgAl_2O_4$–SiO_2, *Neues. Jahrb. Mineral. Abh.*, **147**, 281–303.

Mysen, B.O., Virgo, D., and Seifert, F.A. (1984) Redox equilibria of iron in alkaline earth silicate melts: relationships between melt structure, oxygen fugacity, temperature and properties of iron-bearing silicate liquids. *Amer. Mineral.*, **69**, 834–847.

Mysen, B.O., Virgo, D., Neumann, E.-R., and Seifert, F.A. (1985a) Redox equilibria and structural states of ferric and ferrous iron in melts in the system CaO–MgO–Al_2O_3–SiO_2–Fe–O: relationships between redox equilibria, melt structure and liquidus phase equilibria. *Amer. Mineral.*, **70**, 317–331.

Mysen, B.O., Virgo, D., and Seifert, F.A. (1985b) Relationship between properties and structure of aluminosilicate melts. *Amer. Mineral.*, **70**, 88–105.

Mysen, B.O., Virgo, D., Scarfe, C.M., and Cronin, D.J. (1985c) Viscosity and structure of iron- and aluminum-bearing calcium silicate melts. *Amer. Mineral.*, **70**, 487–498.

Mysen, B.O., Carmichael, I.S.E., and Virgo, D. (1985d) A comparison of iron redox ratios in silicate glasses determined by wet-chemical and ^{57}Fe Mössbauer resonant absorption methods. *Contrib. Mineral. Petrol.*, **90**, 101–106.

Navrotsky, A., Peraudeau, P., McMillan, P., and Coutoures, J.-P. (1982) A thermochemical study of glasses and crystals along the joins silica-calcium aluminate and silica-sodium aluminate. *Geochim. Cosmochim. Acta*, **46**, 2039–2049.

Navrotsky, A., Geisinger, K.L., McMillan, P., and Gibbs, G.V. (1985) The tetrahedral framework in glasses and melts—influences from molecular orbital calculations and implications for structure, thermodynamics, and physical properties. *Phys. Chem. Minerals*, **11**, 284–298.

Nelson, C., and Tallant, D.R. (1984) Raman studies of sodium silicate glasses with low phosphate contents. *Phys. Chem. Glasses*, **25**, 31–39.

Nelson, C., and Tallant, D.R. (1986) Raman studies of sodium phosphates with low silica contents. *Phys. Chem. Glasses*, **26**, 119–122.

Nolet, D.A., Burns, R.G., Flamm, S.L., and Besancon, J.R. (1979) Spectra of Fe–Ti silicates: implications to remote-sensing of planetary surfaces. *Proc. Lunar Planet. Sci. Conf.*, **10**, 1775–1786.

Oestrike, R., Yang, W., Kirkpatrick, R.J., Hervig, R., Navrotsky, A., and Montez, B. (1987) High-resolution ^{23}Na, ^{27}Al and ^{29}Si NMR spectroscopy of framework-aluminosilicate glasses. *Geochim. Cosmochim. Acta*, **51**, 2199–2210.

Ohtani, E., Taulelle, F., and Angell, C.A. (1985) Al^{3+} coordination changes in liquid silicates under pressure. *Nature*, **314**, 31–314.

Osborn, E.F., and Muan, A. (1960a) The system CaO–"FeO"–SiO_2, Plate 7, Phase equilibrium diagrams of oxide systems. American Ceramic Society, Columbus, OH.

Osborn, E.F., and Muan, A. (1960b) The system CaO–"Fe_2O_3"–SiO_2, Plate 10, Phase equilibrium diagrams of oxide systems. American Ceramic Society, Columbus, OH.
Paul, A., and Douglas, R.W. (1965) Ferrous-ferric equilibrium in binary alkali silicate glasses. *Phys. Chem. Glasses*, **6**, 207–211.
Phillips, B., and Muan, A. (1959) Phase equilibria in the system CaO-iron oxide-SiO_2 in air. *J. Amer. Ceram. Soc.*, **42**, 413–423.
Richet, P. (1984) Viscosity and configurational entropy of silicate melts. *Geochim. Cosmochim. Acta*, **48**, 471–483.
Riebling, E.F. (1966) Structure of sodium aluminosilicate melts containing at least 50 mole % SiO_2 at 1500°C. *J. Chem. Phys.*, **44**, 2857–2865.
Rigden, S.M., Ahrens, T.J., and Stolper, E.M. (1984) Densities of liquid silicates at high pressures. *Science*, **226**, 1071–1074.
Rigden, S.M., Ahrens, T.J., and Stolper, E.M. (1985) Shock compression of molten silicate: results for a model basaltic composition. *J. Geophys. Res.*, **93**, 367–382.
Rontgen, H., Winterhager, P., and Kammel, R. (1960) Struktur und Eigenschaften von Schlacken der Metallhuttenprozesse. II. Viskositatsmessungen an Schmelzen der Systeme Eisenoxydul-Kalk-Kiesesaure und Eisen oxydul-Kalk-Tonerde-Kieselsaure. *Z. Erzbergbau. Metallhuttenwes.*, **8**, 363–373.
Rossin, R., Berson, J., and Urbain, G. (1964) Etude de laviscosite de laitiers liquides appertenant au systeme ternaire: SiO_2–Al_2O_3–CaO, *Rev. Hautes Temp. Refract.*, **1**, 159–170.
Ryerson, F.J. (1985) Oxide solution mechanisms in silicate melts: systematic variations in the activity coefficient of SiO_2. *Geochim. Cosmochim. Acta*, **49**, 637–650.
Ryerson, F.J., and Hess, P.C. (1980) The role of P_2O_5 in silicate melts. *Geochim. Cosmochim. Acta*, **44**, 611–625.
Sack, R.O., Carmichael, I.S.E., Rivers, M., and Ghiorso, M.S. (1980) Ferric-ferrous equilibria in natural silicate liquids at 1 bar. *Contrib. Mineral. Petrol.*, **75**, 369–377.
Scarfe, C.M. (1977) Structure of two silicate rock melts charted by infrared absorption spectroscopy. *Chem. Geol.*, **15**, 77–80.
Scarfe, C.M., Mysen, B.O., and Virgo, D. (1987) Pressure dependence of the viscosity of silicate melts. In: *Magmatic Processes: Physicochemical Principles*, edited by B.O. Mysen, pp. 59–68. Geochemical Society, Spec. Publ. No. 1.
Schneider, E., Stebbins, J.F., and Pines, A. (1987) Speciation and local structure in alkali and alkaline earth silicate glasses: Constraints from 29-Si NMR spectroscopy. *J. Noncrystall. Solids*, **89**, 371–383.
Seifert, F.A., and Olesch, M. (1977) Mössbauer spectroscopy of grandidierite, (Ca, Mg) Al_3BSiO_9. *Amer. Mineral.*, **62**, 547–553.
Seifert, F.A., Virgo, D., and Mysen, B.O. (1979) Melt structures and redox equilibria in the system Na_2O–FeO–Fe_2O_3–Al_2O_3–SiO_2. *Carnegie Inst. Washington Year Book*, **78**, 511–519.
Seifert, F.A., Mysen, B.O., and Virgo, D. (1981) Quantitative determination of proportions of anionic units in silicate melts. *Carnegie Inst. Washington Year Book*, **80**, 301–302.
Seifert, F.A., Mysen, B.O., and Virgo, D. (1982) Three-dimensional network melt structure in the systems SiO_2–$NaAlO_2$, SiO_2–$CaAl_2O_4$ and SiO_2–$MgAl_2O_4$. *Amer. Mineral.* **67**, 696–718.
Seifert, F.A., Mysen, B.O., and Virgo, D. (1983) Raman study of densified vitreous silica. *Phys. Chem. Glasses*, **24**, 141–145.
Sharma, S.K., Virgo, D., and Mysen, B.O. (1979a) Raman study of the coordination of aluminum in jadeite melt as a function of pressure. *Amer. Mineral.*, **64**, 779–788.
Sharma, S.K., Hoering, T.C., and Yoder, H.S. (1979b) Quenched melts of akermanite

compositions with and without CO_2—characterization by Raman spectroscopy and gas chromatography. *Carnegie Instn. Washington, Year Book*, **78**.

Shaw, H.R. (1972) Viscosities of magmatic silicate liquids: an empirical method of prediction. *Amer. J. Sci.*, **272**, 870–893.

Shiraishi, Y., Ikeda, K., Tamura, A., and Saito, T. (1978) On the viscosity and density of molten $FeO–SiO_2$ systems. *Trans. Japan Inst. Metals*, **19**, 264–274.

Stebbins, J.F. (1987) Identification of multiple structural species in silicate glasses by ^{29}Si NMR. *Nature*, **330**, 465–467.

Stebbins, J.F. (1989) Effects of temperature and composition on silicate glass structure and dynarnics: Si-29 NMR results. *J. Noncrystall. Solids* (in press).

Stolper, E., Walker, D., Hager, B.H., and Hays, J.F. (1981) Melt segregation from partially molten source regions: The importance of melt density and source region size. *J. Geophys. Res.*, **86**, 6261–6271.

Taylor, M., and Brown, G.E. (1979a) Structure of mineral glasses. I. The feldspar glasses $NaAlSi_3O_8$, $KAlSi_3O_8$ and $CaAl_2Si_2O_8$. *Geochim. Cosmochim. Acta*, **43**, 61–77.

Taylor, M., and Brown, G.E. (1979b) Structure of mineral glasses. II. The $SiO_2–NaAlSiO_4$ join. *Geochim. Cosmochim. Acta*, **43**, 1467–1475.

Thornber, C.R., Roeder, P.L., and Foster, J.R. (1980) The effect of composition on the ferric-ferrous ratio in basaltic liquids at atmospheric pressure. *Geochim. Cosmochim. Acta*, **44**, 525–533.

Tomlinson, J.W., Heynes, M.S.R., and Bockris, J.O'M. (1958) The structure of liquid silicates. Part 2. Molar volume and expansivities. *Trans. Faraday Soc.*, **54**, 1822–1834.

Urbain, G., Bottinga, Y., and Richet, P. (1982) Viscosity of liquid silicates and aluminosilicates. *Geochim. Cosmochim. Acta*, **46**, 1061–1072.

Virgo, D., and Mysen, B.O. (1985) The structural state of iron in oxidized vs. reduced glasses at 1 atm: a ^{57}Fe Mössbauer study. *Phys. Chem. Miner.*, **12**, 65–76.

Virgo, D., Mysen, B.O., and Kushiro, I. (1980) Anionic constitution of silicate melts quenched at 1 atm from Raman spectroscopy: implications for the structure of igneous melts. *Science*, **208**, 1371–1373.

Virgo, D., Mysen, B.O., and Seifert, F.A. (1981) Relationship between the oxidation state of iron and structure of silicate melts. *Carnegie Inst. Washington Year Book*, **80**, 308–311.

Virgo, D., Mysen, B.O., and Danckwerth, P.D. (1983) The coordination of Fe^{3+} in oxidized vs. reduced aluminosilicate glasses: a ^{57}Fe Mössbauer study. *Carnegie Inst. Washington Year Book*, **82**, 309–313.

Visser, W., and Van Groos, A.F.K. (1979) Effect of pressure on the liquid immiscibility in the system $K_2O–FeO–Al_2O_3–SiO_2–P_2O_5$. *Amer. J. Sci.*, **279**, 1160–1175.

Watson, E.B. (1977) Partitioning of manganese between forsterite and silicate liquid. *Geochim. Cosmochim. Acta*, **41**, 1363–1374.

Waychunas, G.A., and Rossman, G.R. (1983) Spectroscopic standard for tetrahedrally coordinated ferric iron: $\gamma LiAlO_2:Fe^{3+}$. *Phys. Chem. Miner.*, **9**, 212–215.

Waychunas, G.A., Apted, M.J., and Brown, G.E. (1983) X-ray K-edge absorption spectra of Fe minerals and model compounds: near-edge structure. *Phys. Chem. Miner.*, **10**, 1–9.

Waychunas, G.A., Brown, G.E., Ponader, C.W., and Jackson, W.E. (1988) Evidence from X-ray absorption for network-forming Fe^{2+} in molten silicates. *Nature*, **332**, 251–253.

Wilson, A.D. (1960) The microdetermination of ferrous iron in silicate minerals by volumetric and a colorimetric method. *Analyst*, **85**, 823–827.

Wones, D.R., and Eugster, H.P. (1962) Stability relations of the ferruginous biotite, annite. *J. Petrol.*, **3**, 82–125.

Chapter 3 Diffusion in Magmatic Melts: New Study

Anatoly S. Chekhmir and Mark B. Epel'baum

Introduction

Studying the physical properties of magmatic melts is one of the basic problems in physico-chemical petrology and geochemistry. Transport properties of magmatic melts and molecular diffusion of components, in particular, may certainly be regarded as the most important physical properties.

It is difficult to indicate a stage of magmatic evolution that does not involve mass-exchange processes. The early stage of magma generation, the process of magmatic intrusion, the interaction between magma and enclosing rocks or transmagmatic fluids and crystallization of magma,—these processes are all controlled, at least to some extent, by a diffusion mechanism.

Mass-exchange processes in magma chambers may proceed by different mechanisms such as convectional mass-exchange or extraction of components by a fluid phase. Nevertheless, molecular diffusion of components in a melt remains one of the elementary and limiting processes.

Molecular diffusion coefficients of the main components of magmatic melts, together with the available data on viscosity and surface tension at the melt-fluid boundary, can help in solution of many problems on magma generation and magmatic evolution to be solved.

Previous Works

The first experimental study of diffusion was made by Bowen as long ago as 1921. All the currently available experimental publications on experiments with diffusion processes in silicate melts may be divided into the following three groups.

The first group includes experiments performed before the early seventies and is concerned, basically with slags and glass systems. Epelbaum (1980) has reviewed the literature on these systems, therefore there is no need to consider them in detail.

In the mid-seventies the transport properties of magmatic melts attracted particular interest among petrologists and geochemists, and experimental studies on the subject appeared. Therefore, this second group involves measurements of ion diffusivities in glasses and melts that approach in composition those of geological interest under normal conditions.

The third group concerns experiments on magmatic melts or their synthetic models at high pressures in the presence of a fluid phase, mostly H_2O.

Experimental results of the last-mentioned two groups are considered here. The present paper is not concerned with the details of the publications available, but considers only the main conclusions.

Jambon & Carron (1976) and Jambon & Delbove (1977) performed a series of experiments to measure the diffusion coefficients of alkali ions (Li, Na, K, Rb, Cs) and Ca in glasses of albite, orthoclase and obsidian. They also discovered that diffusivities of alkali ions are dependent on their ionic radii. Magaritz and Hofmann (1978) showed that diffusivity of alkali ions (Na) is two to three orders of magnitude higher than that of alkaline earths (Ba, Sr). Similar results obtained by Hofmann and Magaritz (1977) and Lowry et al. (1981) regarding diffusion of different components (Ca, Sr, Ba, Co, Li) in basaltic melts confirm the above statement.

Jambon et al. (1978) studied diffusion of Li and Cs in granite melt of pressure of about 3 kbar. Watson (1979a,b, 1982) measured the diffusion coefficients of Ca up to pressures of 20 kbar and between 1100–1400°C and showed that, under isothermal conditions, the diffusion coefficients of Ca decrease with increasing total pressure. Watson also determined the volume activation of diffusion and their relationships with temperature.

Other publications of interest will be discussed as needed in different section of the present paper.

Experimental Investigation of Diffusion Coefficients of the Major Rock-forming Oxides

Experimental Procedures

The geological literature includes descriptions of large numbers of diversified techniques that can be used to determine diffusion coefficients. However, specific peculiarities of high temperature—high pressure experiments restrict the method applicable considerably.

In recent years most investigators have used techniques based on radioactive isotopes. Together with unquestionable merits, i.e. accuracy of determining the matter distribution in a sample, these methods all have restrictions. For example, radioactive isotopes may be inapplicable to processes of multicomponent diffusion and generally do not model the processes occurring in nature very well.

The development and use of microprobe analysis makes it possible to discard the method based on radioactive isotopes in favor of that consisting of making a solution of mineral or a crystalline rock in the melt. The diffusion distribution of the components of dissolving crystalline phases can then be examined. Furthermore, the high resolution obtained from a microprobe (2–3 μm) enables relatively short-term runs to be used, even in acid melts that feature high viscosity and hence thin developing-diffusion zones. On the other hand, the contact of crystalline phase and melt favors non-standard diffusion into a semi-infinite medium, with concentration of the diffusing component at the interface constant. This concentration is identical to the liquidus composition on the diagram of two-component system melt-contact mineral under P–T run conditions. In other words, as long as the crystalline phase exists, the melt composition at its boundary is not changed. The solutions to Fick's equation for a given set of conditions can be found in Crank (1956).

A contact mineral should be stable under the P–T run conditions and have a higher melting temperature. To obtain diffusion profiles of sufficient quality, the water–albite melt, which meets the following requirements, has been chosen:

(i) Its physical (viscosity) and structural properties match those of the granitic melts.
(ii) It is homogeneous in composition
(iii) There are no convective flows
(iiii) It is easily obtained.

Fig. 1 shows a diagram of variations in viscosity of granitic melts versus X_{H_2O}, T, P (according to Persikov, 1984). This diagram is used to characterize the viscosity run conditions resulting from the close relationship between viscosity of melts and diffusivity of components. Under given P–T–X_{H_2O} run conditions (from 0.5 to 6 kbar, 1000–1100°C and 1.7–8.5 wt% H_2O in the melt) the viscosity of albite melt (Persikov and Epelbaum, 1982) is equal to that of granitic magmas in the wide range of thermodynamic parameters (denoted as a shaded region in Fig. 1).

Consequently, the diffusion coefficients measured are characteristic of the transport properties of components in granitic magmas at the most important P–T parameters.

Results and Discussion

The contact minerals used and values of diffusion coefficients of different components in water-albite melt over the temperature and pressure range from 1000 to 1100°C and 0.5 to 6 kbar are listed in Tables 1a–d.

Consider some regularities associated with the absolute values of diffusion

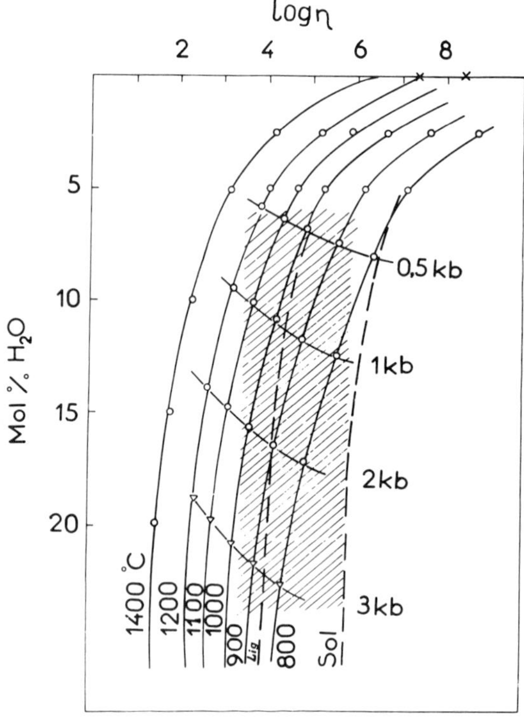

Fig. 1. Viscosity of the system granite-H_2O under equilibrium conditions. (According to Persikov, 1984.)

coefficients of rock-forming oxides. All the components of magmatic melts apparently may be divided into the following three groups:

(i) High-charge network-forming cations whose diffusion coefficients range from about 10^{-11} to 10^{-10} cm^2/s with water content in the melt of 1.7 wt%.
(ii) Oxides of two valence cations-modifiers of the RO (CaO, MgO, FeO) type whose diffusion coefficients are about 10^{-9}–10^{-8} cm^2/s under same conditions.
(iii) The most mobile components of melts, i.e. alkali oxides and H_2O. Their diffusion coefficients are between 10^{-7}–10^{-6} cm^2/s.

In other words, diffusivity of one or another component of magmatic melts is apparently controlled by its structural role.

On the other hand, the diffusion mobility of components depends on the type of contact mineral. For example, the diffusion coefficients of CaO turned out to be two orders of magnitude lower on solution of anorthite than those on solution of wollastonite. In this instance a process of multicomponent diffusion, which will be discussed later, must be dealt with.

Table 1a. Diffusion coefficients of some components in a water–albite melt ($P_{tot} = 0.5$ kbar, $X_{H_2O} = 1.7$ wt%, $T = 1000$–$1100°C$)

Diffusing component	Dissolving mineral	Diffusion coefficients, cm^2/c 1000°C	1100°C
SiO_2	Quartz	8.7×10^{-11}	2.7×10^{-10}
SiO_2	Anorthite	2.5×10^{-11}	4.4×10^{-10}
SiO_2	Nepheline	1.3×10^{-9}	4.1×10^{-9}
Al_2O_3	Cordierite	1.1×10^{-9}	—
Al_2O_3	Anorthite	2.5×10^{-11}	4.4×10^{-10}
CaO	Anorthite	2.5×10^{-11}	4.4×10^{-10}
CaO	Wollastonite	1.3×10^{-8}	2.0×10^{-8}
MgO	Enstatite	2.5×10^{-9}	5.0×10^{-8}
MgO	Cordierite	1.0×10^{-8}	—
FeO	Fayalite	6.8×10^{-9}	2.0×10^{-8}
Na^+–K^+	Ab_L–Or_L	6.0×10^{-7}	1.7×10^{-6}
Na_2O	Nepheline	1.3×10^{-9}	4.1×10^{-9}
P_2O_5	Apatite	3.4×10^{-11}	5.3×10^{-11}
WO_3	Scheelite	5.3×10^{-11}	1.8×10^{-10}
MoO_2	Molybdenite	2.5×10^{-10}	3.6×10^{-9}
TiO_2	Sphene	1.4×10^{-10}	8.0×10^{-10}
Fe_2O_3	Hematite	2.2×10^{-10}	5.3×10^{-10}
SnO_2	Cassiterite	4.4×10^{-9}	1.3×10^{-8}
Cu_2O	Cuprite	6.1×10^{-8}	—
H_2O	H_2O	9.1×10^{-7}	1.6×10^{-6}
H_2	H_2	8.0×10^{-4}	—

Table 1b. Diffusion coefficients of some components in water–albite melt ($P_{tot} = 2$ kbar, $X_{H_2O} = 3.8$ wt%, $T = 1000°C$)

Diffusing component	Dissolving mineral	Diffusion coefficient, cm^2/c 1000°C
SiO_2	Quartz	3.4×10^{-10}
SiO_2	Anorthite	4.1×10^{-10}
Al_2O_3	Anorthite	2.4×10^{-9}
CaO	Anorthite	8.5×10^{-8}

Table 1c. Diffusion coefficients of some components in water–albite melt ($P_{tot} = 4$ kbar, $X_{H_2O} = 6.5$ wt%, $T = 1000$–$1100°C$)

Diffusing component	Dissolving mineral	Diffusion coefficient, cm^2/c 1000°C	1100°C
SiO_2	Quartz	2.8×10^{-9}	—
SiO_2	Anorthite	2.5×10^{-9}	4.4×10^{-9}
SiO_2	Nepheline	6.4×10^{-9}	8.6×10^{-9}
Al_2O_3	Nepheline	6.4×10^{-9}	8.6×10^{-9}
Al_2O_3	Anorthite	6.5×10^{-9}	9.5×10^{-9}
CaO	Anorthite	4.0×10^{-7}	5.1×10^{-7}
MgO	Enstatite	7.0×10^{-8}	9.5×10^{-8}
Na_2O	Nepheline	6.4×10^{-9}	8.6×10^{-9}

Table 1d. Diffusion coefficients of some components in water–albite melt ($P_{tot} = 6$ kbar, $X_{H_2O} = 8.5$ wt%, $T = 1000°C$)

Diffusing component	Dissolving mineral	Diffusion coefficients, cm^2/c 1000°C
SiO_2	Anorthite	3.4×10^{-9}
Al_2O_3	Anorthite	9.5×10^{-9}
CaO	Anorthite	9.1×10^{-7}
MgO	Enstatite	1.1×10^{-7}

Experimental Investigation of Diffusion Coefficients of Components of a Fluid

Experimental Procedure and Results

Knowledge of the diffusivities in magmas of aqueous fluid components is also of great importance. Therefore, the diffusion coefficients of HCl and HF have been determined.

For this determination, a polished glass cylinder was placed in a Pt tube that was then filled with a 5 *N* solution of HF or HCl. Because only a small amount of HCl (HF) is dissolved in the melt during the short-term experiments, changes in its activity in the fluid may be neglected and consideration given to the HCl (HF) content on the sample surface constant. Consequently, it is possible to consider diffusion into a semi-infinite medium with a constant concentration of the diffusing component at the interface.

After the experiment the distribution of Cl and F along the sample were determined with a microprobe. The results are shown in Table 2.

Table 2. Diffusion coefficients of the components of a fluid ($P_{tot} = 0.5$ kbar, $X_{H_2O} = 1.7$ wt%, $T = 1000$–$1100°C$)

Diffusing component	Source	Diffusion coefficient cm^2/c 1000°C	1100°C
HF	5 *N* sol HF	$\sim 10^{-7}$	$\sim 10^{-7}$
HCl	5 *N* sol HCl	1.4×10^{-9}	5.1×10^{-9}

Experimental Study of the Hydrogen Transport Through Magmatic Melts

All the experimental publications now available on the problem of hydrogen transport are concerned with systems grossly different in composition from those of geological interest. To identify a relative mobility of hydrogen and peculiarities of its transport through magmatic melts out a series of experiments applying a specially developed method were performed, using the redox color indication technique.

The method is as follows. A specific amount of variable-valence ions (so-called traps) is introduced in a sample. Interaction with a redox agent causes these ions to change the color of that zone of the sample in which the redox reaction takes place. Measuring the thickness of the above zone allows calculation the of diffusion coefficients of the corresponding agents (or diffusing species). It is logical that the rate of advance of this changed zone is dependent on solubility of hydrogen and concentration of traps in the sample. From knowledge of the solubility of hydrogen and concentration of traps a value for the "intrinsic" diffusion coefficients of the redox agent (in this example, hydrogen), i.e. the rate of its penetration in the sample that contains no traps can be calculated. Ignoring concentration of traps permits measurement of the value of the apparent diffusion coefficients. The present method is fairly well developed theoretically (Crank, 1956).

It is obvious that with decreasing concentration of traps the apparent diffusion coefficient value of the redox agent will approach the "intrinsic" one. By contrast, with increasing concentration of traps in the sample the rate of advance of the chemical reaction front decreases, i.e. there is a "delay" caused by consumption of the redox agents in the chemical reaction.

The authors performed experiments with albite glass containing CO_2 (0.05 wt%) as traps. The diffusion runs were conducted in a gas-pressure vessel equipped with a special levelling-partitioner that allowed regulation of the pressure of a pure H_2 up to 4 kbar. The experiments were carried out at $P_{H_2} = P_{tot} = 1.8$ kbar in the temperature range 750–880°C. After the experiments, the surfaces of the samples darkened. The average thickness of the zone where the redox reaction takes place was measured by optical methods and equaled 1–1.3 mm at different temperatures. (For further details see Chekhmir et al., 1985).

Table 3. Diffusion coefficients of hydrogen in an albite glass (P_{H_2} = 1.8 kbar)

T°C	750	820	880
D cm^2/c	1.7×10^{-6}	5.0×10^{-6}	7.6×10^{-6}

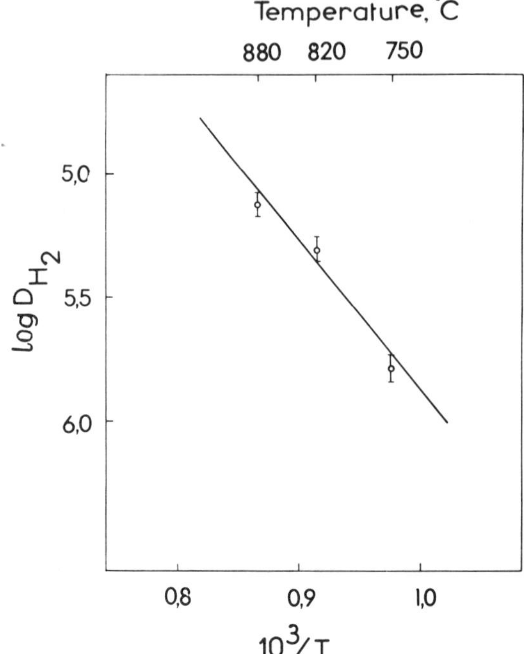

Fig. 2. Arrhenius-type plot of diffusion coefficients of the hydrogen calculated for albite glass.

The values of "intrinsic" diffusion coefficients of hydrogen in albite glass are listed in Table 3. Fig. 2 shows the relations between $\log D$ and $1/T$.

The apparent diffusion coefficients calculated from the thickness of the reaction zone (neglecting concentration of traps) and "intrinsic" diffusion coefficient value actually turned out to be the same due to low solubility of CO_2 in the sample. Consequently, CO_2 cannot be used as traps in experiments on melts, because diffusion rates are higher in melts than in glasses.

To measure the diffusion coefficients of H_2 in albite melts Mn_2O_7 was used as traps. This color indicator was chosen because of its high solubility in melts in addition to the clearly defined colors of reducing and oxidizing Mn forms in silicate glasses. These experiments were carried out at water pressure of 0.5 kbar and 1000°C. Due to the Ni–NiO buffer the partial hydrogen pressure was about 5 bar. The apparent diffusion coefficient value was estimated to be $2 \cdot 10^{-7}$ cm^2/s, whereas the "intrinsic" one equals $8 \cdot 10^{-4}$ cm^2/s.

With greatly increasing concentration of traps in the sample the rate of advance of a front of chemical reaction is seen to be 3 orders of magnitude lower than "intrinsic" mobility of H_2. In other words, the components available in the melt and capable of interacting with hydrogen may slow the rate of hydrogen penetration into the melt considerably, regardless of its extremely high "intrinsic" mobility.

The Effect of H_2O Concentration in the Melt on the Diffusion Mobility of Components

To the authors' knowledge no systematic experimental investigations have been conducted on the H_2O effect on diffusivity of components. Of all the available publications, only the experimental studies of Jambon (1978) and Watson (1979a) on Cs diffusion in granitic melts between 700° and 900°C and H_2O content in the melt of 0, 1, 2 and 6 wt% (see Fig. 3) may be cited. Although the authors performed the experiments with almost identical melts, their results agree only qualitatively.

Karsten et al. (1982) studied the effect of water content in rhyolitic melt on

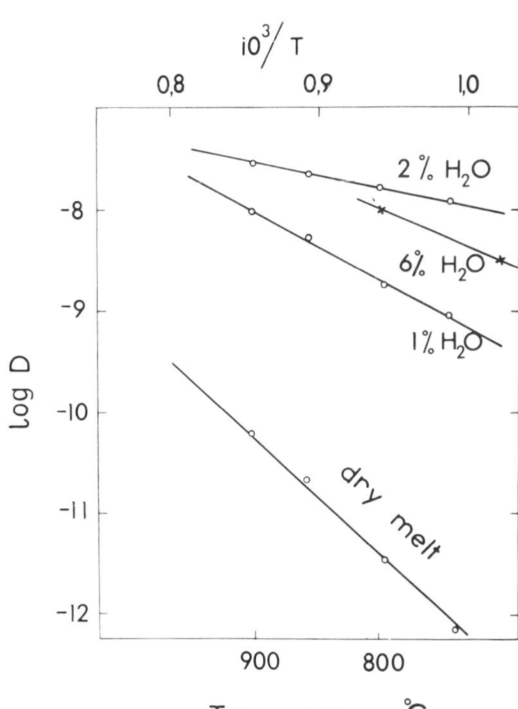

Fig. 3. Diffusivity of Cs in granite melt at different H_2O content (open circle, Jambon, 1978; solid circle, Watson, 1979a).

Table 4. Diffusion coefficients of SiO_2 in a water-albite melt at 1000°C and different pressures and water content

Diffusing component	Total pressures	Water content wt%	Diffusion coefficient
SiO_2	1 atm	0	4.1×10^{-12}
SiO_2	0.5 kb	1.7	8.7×10^{-11}
SiO_2	2.0 kb	3.8	3.4×10^{-10}
SiO_2	4.0 kb	6.5	2.8×10^{-9}
SiO_2	6.0 kb	8.5	3.4×10^{-9}

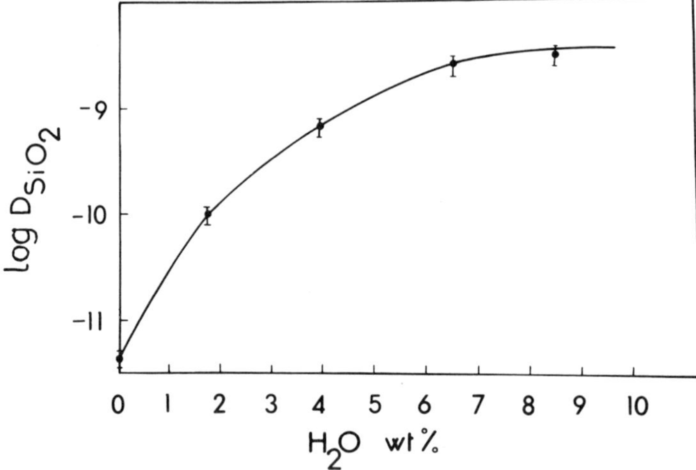

Fig. 4. Diffusivity of SiO_2 in the water–albite melt at 1000°C plotted against the total water content (wt%) in the melt.

the diffusion mobility of H_2O between 650 and 950°C and at water pressure of 0.7 kbar. It should be noted that the activation energy of the water diffusion is constant regardless of the content (1–3 wt%) in the melt. The results were rather unexpected in view of the conclusions of Jambon et al. (1978) regarding Cs diffusion. These authors concluded that the activation energy for Cs diffusion varies from 70 to 17 kcal/mol as H_2O content increases from O to 2 wt%.

Table 4 and Fig. 4 represent the author's measurements on, for example, D_{SiO_2} in water-albite melts in a wide range of H_2O concentrations. On the assumption that the effect of total pressure (ranging from 0.5 to 6 kbar) is negligible in comparison to that of H_2O, $\log D - X_{H_2O}$ relationships may be described by the equation:

$$\log D_{SiO_2} = \frac{-20.879}{X_{H_2O} + 4.633} - 6.839 \tag{1}$$

It is evident from Fig. 4 that initial portions of the dissolved H_2O (up to 3 wt%) have the most marked effect on the diffusion coefficient value of components and the curve becomes less steep on further solution. The acid melt viscosity is affected in a similar way (see Fig. 1). The mechanism of influence of H_2O and some other factors (T, P_{tot}) on diffusivity of components will be given below.

Characteristic Features and Regularities of Molecular Diffusion Processes in Magmatic Melts

Characteristic Features of the Diffusion Distribution of Components Near Contact Zones

Analyzing the diffusion distribution patterns of components near contact zones reveals some features of mineral dissolution. In sum, consideration of solution of simple one-component minerals—oxides, such as SiO_2, the distribution curves are typical of diffusion (see Fig. 5), indicating that there is no interaction between dissolving silica and the albite network. This is the simplest condition to the authors' knowledge.

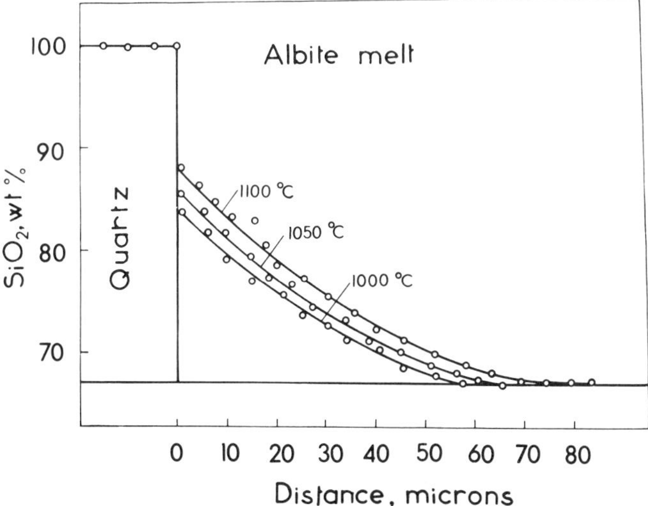

Fig. 5. Electron microprobe traverses across boundary layers of water–albite melt in contact with dissolving quartz crystal. ($P_{H_2O} = 0.5$ kbar.)

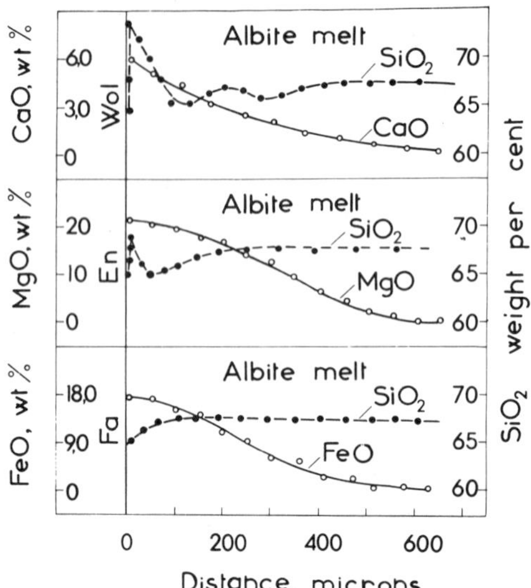

Fig. 6. Electron microprobe traverses across boundary layers of water–albite melt in contact with dissolving wollastonite, enstatite and faylite crystals. ($P_{H_2O} = 0.5$kb.)

Distribution of components of the silicates of CaO, MgO and FeO (wollastonite, enstatite, fayalite) is more complicated (Fig. 6). The distribution curves of CaO, MgO and FeO are typical of diffusion, but those of SiO_2 on solution of wollastonite and enstatite are not. If solution of mineral is not accompanied by other processes, the distribution of its components must become similar to that available at the contact of fayalite with albite melt. As it passes into the melt, the mineral is decomposed and its components move with their own diffusion coefficients. If the latter are different ($D_{Feo} \approx 10^{-9}$ cm^2/s and $D_{SiO_2} \approx 10^{-10}$ cm^2/s), the thickness of the diffusion zones is also different (600 and 100 μm, respectively).

The situation is quite different if solution is accompanied by some other processes. Fig. 6 shows that SiO_2 entering the melt on solution of wollastonite and enstatite is concentrated near the contact zone. Moreover, some amounts of silica in albite are also concentrated there, leading to the appearance of a "cavity" on the SiO_2 distribution curve at a distance of about 200 μm from the contact. It is inferred that this phenomenon results from acid-basic interaction between components of the melt. Hence, an influx of the basic components (for example, CaO) in the melt, involves a decrease in the activity coefficients of acid components such as silica. Consequently, there appears a SiO_2 chemical potential gradient that inhibits the movement of silica from wollastonite and gives rise to the counter flow of SiO_2.

This idea is confirmed by the peak height that corresponds to basicity of the diffusion oxides (minerals). When CaO is introduced, the concentration of silica increases to about 8 wt%, MgO to 4 wt%, and silica is not redistributed on diffusion of FeO (see Fig. 6).

Experiments carried out in gabbro-norite melts are strong evidence for valid-

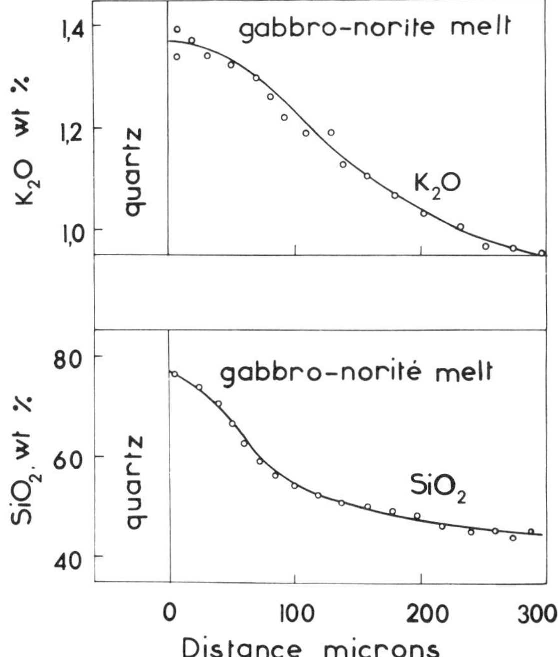

Fig. 7. Electron microprobe traverses across boundary layers of gabbro-norite melt in contact with dissolving quartz crystal. ($P = 1$ atm, $T = 1300°C$.)

ity of our interpretation pertaining to this phenomenon. In this example the solution of an acid component (SiO_2) gives rise to a similar decrease in activity coefficients of the basic components, namely K_2O and Na_2O. Chemical potential gradients of K_2O and Na_2O, formed in the melt, lead to their re-distribution and an increase in their concentration at the contact zone (see Fig. 7).

Sato (1975) reported a similar K_2O distribution in basaltic melt. He described diffusion coronas around xenocrysts in andeside and basalt from the Tertiary volcanic region in Northeastern Shikoku, Japan. Microprobe analyses of these samples have shown that concentration of K_2O in basalt increases from 1.4 to 6.2 wt% along the diffusion zone toward the contact. Moreover, Sato's experiments are similar to the authors', i.e. quartz crystals have been dissolved in basaltic melt at 1400°C for 7 hr. The K_2O and Na_2O concentrations in the diffusion zone of about 120–150 μm length are shown to increase towards the contact with quartz crystal.

The solution of silicates of CaO, MgO, and FeO in the melt causes severance of the chemical bonds between their constituent oxides (RO and SiO_2). The latter, for example FeO and SiO_2, therewith, move along the contact zone with individual diffusion coefficients varying by almost two orders of magnitude. This movement leads to different thicknesses of the diffusion zones.

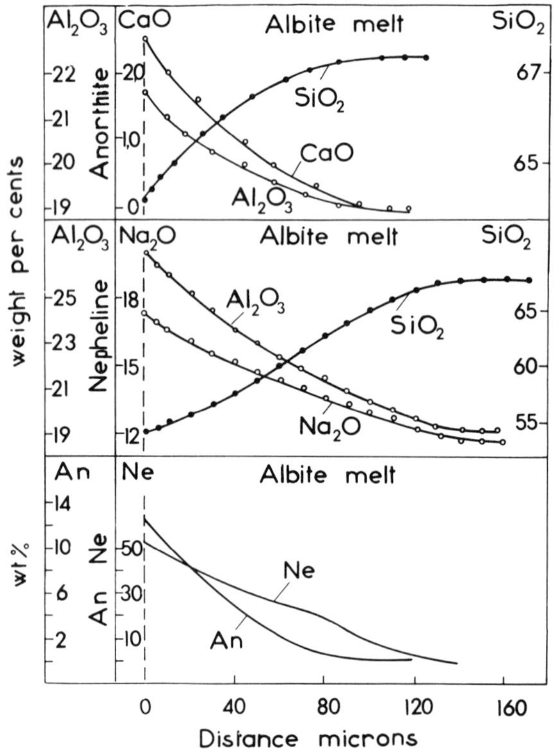

Fig. 8. Electron microprobe traverses across boundary layers of water–albite melt in contact with dissolving anorthite and nepheline crystal. ($P_{H_2O} = 0.5$ kbar, $T = 1100°C$.)

However, on solution of some aluminosilicates, in part anorthite and nepheline, differential mobility of their constituent oxides (for example, CaO, Al_2O_3, SiO_2) in the contact zone is lacking. The oxides mentioned move with one and the same diffusion coefficient (see Fig. 8 and Table 1a).

Moreover, quantitative relationships between the oxides that compose anorthite and nepheline are consistent with stoichiometric mineral composition in each section of the diffusion zone, i.e. the movement of anorthite and nepheline molecules (or minals) is fixed. Therefore, the values of diffusion coefficients of such normally mobile oxides as Na_2O and CaO are two to three orders of magnitude lower than those obtained on solution of wollastonite or in ionic exchange between albite and orthoclase.

Similar results were obtained by Yoder (1973) in his experiment on basanitic-rhyolitic melts interaction. The thickness of the diffusion zones between the rhyolitic and basanitic melts for all their componens were the same. Although Yoder did not list the diffusion coefficients' values they are readily calculated, and turn out to be equal. However, it is not yet clear from these experiments whether the author excluded the possibility of the melt mixing through convection.

Aliberty and Carron (1980) reported diffusion coefficient measurements for SiO_2, TiO_2, Al_2O_3, CaO, MgO, K_2O, Na_2O at ionic exchange between couples of glasses and melt of basaltic and rhyolitic composition. In this work, Yoder's experiment was actually reproduced, but with a view to measuring the diffusion constants. The procedure involves heating of co-axial cylinders from the mentioned glasses between 900 and 1300°C at 1 atm. Distribution of components in the diffusion zone was determined with a microprobe. The results showed that the thicknesses of diffusion zones and values of diffusion coefficients are equal. The fact that the components in glasses have also a tendency to coherent movement seems particularly important because convection in these conditions is impossible.

Simultaneous movement of the dissolving mineral components in the melt is unequivocal evidence of the chemical bonds and molecules available. The latter are consistent with a crystalline state of the system.

In addition to its structural interest, this phenomenon is of great importance when calculating all the possible mass-exchange processes in magmatic melts. It should be noted that using only "individual" diffusion coefficients involves considerable errors in calculation, for example, of the thickness of diffusion zones.

It is worth noting that not all aluminosilicates are dissolved as molecules in the albite melt. Thus, on solution of cordierite and gehlenite there takes place differentiation of the oxides on mobilities. This differentiation causes a loss of stoichiometry in the sections parallel to the interface mineral-melt and results in different thicknesses of diffusion zones (see Fig. 9).

To ascertain the H_2O effect on the movement of anorthite and nepheline components these minerals were dissolved in the water-albite melt at varying

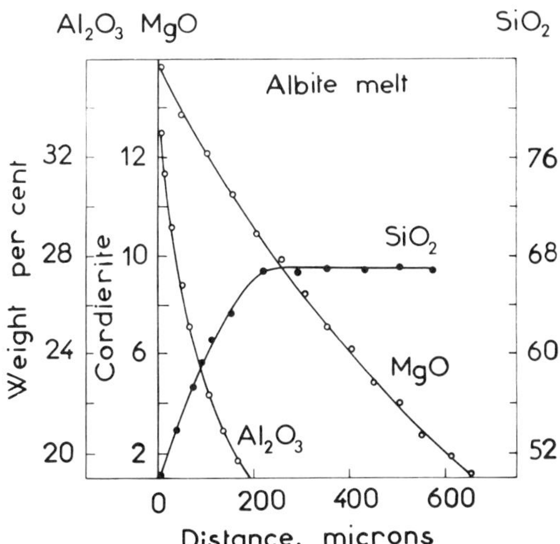

Fig. 9. Electron microprobe traverses across boundary layers of water–albite melt with dissolving cordierite crystal in contact. ($P_{H_2O} = 0.5$ kbar, $T = 1000°C$.)

water contents. Nepheline tends to stoichiometrically move in the whole range of H_2O contents (up to 8.5 wt%). However, it is quite natural that diffusion coefficients of all the nepheline oxides increase under the other conditions. As for anorthite, at a pressure of 1.5 kbar ($\sim$3 wt% H_2O in melt) its molecules start to be broken and constituent oxides to move with the diffusion coefficients typical of the free state (see Table 1b–d).

Relations Between Diffusivity of the Melt Components, Viscosity and the Nature of Diffusing Oxides

As mentioned above, the values of diffusion coefficients obtained for some components of the melt are typical of granitic magmas.

However, it is also necessary to have data on the diffusivity of various components in the melts of other compositions (for example, basalt, andesite). In addition, a great deal of the components of interest as well as labor-consuming experiments require a special method for a quick and sufficiently accurate estimation of the diffusion coefficients of any component in different melts. The following main principles form the basis of the method suggested:

(i) linear relations between $\log D$ and $\log \eta$
(ii) the diffusivity of components as a function of their charge and radius.

The linear relations between $\log D$ and $\log \eta$ follow from the well known Stokes–Einstein relations:

$$D = \frac{k \cdot T}{6\pi R \eta},$$

where k = the Boltzmann constant, R = the radius of the diffusing cation, η = the viscosity of melt, T = the absolute temperature, whose taking logs is as follows:

$$\log D = A - B \log \eta.$$

A great body of experimental evidence on diffusion of various components in glasses, their melts and melts of metallurgical slags confirms the above linear relation (Epel'baum, 1980).

Fig. 10 shows that for a wide viscosity range a linear relationship is seen between $\log D$ and $\log \eta$. This view enables the D values of components both in granitoid melts (in the wide P-T-X_{fl} conditions) and in the melts of the other compositions to be estimated.

The relationships considered may be described by the following equations:

$$\log D_{SiO_2} = -0.580 \log \eta - 6.633 \quad (2)$$

$$\log D_{Al_2O_3} = -0.352 \log \eta - 6.927 \quad (3)$$

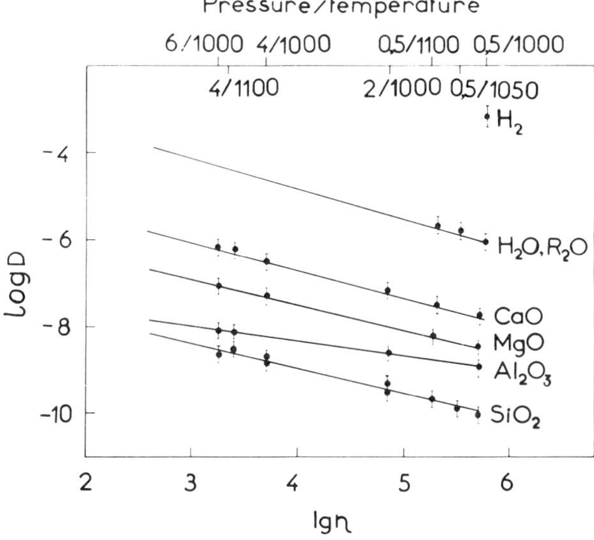

Fig. 10. Relationships between diffusivity of components and viscosity of metls (experimental data).

$$\log D_{\mathrm{MgO}} = -0.607 \log \eta - 5.107 \tag{4}$$

$$\log D_{\mathrm{CaO}} = -0.671 \log \eta - 4.048 \tag{5}$$

Using viscosity as a parameter for comparison allows one to ignore a set of factors that can act on diffusion mobility of the melt components. These factors are temperature, total pressure, volatile content and composition of melt.

To estimate the values of diffusion coefficients for various components in the melt of a fixed composition one way use experimental relations between $\log D$ and ionic potential components (or Cartly's potential), Z/R, where Z is the charge of the cation and R its radius.

Fig. 11 shows such relations for diffusion of 11 components in the water-albite melt at pressures of 0.5; 2 and 4 kbar.

The 0.5-kbar curve, which contains the largest amount of information, may be described by the following equation:

$$\log D = \frac{10.561}{Z/R + 1.439} - 11.161.$$

A very good relationship between $\log D$ and Z/R allows estimation of the possible diffusion coefficient values of rock-forming or ore components of interest without additional experiments.

Relationships are thus obtained that can be used not only to estimate the values of diffusion coefficients for completely different components in the water-

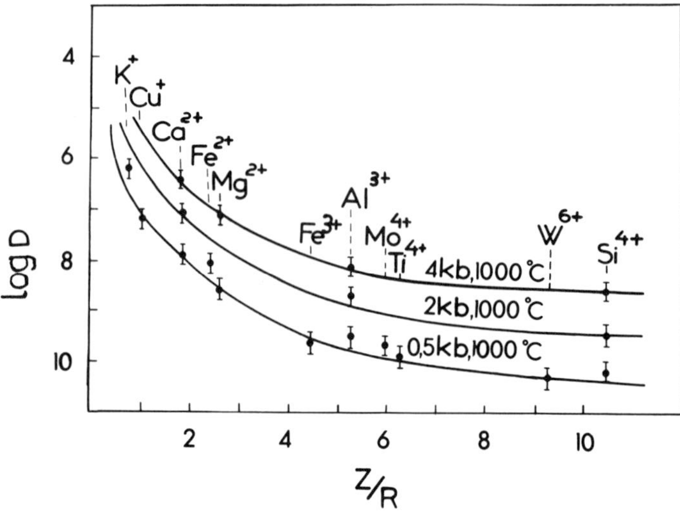

Fig. 11. Relationships between diffusivity of components and Z/R (Z, charge ions, R, their raidus).

albite melt, but also to apply these data to melts of different composition, using the $\log D - \log \eta$ relationships.

Applications

Inasmuch as this paper is dedicated mainly to experimental results, only the data of greatest interest will be considered here.

In the previous section of the paper were described experiments in which the components of the melts concentrated at the contacts of interfaces and might be found both in experimental and natural systems (Sato, 1975). Although the above instances are characteristic of small-scale processes, the "uphill" diffusion may occur in large-scale processes, such as interaction of magmatic melts with enclosing rocks.

For example, at the contact of the Chiney gabbro-noritic pluton (North Baikal) with sedimentary terrigenous rocks there is a narrow zone enriched in Ksp and Qz in contrast to the gabborid massif under consideration. The curves of component distribution in the zone are characteristic of the diffusion processes and likely to be formed during hybridization.

However, lower contents of K_2O in enclosing rocks (sandstones) compared to those in gabbro-norites suggest that re-distribution of K_2O is caused by acid-base interaction of components on solution of SiO_2 from the sandstone.

This sequence was predicted from experiments (Fig. 7). For further details, see Konnikov, Epel'baum, Chekhmir, 1981.

The "Diffusion" Technique for Studying Liquidus Relations Under Hydrothermal Conditions

The compositions of melt at the contact with quartz have been analyzed with a microprobe in a series of experiments (for example, for system quartz-albite) at $P = 0.5$ kbar and $T = 1000$, 1050 and 1100°C. The compositions analyzed are consistent with those obtained by Tuttle and Bowen (1958) for this system. One of the main advantages of the proposed method is that there is no need to attain equilibrium and, as a consequence, the experiments are less labor-intensive. For example, the run duration of Tuttle and Bowen is 5–7 days, whereas the authors' is 2–3 hr.

Moreover, the solubility of minerals (in part ore) as well as investigations on distribution of accessory components (Watson, 1984) may be determined by this method.

Conclusions

Experimental study of processes pertaining to molecular diffusion of components in magmatic melts is now in progress. Different investigators have measured the diffusion coefficients for a wide range of components of magmatic melts at high pressures and temperatures.

The H_2O influence upon the increase in diffusivity of components is revealed.

Various methods have been developed to estimate the values of diffusion coefficients in magmatic melts over a wide range of temperatures and pressures.

However, the molecular diffusion studies should not be considered complate. The diffusion processes of redox agents followed by chemical reaction seem to be worthy of further study and look promising for investigating evolution processes of natural magmas. Processes similar to those mentioned are responsible for regulation of redox potentials in magmas and oxidation states of variable valence ions as they occur in magmas. To date, the authors' experiments on hydrogen transport through model magmatic melts appear to be unique in this respect. However, it is unlikely that hydrogen is the only component responsible for redox conditions in melts.

The present results certainly must be used in corrected calculations regarding various mass-exchange processes. The first such calculations have been done already. For example, evaluation of the time needed for complete dissolution of xenoliths in melts (Watson, 1982), a quantitative approach to the explanation of zoned crystal formation (Simakin, 1983), estimations of true rates of crystal growth in melts (Simakin, Epel'baum, 1984) and other calculations.

Acknowledgements

The authors wish to express their sincere thanks to Mrs. G.G. Gonchar for translating this manuscript from Russian into English. The authors have greatly benefited from the critical review and helpful suggestions of Dr. E.B. Watson. Their thanks are also due to Mrs. T.V. Prasol for her assistance during the preparation of the manuscript.

References

Aliberty, C., and Carron, J-P. (1980) Donnes experimentales la diffusion des elements majeurs entre verres on liquides de composition basaltique rhyolitique et phonolitique entre 900°C et 1300°C, a pression ordinate. *Earth Planet. Sci Lett.*, **47**, 294–306.

Bowen, N.L. (1921) Diffusion in silicate melts. *J. Geol.*, **29**, 295–317.

Chekhmir, A.S., Epelbaum, M.B., and Lytov, V.S. (1979) The diffusion coefficients of some components of fluid in water-albite melt at high pressures. *Acad. Nauk USSR Dokl.* **246**, No.3, 698–701 (in Russian).

Chekhmir, A.S., and Epelbaum, M.B. (1982) Experimental study of diffusion processes in magmatic melts. In: *Contributions to Physico-Chemical Petrology, Vol X*, pp. 204–215. Nauka, Moscow (in Russian).

Chekhmir, A.S., Persikov, E.S., Epelbaum, M.B., and Bukhtiarov, P.G. (1985) Experimental study of the hydrogen transport through model magmatic melts. *Geochimia*, No. 594–598 (in Russian).

Crank, J. (1956) *The Mathematics of Diffusion.* Clarendon Press, Oxford.

Epelbaum, M.B. (1980) *Silicate Melts with Volatiles Components.* Nauka, Moscow (in Russian).

Harrison, T.M., and Watson, B.E. (1984) The behavior of apatite during crustal anatexis: Equilibrium and kinetic considerations. *Geochem. Cosmoch. Acta*, **48**, 1467–1477.

Hofmann, A.W., and Magaritz, M. (1977) Diffusion of Ca, Sr, Ba and Co in a basalt melt: Implication for the geochemistry of the Mantel. *Jour. of Geophy. Res.* **82**, No. 33, 5432–5440.

Jambon, A., and Carron, J-P. (1976) Diffusion of Na, K, Rb, Cs in glasses of albite and orthoclase compoisition. *Geochem. Cosmochem. Acta*, **40**, 897–903.

Jambon, A., and Delbove, F. (1977) Etude a 1 bar, entre 600 et1100 de la diffusion du calcium dans les verres feldspathiques. *C.R. Acad. Sci. Paris (Ser. D)*, **284**, 2191–2194.

Jambon, A., and Semet, M.P. (1978) Lithium diffusion in silicate glasses of albite, orthoclase and obsidian composition: an ion-microprobe determination. *Earch Planet. Sci. Lett.*, **37**, 445–450.

Jambon, A., Carron, J-P., and Delbove, F. (1978) Donnes preliminaires sur la diffusion dans les magmas hydrates: le cesium dans un liquide granitique à 3 kbar. *C.R. Acad. Sci. Paris (Ser. D)*, **287**, 403–406.

Karsten, J. L., Holloway, J.R., and Delaney, J.R. (1982) Ion-microprobe studies of water in silicate melts: temperature-dependent water diffusion in obsidian. *Earth Planet. Sci. Lett.*, **59**, 420–428.

Konnikov, E.G., Epelbaum, M.B., and Chekhmir, A.S. (1981) The causes of potassium concentration in endocontact zone of the Chiney gabbro-norite pluton. *Geochimia*, **2**, 257–263 (in Russian).
Lowry, R.K., Reed, S.J.B., Nolan, J., Hederson, P., and Long, J.V.P. (1981) Lithium tracer-diffusion in an alkali-basaltic melt—an ion microprobe determination. *Earth Planet. Sci. Lett.*, **59**, 36–40.
Magaritz, M., and Hofmann, A.W. (1978) Diffusion of Sr, Ba and Na in obsidian. *Geochem. Cosmochim. Acta*, **42**, 595–605.
Persikov, E.S., and Epelbaum, M.B. (1980) The experimental investigation of the viscosity water-albite melt at high pressures. In *Contributions to Physico-Chemical Petrology, Vol IX*, pp. 134–141. Nauka, Moscow (in Russian).
Persikov, E.S. (1984) *Viscosity of Magmatic Melts.* Moscow, Nauka (in Russian).
Sato, H. (1975) Diffusion coronas around quartz xenocrysts in andesite and basalt from tertiary volcanic region in Northeastern Shikoku, Japan. *Contrib. Mineral. Petrol.*, **50**, No. 1, 275–286.
Simakin, A.G. (1983) Rhythmical zonality of crystals: simple quantitative model. *Geochmia*, **12**, 1720–1729 (in Russian).
Simakin, A.G., and Epelbaum, M.B. (1984) Some aspects of the crystalization dynamic of supercooling silicate melts. *Geochimia*, **5**, 756–760 (in Russian).
Tuttle, O.F., and Bowen, N.L. (1958) Origin of granite in light of experimental studies in the system $NaAlSi_3O_8$–$KAlSi_3O_8$–SiO_2–H_2O. *Geol. Soc. Amer., Memoir*, **74**, 153 p.
Watson, E.B. (1979a) Diffusion of cesium ions in H_2O-saturated granitic melt. *Science*, **205**, 1259–1260.
Watson, E.B. (1979b) Calcium diffusion in a simple silicate melt to 30 kbar. *Geochemic. Cosmochim. Acta*, **43**, 313–322.
Watson, E.B., and Harrison, T.M. (1984). Accessory minerals and the geochemical evolution of crustal magmatic systems: a summary and prospectus approaches. *Phys. Earth Planet. Int.* **35**, 19–30.
Yoder, H.S., Jr. (1973) Contemporaneous basaltic and rhuolitic magmas. *Amer. Mineral.*, **58**, 153–171.

Chapter 4
Chemical Diffusion in Magmas: An Overview of Experimental Results and Geochemical Applications

E. Bruce Watson and Don R. Baker

Abstract

The subject of chemical diffusion in magmas has attracted the interest of petrologists and geochemists seeking to place time constraints on phenomena ranging from magma mixing to crystal growth. Experiments have been devised to examine chemical diffusion effects during such processes as interdiffusion of two liquids, growth and dissolution of crystals, exchange of halogens with oxygen in the air, reduction or oxidation of dissolved iron oxide, and introduction of dissolved volatiles. A few experiments have even been done using a temperature gradient to induce thermal migration.

Many of the studies carried out to date have incorporated variations in temperature, pressure, and dissolved H_2O content, so the collective results allow diffusivities in magmas to be estimated quite well for most geologically-realizable conditions. In general, the following major characteristics appear to hold:

(1) Network-forming species, most notably SiO_2, are the slowest-moving magmatic components, although network-modifiers that form stable complexes in the melt may be equally sluggish;
(2) alkalies, divalent cations, oxygen, and fluorine are the most mobile magmatic components when their transport is not rate-limited by counterdiffusion of slower species; and
(3) the effect of H_2O content on chemical diffusion of most components (including H_2O itself) is extremely large, sometimes amounting to several orders of magnitude at crustal melting conditions.

Introduction

Geochemists' Perspective on Diffusion in Molten Silicates

Diffusion in molten silicates has been a topic of major interest in several scientific disciplines, including physical chemistry, materials science, and geology. The underlying reasons for this interest vary considerably: Some chemists may be concerned with fundamental properties of metal oxides at high temperature, and other researchers in materials science may be motivated by more practical considerations such as the rate of melt homogenization during glass making.

Geoscientists working on diffusion in molten silicates seek answers to questions regarding the internal processes of the Earth and terrestrial planets. Petrologists and geochemists have recognized for some time that the attainment of equilibrium in igneous systems can be hindered by slow diffusion of some species in the melt, even in situations that involve time scales far exceeding those of human experience. This realization has prompted numerous laboratory measurements of chemical diffusion. The common objective of these studies has been to provide the requisite data for (1) constraining the rates at which certain natural processes can occur, or (2) extracting information on the thermal history experienced by a given system that preserves a disequilibrium state. Basically, geologists want to know "How fast can (or did) a specific magmatic transport process take place?"

Scope of this Chapter

For the sake of managability, the scope of this paper is restricted in several ways. First, it is intended as an overview of *chemical diffusion* only, by which is meant element migration in response to a chemical (potential) gradient, resulting in a net flux of that element and necessitating simultaneous motion of charge-balancing species in the opposite direction. No consideration is given here to *self-, or tracer, diffusion*, which specifically characterizes atomic migration under circumstances of effectively no potential gradient and therefore results in no net flux. Self- and tracer-diffusion measurements are made by use of a radioactive (traceable) isotope that intermingles with other isotopes of the element of interest. Although the strict relevance of tracer diffusion is limited to the specific fields of isotope geochemistry and geochronology, for practical purposes the data can also be applied to problems involving trace element transport. Chemical diffusion in magmas has somewhat broader applications in geochemistry and petrology, as should be apparent in the "applications" section. For a review that includes a discussion of tracer diffusion measurements, the reader is referred to Hofmann (1980).

The second deliberate limitation of this chapter is that it summarizes contributions from the geologic community only. A good deal of research highly germane to diffusion in magmas has been done by glass scientists and ceramists; for the present purposes, however, these contributions are regarded as belonging to a separate, though no less important, literature. The compilation by Freer (1981) is a good starting point from which to enter the glass and materials science literature.

Considering that the overall objective of this paper is to provide the reader with a practical overview of experiments, data, and applications of chemical diffusion in magmas, relatively little space is devoted to theoretical and mathematical considerations. This shortcoming is perhaps most significant in interdiffusion of two complex liquids, which has been the subject of detailed mathematical treatments. For a recent paper illustrating the ion flux approach of Cooper (1965), the reader is referred to Oishi et al. (1982). Some discussion on this subject is also given by Hofmann (1980). In the near future, contributions on this topic can be anticipated from Lasaga and co-workers (Muncill and Lasaga, 1984). The only discussion of theoretical considerations in this paper is the presentation of a general method for estimating chemical diffusivities by use of transition state theory.

It will be immediately apparent that this paper deals almost exclusively with cation diffusion, with only brief discussion of the anions F^- and O^{2-}. This limitation reflects simply the availability of experimental data. Studies of chemical diffusion in magmas have focused primarily upon major elements of the Earth, which include Si, Ti, Al, Mg, Fe, Ca, Na, K, Mn, and P. Oxygen, of course, is also a major component, but has received relatively little attention due to inherent analytical difficulties.

Summary of Experimental Approaches and Results

In this section, the results of experimental work by petrologists and geochemists on the topic of chemical diffusion in magmas are condensed into a few pages. For simplicity in presentation, the research efforts are categorized either by experimental approach (e.g., interdiffusion of two liquids) or by the principal objective of the experiments (e.g., to evaluate the effect of pressure), whichever seems to take precedence. Thus, for example, although Kushiro (1983) reports the results of experiments involving liquid/liquid interdiffusion, his paper is discussed under the heading "Effect of Pressure," because pressure was the focus of the study.

Interdiffusion of Two Liquids

The earliest attempt to characterize chemical diffusion in magmas was made by Bowen (1921), who examined interdiffusion of two magma analogs, molten diopside ($CaMgSi_2O_6$) and molten plagioclase [$(Ca, Na)Al_{1-2}Si_{3-2}O_8$]. In experiments remarkably similar to those still conducted today, Bowen prepared glasses of the two silicates, placed them in gravitationally-stable contact in a platinum crucible, and allowed them to interdiffuse at 1500°C and 1 atmosphere pressure for 17 to 48 hours. Without recourse to *in situ* microanalytical techniques (i.e., the electron microprobe), Bowen determined the extent of diffusional intermingling by measuring refractive indices along the interdiffusion zone. An independent calibration of refractive index vs. glass composition across the diopside-plagioclase join then allowed him to draw compositional profiles, which led, through an error-function solution of Fick's second law, to bulk diffusivities.

Implicit in Bowen's technique was the assumption of "linear" interdiffusion, i.e., that the composition of any liquid in the interdiffusion zone could be characterized in terms of the two intermingling endmembers, diopside and plagioclase. Bowen had, in fact, no means by which to characterize the gradients in individual elements or oxides (i.e., the "chemical path"). Despite this necessary oversimplification, the diffusivities obtained (10^{-6} to 10^{-7} cm^2/sec) are in good agreement with much more recent studies made after the advent of microanalytical techniques.

Following Bowen's classic study, no further work on diffusion in magmas was done until 1973, when the results of two studies were published—one by Yoder (1973) on interdiffusion of molten basalt and rhyolite, the other by Medford (1973), aimed specifically at characterizing Ca transport in mafic magmas. Yoder's objective was to show that melts of rhyolitic and basaltic composition are in fact miscible. In experiments broadly similar to those of Bowen, he placed powders or cylinders of rhyolitic obsidian and crystalline basalt in opposite ends of small Pt capsules. Sufficient water was added to saturate the melts at run conditions, and the capsules were sealed by welding. The resulting diffusion couples were run in an argon-medium vessel at 1 kbar pressure and 1200 or 950°C for 1 and 2 hours, respectively. Inasmuch as they produced interdiffusion zones easily resolvable with the electron microprobe, the experiments were a convincing demonstration that the two magma types are miscible at the specific conditions tested. Although Yoder made no attempt to compute actual interdiffusivities from his analytical profiles, apparent "bulk" values on the order of $1–5 \times 10^{-9}$ cm^2/sec at 1200°C are readily calculable from the smoothed curves presented. The data are not sufficiently detailed to reveal diffusivity differences among the major oxides analyzed, but there was some suggestion of anomalous behavior with potassium.

Medford's (1973) objective of obtaining a diffusivity for Ca in generally basaltic magma (specifically, a mugearite) was pursued by preparing glass-in-Pt-tube diffusion couples in which one half consisted of Ca-enriched material (10 wt% CaO relative to 7.5 wt% in the unadulterated rock composition). In this respect, Ca was the only variable component across the diffusion couple; the work still amounted to an experiment in chemical diffusion, however, because a gradient in Ca concentration was set up, and any net flux of Ca would have to be charge-compensated by an opposing flux of some other element. Through Boltzmann-Matano analysis, Medford's microprobe scans for Ca on samples run at 1230 to 1423°C produced the following well-defined Arrhenius relationship:

$$D_{Ca} = 0.0023 \exp(-29{,}500/RT) \qquad (1)$$

where D_{Ca} is calcium diffusivity, R is the gas constant (cal/deg-mole), and T is absolute temperature. Interestingly, the activation energy of ~30 kcal/mol is similar to published values for Ca tracer diffusion in simple melts at low pressures (e.g., Watson, 1979). The nature of Medford's experiments precluded measurements of diffusivities for elements other than Ca.

The first systematic and general study of chemical diffusion in magmas was that of Smith (1974), who examined interdiffusion of alkalies (Na, K, and Rb) in basaltic, andesitic, and rhyolitic melts. His experiments again involved the use of diffusion couples, the two halves of which initially differed only in the identities of the alkalies present—one half was artificially enriched in one alkali, the other half in another alkali. Smith's results are described in some detail by Hofmann (1980); the salient conclusions are the following: (1) interdiffusion of alkalies is governed by diffusivities in the range $\sim 10^{-5}$ to 10^{-7} cm^2/sec ($T \sim$ 1200–1500°C); and (2) the larger alkalies are generally somewhat slower than smaller ones with which they interdiffuse, so in order to maintain charge balance, motion of another cation (usually Ca) is required. In an experiment similar to that of Yoder (1973) on basalt/rhyolite interdiffusion, Smith confirmed a bulk interdiffusivity of $\sim 10^{-9}$ cm^2/sec.

From experiments designed primarily to examine differentiation processes in basaltic magma, Dowty and Berkebile (1982) were able to extract some information on chemical diffusion. These authors melted coarsely-crushed natural diabase and found that, early in the melting process, plagioclase crystals floated and pyroxene crystals sank. Upon completion of melting, vertical composition gradients in some elements were thus established, and the decay of these gradients over time could be used to compute diffusivities. Sodium, Fe^{2+}, Mg, and Al were found to have approximately the same diffusivity of $10^{-6.6}$ cm^2/sec at 1300°C. These numbers are similar to Smith's (1974) values for alkalies and those of Medford (1973) for Ca. Although the Dowty and Berkebile diffusivities were obtained in a sense as by-products, they are interesting because they pertain to a situation of no initial gradient in SiO_2 concentration.

As part of a larger study of the interaction between mafic magmas and felsic materials of the continental crust, Watson (1982) carried out several liquid/liquid interdiffusion experiments in which molten basalt was juxtaposed in diffusion couples with molten feldspar or granite. These studies were sufficiently detailed to reveal very clearly the phenomenon of 'non-linear' or 'selective' interdiffusion. For all conditions investigated (0.001 and 6 kbar; 1200 and 1300°C), the following general conclusions apply: (1) diffusional mixing of mafic and felsic liquids with respect to SiO_2 is quite slow, the interdiffusion profiles giving Boltzmann-Matano diffusivities on the order of 10^{-9} to 10^{-10} cm^2/sec; (2) interdiffusion of alkalies can be extremely fast (up to 10^{-5} cm^2/sec) and may result in net fluxes either down or *up* the initial concentration gradient, whichever tends to establish a felsic liquid/basaltic liquid concentration ratio of ~2:1–3:1; and (3) if the diffusion reservoirs are of limited dimension, the initial rapid transfer of alkalies ceases,

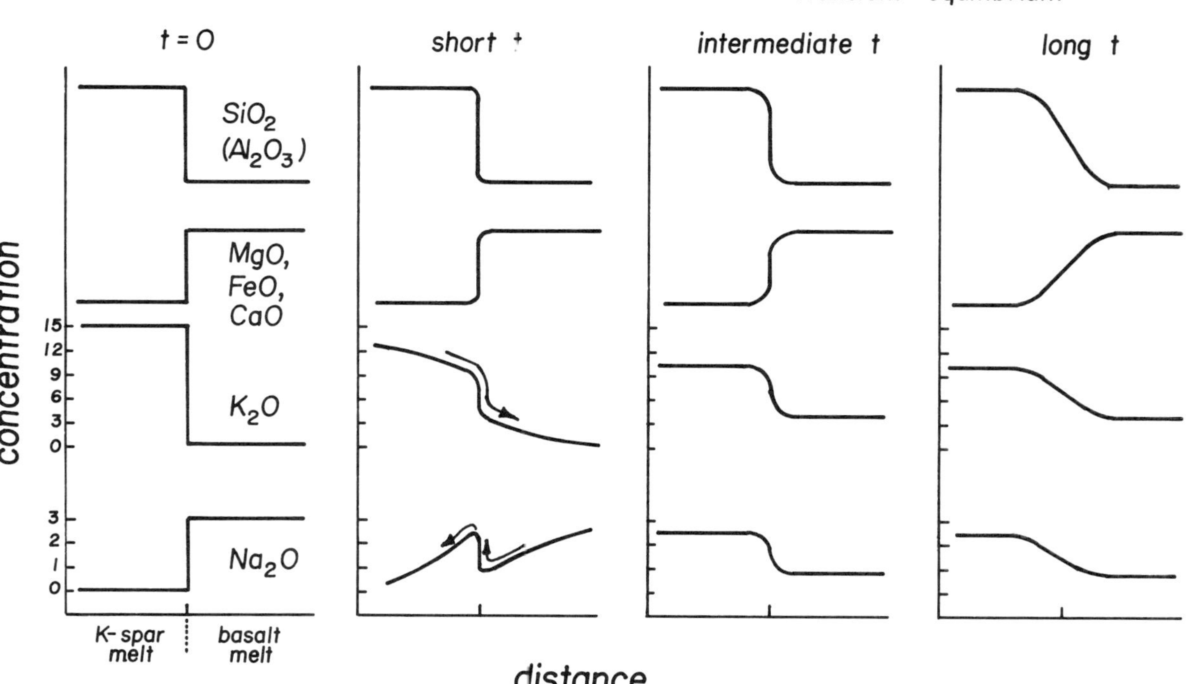

Fig. 1. Series of schematic diagrams illustrating the behavior through time of a basalt melt/K-feldspar melt diffusion couple. The alkalies show the greatest mobility, diffusing quickly to establish a K-feldspar/basalt concentration ratio of 2–3:1, potassium moving 'downhill' and sodium 'uphill'. Once this 'transient partitioning equilibrium' is reached, all concentration gradients are of the same length and the two liquids further intermix at a much slower rate determined by the diffusivity of SiO_2. (See text).

and the concentration gradients assumed by all elements become equal in length to (though possibly in the opposite direction of) that of SiO_2. (See schematic representation of this process in Fig. 1). From these observations, Watson (1982) concluded that interdiffusion of all magma components is regulated by diffusion of the principal melt structure-controlling species, SiO_2. Some melt components, most notably the alkalies, have lower activities (at a given concentration) in melts of high SiO_2 content, while for other components (e.g., CaO, MgO, FeO) the opposite is true. The response of all components can thus be viewed as an attempt to establish a "transient partitioning" equilibrium between the high- and low-Si melts, an equilibrium that persists as long as the two melts remain structurally distinct due to sluggish SiO_2 transport. To reiterate the proviso of point No. 3 above, it should be emphasized that the attainment of transient equilibrium is possible only when the diffusion reservoirs are limited in size, that is, finite with respect to transport of the fastest-moving species.

The concept of transient equilibrium is entirely consistent with observed partitioning between melts of differing SiO_2 content that are demonstrably immiscible (Watson, 1976; Ryerson and Hess, 1978). A possibly significant implication is that apparent diffusion gradients for most magma components may contain no diffusivity information at all—they may simply represent the temporarily "stable" distribution of components in an activity gradient imposed by the slow-decaying gradient in SiO_2 content. This interpretation needs further support before it gains wide acceptance; however, it has already been shown to be consistent with diffusion phenomena occurring in an imposed temperature gradient (Walker et al., 1981; Walker and DeLong, 1982; Lesher and Walker, 1985; See discussion of Soret diffusion below). The advantage of viewing liquid/liquid interdiffusion in this way is that it allows simple predictions to be made regarding the direction and extent of diffusional exchange of components between contacting magmas of differing SiO_2 content. This may be helpful when a mathematical analysis of the type presented by Oishi et al. (1982) is impractical or impossible.

Watson and Jurewicz (1984) followed up the initial paper by Watson (1982) with a more narrowly-defined study at 10 kbar and ~1250°C of interdiffusion processes resulting from contact of molten basalt with partially-molten granite. The results of this project confirmed an earlier prediction that "uphill" diffusion of Na from basalt to granite would occur, and also allowed a value of $\sim 3 \times 10^{-8}$ cm^2/sec to be assigned to K chemical diffusivity in molten basalt. Because of the complexity of the system, the charge-balancing (and possibly rate-limiting) species could not be identified.

To better understand chemical diffusion between intermediate and silicic melts, Baker (1990) studied chemical interdiffusion between metaluminous melts at 1 atm and 10 kbar, and between peralkaline and peraluminous melts at 10 kbar. As found by Watson (1982), the diffusivities of all non-alkali cations, even those found at only parts-per-million concentrations, are similar to, and appear controlled by, the diffusivity of the network formers Si and Al. Alkalies (Na, K, and Rb) appear to diffuse much more rapidly, however. Diffusivities are only moderately affected by changes in the SiO_2 content of the melt (Fig. 2).

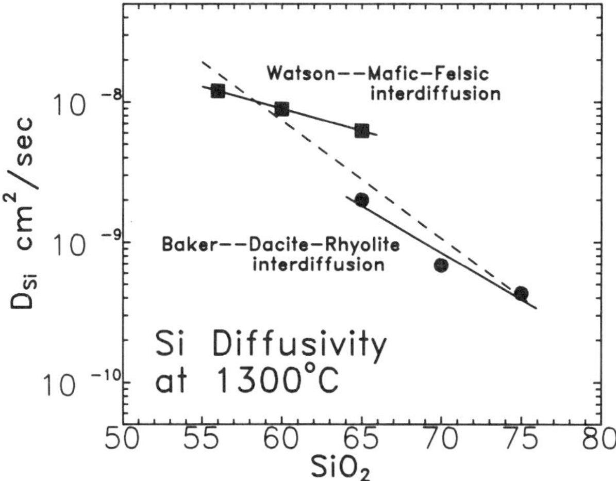

Fig. 2. Silicon diffusivity as a function of melt SiO_2 content during interdiffusion at 1300°C. Squares are mean Si diffusivities determined by Watson (1982) during interdiffusion of felsic and basaltic melts at 6 kbar. Circles are mean Si diffusivities determined by Baker (1990) during interdiffusion of dacitic and rhyolitic melts at 10 kbar. The solid lines are best fits to the individual data sets and the dashed line is the best fit line to the combined data set. The similarity of the results of the two studies at 65% SiO_2 suggests that diffusivities are relatively independent of differences in initial bulk compositions, but moderately dependent upon the SiO_2 content of the melt.

Because chemical diffusivities are dependent upon the specific chemical potential gradient during diffusion, there is no *a priori* reason to expect similar diffusivities in the 10-kbar dacite/rhyolite interdiffusion experiments of Baker (1990) and the 6-kbar felsic melt/basalt interdiffusion experiments of Watson (1982). Nevertheless, at 65 wt% SiO_2 the Si diffusivities measured in the two studies (Fig. 2) differ by a factor of only 3 at 1300°C. (The pressure difference between the two studies is considered insignificant; see below.) Combining the data sets of Watson (1982) and Baker (1990) reveals that from 56 to 75 wt% SiO_2 the Si diffusivity decreases by only about 1.5 orders of magnitude at these pressures (Fig. 2).

Changes in the alkali/alumina ratio of silicic melts were shown by Baker (1990) to result in a diffusivity minimum and an activation energy maximum at $(Na + K)/Al \sim 1$ (i.e., in metaluminous melts). These results are consistent with viscosity measurements made on simple systems (Riebling, 1966) and suggest that similar mechanisms control viscosity and interdiffusion rates in silicate melts.

Crystal Dissolution

The dissolution of a crystalline phase into a melt or magma undersaturated in that phase is an obvious means of introducing chemical heterogeneity and consequent chemical diffusion. In recent years, several experimental studies have been undertaken along this line, generally with the ultimate objective of characterizing crystal dissolution rates.

Even 50 years ago, at least one perceptive geologist (Holmes, 1936) was aware that the interfacial region between unstable minerals and magma could be one of interesting chemical diffusion effects. In examining natural basic rocks containing quartzose xenoliths, Holmes was able to document that the zone of glass separating quartz from engulfing basic lava was enriched in alkalies, particularly potassium. Potassium had in fact diffused "uphill" into the silica-rich melt immediately surrounding the dissolving quartz. The same phenomenon was described years later by Sato (1975), who was the first to confirm by experiment the occurrence of chemical diffusion effects inferred initially from examination of natural samples. Sato immersed 2–5 mm diameter quartz crystals in molten basalt (1400°C) for a period of 7 hours. Microprobe traverses on the quenched and sectioned samples revealed ~200-micron compositional gradients in the glass against the dissolving quartz, in which SiO_2 content decreased monotonically from the crystal/glass interface. Concentration profiles of K_2O and Na_2O mimicked the SiO_2 gradient, showing similar relative decreases away from the interface. Unlike SiO_2, however, the alkalies could not have come from the quartz—rather, they were enriched in the high-SiO_2 melt around the quartz by uphill diffusion from the basalt.

Watson (1982) carried out quartz dissolution experiments under both static and dynamic conditions that confirmed and extended Sato's early results and served as the impetus for the two-liquid interdiffusion experiments described in the preceding section. The idea was to test the hypothesis that alkalies always migrate up a concentration gradient from a relatively SiO_2-poor melt (i.e., basalt, ~50% SiO_2) into an SiO_2-rich one. When differences in temperature are taken into account, the quartz dissolution experiments of Sato (1975) and Watson (1982) are in good agreement, indicating bulk diffusivities in the range of $\sim 10^{-9}$ to 10^{-10} cm^2/sec at 1200 to 1300°C. (A diffusion model involving constant surface concentration and semi-infinite diffusion medium can be used to obtain these values). These diffusivities are also consistent with basaltic liquid/felsic liquid bulk interdiffusion constants obtained from the experiments of Watson (1982) and Smith (1974). As noted previously, there is good reason for regarding these as diffusivities of the melt structure-controlling component, SiO_2.

More detailed work on chemical diffusion processes associated with mineral dissolution has recently been reported by two principal research groups—that of E.B. Watson and T.M. Harrison, and that of A. Chekhmir and M. Epelbaum at the U.S.S.R. Academy of Sciences Institute for Experimental Mineralogy. At Rensselaer, the authors have been concerned specifically with the dissolution kinetics of accessory minerals (e.g., apatite [$Ca_5(PO_4)_3(F,Cl,OH)$] zircon [$ZrSiO_4$], sphene [$CaTiSiO_5$], and monazite [$(Ce,La,Th)PO_4$]) in granitoid magmas, in particular as they might bear on inheritance of radiogenic isotopes by anatectic magmas forming in old continental crust. (i.e., the authors have considered it possible that these phases do not dissolve under natural conditions

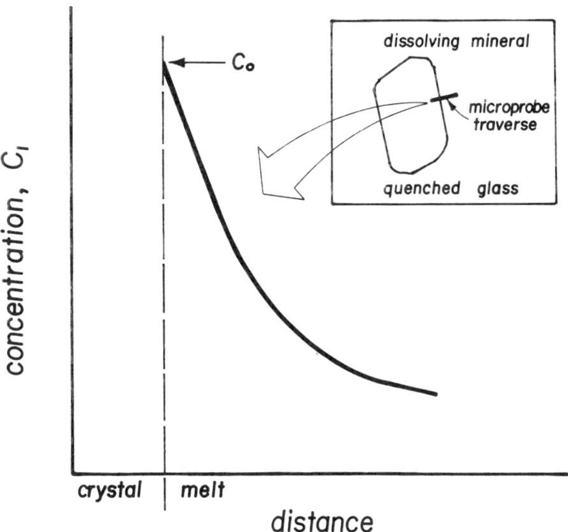

Fig. 3. Schematic representation of the concentration gradient of a major component of a mineral that is dissolving in a melt undersaturated in that mineral. C_0 represents the concentration required for saturation. If diffusion in the melt limits the rate of dissolution, the concentration gradient should generally conform to an ideal error-function diffusion model. See text for further explanation.

because the reaction kinetics are too slow). The approach is to surround a cut-and-polished, 1–2 mm diameter piece of gem-quality accessory mineral in finely-ground granitic obsidian contained in a small Pt capsule. When this assembly is subjected to high pressure and to temperatures well above the saturation surface of the accessory mineral, it begins to dissolve, producing diffusion gradients in the melt that are readily characterized in the quenched glasses by electron microprobe (Fig. 3).

It was found that diffusion profiles generated in this way usually conform very well to an ideal model that assumes one-dimensional diffusion into a semi-infinite medium from a planar surface at which the diffusant concentration is held constant; i.e.

$$C_1/C_0 = \mathrm{erf}\{x/(4Dt)^{1/2}\} \tag{2}$$

(Crank, 1975). An additional imposed condition implicit in the use of this model is that the surface from which the diffusant originates is stationary in space. Owing to the very low solubilities of accessory minerals in granitoid melts, this condition is met for all intents and purposes: The retreat of the crystal surface due to dissolution is minimal relative to the diffusion distance. The assumption of one-dimensional diffusion is valid as long as the crystal/liquid interface is planar and the diffusion distance is small relative to the dimensions of the interface. The concentration of diffusant at the surface has relatively little relevance to the diffusion measurements themselves, but is nevertheless of great interest because it represents the saturation level of that component with respect to the accessory mineral stability (See Harrison and Watson, 1983; 1984; Fig. 3).

The above method is readily amenable to investigating the effects on diffusion of variable water content in the melt (H_2O can be added to the Pt capsule in any amount desired), and works well as long as the temperature is sufficiently high to generate resolvable diffusion profiles in a laboratory time frame. Experiments generally cannot be done at the temperatures of interest to accessory mineral behavior in nature (i.e., 650 to 850°C), but Arrhenius relations that have been well defined at higher temperatures allow diffusivities at crustal melting conditions to be estimated with some confidence.

To date, work has been completed on zircon (Harrison and Watson, 1983), apatite (Harrison and Watson, 1984) and monazite (Rapp and Watson, 1985), giving chemical diffusion information for Zr, P, Ca, and several rare earth elements (REE). The data, summarized in Fig. 4, cover the temperature interval ~1100 to 1500°C and include dissolved water contents between 0 and 6 wt%. The salient features are: (1) remarkably low diffusivities and high activation energies for diffusion of highly-charged ions in dry granitic melts (~100 kcal/mole for Zr^{4+}; ~140 kcal/mol for P^{5+}); and (2) major reduction in activation energy and increase in diffusivity (below ~1500°C) as water content increases. The increase in diffusivity due to addition of water may amount to several orders of magnitude, depending upon temperature. It also seems clear that during monazite dissolution, *P* and the REE diffuse away from the crystal at the same rate, perhaps indicating a charge-balance or complexing process in the melt [as suggested on different grounds by Ryerson and Hess (1978)]. The behavior of P and Ca during apatite dissolution is quite different, inasmuch as P is far outdistanced by Ca (see Harrison and Watson, 1984).

Mineral dissolution studies of even broader scope than the Harrison-Watson accessory mineral experiments were described by Chekhmir (1984). Using similar but independently-developed techniques Chekmir performed dissolution experiments on numerous minerals to obtain chemical diffusivities for most major and some trace elements in a variety of silicate liquids. The range of conditions and compositions covered by Chekhmir's experiments was sufficiently extensive to enable development of an empirical model for prediction of the diffusivity of virtually any species in any assumed conditions. No effort is made here to summarize these results because this is done by Chekhmir and Epelbaum elsewhere in this book.

The mineral-dissolution approach to studies of chemical diffusion in magmas has considerable potential for illuminating mineral/melt interaction processes in nature and has recently been taken up by additional research groups, including Kuo and Kirkpatrick (1985) and C.M. Scarfe and co-workers (e.g., Brearley and Scarfe, 1984). The focus of both these research efforts is upon general dissolution kinetics of silicates, and upon the interaction between alkalic basalt magma and entrained, mantle-derived xenoliths and xenocrysts, respectively.

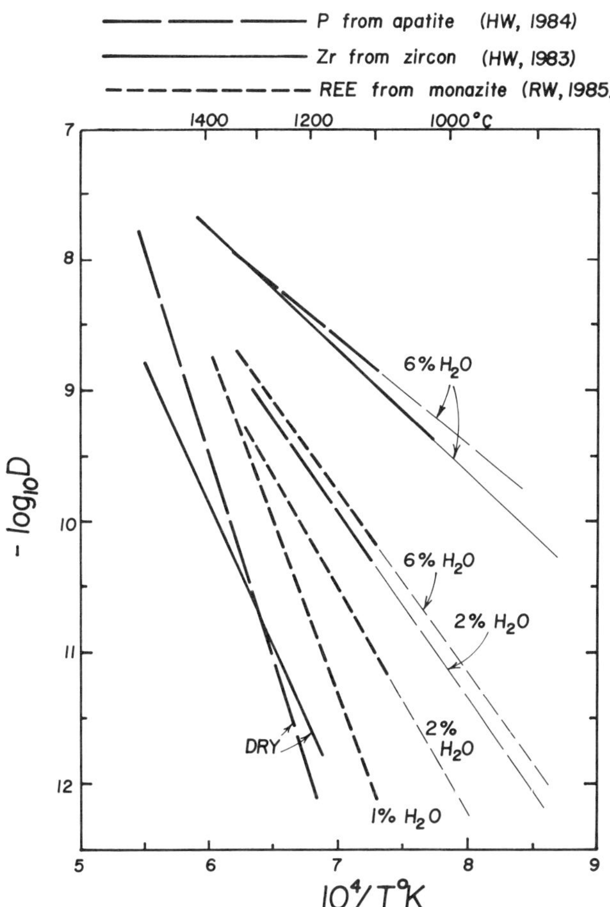

Fig. 4. Summary of chemical diffusion data obtained by partial dissolution of accessory minerals in granitic melts of variable water content. The heavy lines define the range over which actual data were acquired; the light lines are extrapolations. Note the extreme enhancement effect of even small amounts of H_2O. Sources: HW = Harrison and Watson; RW + Rapp and Watson. See text and Fig. 4.

Crystal Growth

Like crystal dissolution, *growth* of crystals under the appropriate conditions can result in chemical diffusion effects in the contacting liquid. Outside geology, this fact has been recognized for many years, receiving a good deal of attention during the 1950s from materials scientists concerned with the behavior and fabrication of semiconductors (e.g., Smith et al., 1955). Simply stated, diffusion gradients in the liquid can be expected to develop near a rapidly-growing crystal of composition different from that of the host melt—incompatible elements will "pile up" against the crystal and strongly compatible elements will be "drawn down" if diffusion in the melt cannot keep pace with crystal growth. Albarede and Bottinga (1972) were the first to emphasize the potential significance of crystal growth-induced diffusion phenomena to crystal/liquid disequilibrium in magmas. These effects were documented by Grove and Raudsepp (1978) with major-element diffusion profiles generated in controlled-cooling experiments on a lunar basalt. Chemical gradients up to 50 microns long were observed around pyroxenes that had been grown at cooling rates of 30, 60, and 150°C/h (quench temperature ~1000 to 1200°C). Due to the large number of system variables, extracting actual diffusivities from crystal-growth studies is not straightforward, although Powell et al. (1980) were able to calculate diffusivites ranging between 10^{-9} cm^2/sec for Si and 10^{-7} cm^2/sec for Ca from controlled-cooling experiments (1150 to 1050°C) on a eucrite meteorite.

Chemical Diffusion of Anions (O, F) and Dissolved Volatiles (H_2O, CO_2)

The relatively few publications on chemical diffusion of magma components other than cations have focused mainly upon oxygen, fluorine, and the principal magmatic volatiles, H_2O and CO_2.

Water. Because of its important role in phase equilibria and its pronounced effect on melt viscosity (Shaw, 1963), diffusive transport of dissolved water has received by far the most attention. Shaw (1974a) made the first measurements of water diffusion by performing hydration experiments on 2.5 mm diameter cylinders of obsidian (i.e., natural glass of granitic composition). These cylinders were sealed in gold capsules with more than enough water for saturation of the obsidian melt at the elevated P–T conditions of the experiments. Conventional cold-seal pressure vessels were used to subject the encapsulated obsidian cylinders to conditions of 100–2000 bars and 750 to 850°C for durations ranging up to 6 days. Diffusional uptake of water was gauged by weight gain of the cylinders during the experiments and by weight loss upon heating the hydrated cylinders at atmospheric pressure. Diffusivities extracted from these experiments proved to be strongly dependent upon H_2O concentration, ranging from $\sim 10^{-9}$ cm^2/sec at vanishingly small H_2O content to $\sim 10^{-7}$ cm^2/sec at 6 wt% dissolved water. On the other hand, the temperature dependence was found to be small, with a calculated activation energy of 15 $\pm$ 5 kcal/mole. The experimental technique did not allow recovery of information on the nature of the diffusing species or the existence of counter-diffusing components.

Jambon (1979) performed dehydration and hydration experiments on a natural obsidian to determine H_2O diffusivities at low ($<$0.4 wt%) and high (1–5 wt%) water concentrations, respectively. The dehydration runs consisted of simple heating of obsidian slabs in air at 500 to 980°C; diffusivities were computed from weight-loss measurements. In contrast the hydration runs were made in cold-seal pressure vessels (1 kbar, 800 to 900°C) and actual diffusional uptake profiles were characterized by weight-loss measurements on micro-sectioned wafers. An important innovation in Jambon's study was the addition of obsidian powder to the water reservoir in the hydration experiments, which facilitated attainment of cationic equilibrium between the diffusion sample and the surrounding water. The overall results of the study reveal a 3-order-of-magnitude difference in H_2O diffusivity between obsidians with low and high water contents. The "water-rich" values resemble those of Shaw (1974a; $\sim 10^{-7}$ cm^2/sec), and the "water-poor" data define an Arrhenius line giving much lower diffusivities:

$$D_{\text{water}} = 1.5 \times 10^{-8} \exp(-11{,}250/RT)$$

The activation energy for H_2O diffusion in the water-poor samples is within the overall range suggested by Shaw.

In experiments very similar to those of Shaw, Delaney and Karsten (1981) and Karsten et al. (1982) generated diffusional uptake gradients of water in obsidian that were subsequently characterized by ion-microprobe analysis of H concentration. This direct analysis method resulted in diffusivities qualitatively similar to those of Shaw: Over the range 650 to 950°C at 700 bars pressure, H_2O diffusivity was found to increase exponentially with increasing water concentration. At constant water content of 2 wt%, the diffusivity ranged from $\sim 10^{-8}$ cm^2/sec at 650° to $\sim 10^{-7}$ cm^2/sec at 950°C, giving an activation energy of ~ 20 kcal/mol.

Another study leading indirectly to information on the chemical diffusivity of H_2O is that of Arzi (1978), who examined the melting kinetics of water-bearing granitic rocks. Although Arzi's estimated water diffusivities are generally similar to those of Shaw (1974a) and Karsten et al. (1982) at the temperatures of geologic interest, he inferred that H_2O diffusivity is concentration-independent and governed by a considerably higher activation energy, i.e., 30–40 kcal/mol. Due to the indirect nature of Arzi's deductions, however, it seems appropriate to accept for the present the general conclusions of the other investigators.

It was recently shown by Zhang et al. (1989) that, although water-bearing silicate melts contain both hydroxyl and molecular water, the diffusivity of molecular water is much higher than that of OH^-. Consequently, the observed dependence of $D_{\text{"water"}}$ upon bulk water content of a melt (Shaw, 1974a; Karsten et al., 1982; Jambon, 1979) may actually be due to the increasing proportion of molecular water relative to hydroxyl (see, for example, Stolper, 1982) as total dissolved water increases.

In closing this section on water diffusion, it is interesting to note that the exponential dependence of D_{water} upon dissolved H_2O content that was documented by Shaw (1974a) and Karsten et al. (1982) is qualitatively similar to the dependence of cation diffusivities (e.g., Zr^{4+}, P^{5+}, REE^{3+}) upon melt water content (Harrison and Watson, 1983; 1984; Rapp and Watson, 1985; See Fig. 3).

Carbon dioxide. The diffusive behavior of the other magmatic volatile of widespread importance—carbon dioxide—has received relatively little attention in comparison with H_2O transport. The only available information is the high-pressure study by Watson et al. (1982), who used $Na_2{}^{14}CO_3$ to introduce traceable dissolved carbonate at one end of a Pt capsule containing synthetic, iron-free basalt glass or sodium-calcium aluminosilicate. It should be noted that although this technique involved the use of a radioisotope, the experiments were nevertheless *chemical* diffusion experiments, because a gradient in dissolved carbonate content was set up. The radiotracer was used simply to monitor chemical transport of carbonate, which could not be detected by more conventional techniques such as the electron microprobe.

The study revealed a considerably higher activation energy for diffusion of dissolved carbonate than for water, the Arrhenius equation for the simple aluminosilicate being

$$D_{\text{carbonate}} = 3.5 \exp(-46{,}600/RT)$$

This relationship was shown to hold over the temperature interval 800 to 1350°C and to be relatively insensitive to pressure (At a given temperature, $D_{\text{carbonate}}$ decreases by about an order of magnitude as pressure increases from 1 bar to 25 kbar). Although the experiments on synthetic basalt covered a much smaller temperature interval (1350 to 1500°C), the Arrhenius parameters appeared to be similar to those of the simple aluminosilicate. Watson et al. (1982) concluded on the basis of indirect reasoning that dissolved carbonate diffuses as ionic CO_3^{2-}.

Current work at Rensselaer includes re-examining "CO_2" diffusion in magmas with a view toward broadening the data base to other melt compositions and including the effect of concurrently dissolved H_2O. It is already clear that water has a pronounced effect on CO_2 diffusion, causing increases in $D_{\text{carbonate}}$ similar to those documented for other elements (i.e., orders of magnitude at crustal melting temperatures).

Oxyen. Although several studies of oxygen tracer diffusion have been reported [mostly in the materials science literature; see Dunn (1982) for a summary], only Dunn (1983) and Wendlandt (1980) have undertaken measurements of oxygen chemical diffusion in geological melts. The experiments of these two workers were similar in that both monitored redox changes in spherical droplets of iron-bearing melts. Wendlandt used a low-pressure furnace in which the ambient oxygen pressure could be regulated with gas mixtures. A diffusion experiment was done by equilibrating a droplet with a specific atmosphere and then abruptly changing the oxygen fugacity. Uptake or loss of oxygen was quantified by monitoring the mass of the droplet through time. Dunn's spherical glass beads were synthesized under oxidizing conditions at low pressure, then placed in a graphite container in a piston-cylinder apparatus, where they could be subjected for pre-determined durations to simultaneous high pressure and temperature under relatively reducing conditions. The extent of oxygen loss by diffusion from the sample was determined by measuring the bulk ferrous/ferric ratio of the quenched droplet after the experiment.

The experiments of Wendlandt (1980) and Dunn (1983) are in general agreement, giving diffusivities of 10^{-7} to 4×10^{-6} cm^2/sec at temperatures of 1160 to 1450°C; activation energies were generally in the 50–70 kcal/mol range. Wendlandt found a lower diffusivity and slightly higher apparent activation energy in

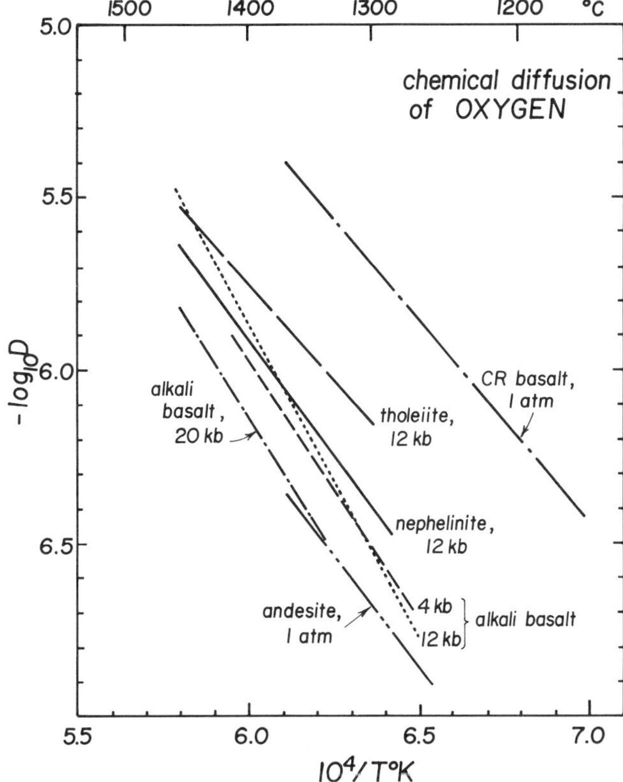

Fig. 5. Summary of experimental data on chemical diffusion of oxygen in basic to intermediate magmatic liquids. The 'Columbia River (CR) basalt' and 'andesite' lines are from Wendlandt (1980) and are based on only two data points each. All other lines represent results of Dunn (1983) and are based on at least three measurements each.

molten andesite than in basalt melt. Although Dunn's melt compositions were all broadly basaltic, the wider range of conditions covered by his experiments led him to the interesting conclusion that the pressure dependence of oxygen diffusion (which may be positive or negative) is correlated with the stability regions of liquidus minerals.

Oxygen chemical diffusion data for basaltic and andesitic melts are summarized in Fig. 5.

Fluorine. Two recent studies were devoted to chemical diffusion of fluorine, another volatile, anionic component of considerable significance in some magma types (Dingwell and Scarfe, 1984; 1985). The first of these two studies focused on jadeite melt at elevated pressure, and incorporated diffusion-couple experiments in which one half of the couple was initially fluorine-free, and the other contained 6 wt% fluorine. At pressures of 10–15 kbar and temperatures of 1200 to 1400°C, the diffusion process was found to be characterized by concentration-independent interdiffusion of fluorine and oxygen at diffusivities ranging from ~ 1 to 7×10^{-7} cm^2/sec. The interdiffusion process has an activation energy of 36–39 kcal/mol and is insensitive to pressure. The authors made the additional important points that (1) fluorine diffusion is as fast as or faster than any magmatic component other than the alkalies; and (2) dissolved fluorine is similar to dissolved H_2O in its strong enhancement effects on cation diffusion. (This last conclusion was based not upon Dingwell and Scarfe's own data, but upon experiments done by materials scientists.)

The second contribution of the above authors to the data base on chemical diffusion of fluorine (Dingwell and Scarfe, 1985) involved l-atmosphere devolatilization experiments. Fluorine-enriched spheres of albite, jadeite, and peraluminous glass (75 wt% SiO_2) were suspended from Pt loops in an oxygen-flow furnace, such that at high temperature (1200 to 1400°C) the fluorine in the spheres exchanged with ambient oxygen. Concentration-independent F–O interdiffusion was again documented, and was found to increase in rate in the order albite < peraluminous melt < jadeite. For the range in temperature covered, all interdiffusivities fall between 10^{-7} and 10^{-9} cm^2/sec, and activation energies between 29 and 43 kcal/mol. Interestingly, there was no suggestion of a reciprocal relationship between melt viscosity and F–O interdiffusion rate.

Baker and Watson (1988) investigated interdiffusion between a pantellerite melt containing ~ 0.5 wt% Cl and a rhyolitic melt containing ~ 0.5 wt% F. As found in the anhydrous diffusion studies, the diffusivities of all non-alkali cations are similar to one another at any given P and T. Once again it was deduced that, during interdiffusion between two melts, the diffusion of Si and Al controlled the diffusivities of all non-alkali cations. When compared with Zr diffusivities obtained by Harrison and Watson (1983) for halogen-free rhyolitic melts, the interdiffusion data of Baker and Watson (1988) on halogenated rhyolite indicate that, at 1200°C, Zr diffusion is enhanced by 2 orders of magnitude by the presence of small amounts of halogens.

The Zr diffusivity measured by Baker and Watson (1988) is similar to that reported by Harrison and Watson for rhyolite melt containing 1 wt% H_2O. The addition of halogens was also demonstrated to decrease the activation energy for Zr diffusion from 98 (Harrison and Watson, 1983) to 63 kcal/mol in a manner analogous to the effect of dissolved H_2O.

Effect of Pressure

Although many of the experimental studies described thus far included some high-pressure runs, the effect of pressure on chemical diffusion was seldom the principal focus of the work. Variable-pressure, *tracer*-diffusion experiments were conducted by Watson (1979; 1981), Watson and Bender (1980), Ross (1982), and Shimizu and Kushiro (1984), but only Fujii (1981), Kushiro (1983) and Baker (1990) were concerned with pressure effects on chemical diffusion.

Fujii's study involved interdiffusion of Ca and Sr in albite melt diffusion couples at 1250 to 1450°C and 2.5–20 kbar. Although the effect of pressure on Ca–Sr interdiffusion was found to be negative over the entire range of pressures investigated, an abrupt decrease in the magnitude of the pressure dependence was noted at ~7 kbar. At 1400°C, the overall decrease in diffusivity was from 3×10^{-7} cm^2/sec at 2.5 kbar to 9×10^{-8} cm^2/sec at 20 kbar. In addition to varying pressure, Fujii also added 4 wt% H_2O to one of his high-pressure runs and recorded a 6-fold increase in the rate of Ca–Sr interdiffusivity relative to the "dry" condition.

Kushiro's (1983) thorough study focused on interdiffusion of network-formers, specifically Ge–Si and Ga–Al, in diffusion couples of jadeite-analog melt at 6–20 kbar and 1400°C. He found not only significant pressure effects—roughly an order-of-magnitude increase in interdiffusion rate over a 20-kbar increase in pressure—but also pronounced dependence upon composition, especially in the Ge–Si experiments. These experiments revealed decreases in interdiffusion rate by as much as 100X as the composition changed from Ge- to Si-endmember jadeite; in the Ga–Al case, up to 50-fold decreases were recorded as composition varied from pure Ga endmember to pure Al endmember. The total range in interdiffusivities for both elements was about 10^{-7} to 10^{-9} cm^2/sec.

In their study of oxygen diffusion in jadeite and diopside melts, Shimizu and Kushiro (1984) demonstrated a positive correlation between oxygen diffusion and pressure for jadeite melts at 1400°C and a negative correlation for diopside melts at 1650°C. The pressure effects on oxygen diffusion are small in both cases, however: for jadeite melt, D increases by a factor of only 2 between 5 and 20 kbar, and for diopside melt D decreased by ~30% as pressure was increased from 10 to 17 kbar. Apparent negative activation volumes for oxygen diffusion in jadeite melts were interpreted as due to local collapse of the network structure during diffusion. Positive activation volumes for oxygen diffusion in diopside melts were correlated with the molar volume of oxygen and interpreted to indicate that individual oxygen ions were the diffusing units. Furthermore, Shimizu and Kushiro (1984) demonstrated the utility of the Eyring equation in the prediction of oxygen diffusivities from melt viscosities (discussed in more detail below).

Baker's (1990) study of interdiffusion between metaluminous dacitic and rhyolitic melts at 1 atm and 10 kbar also demonstrated a positive correlation between pressure and diffusion. The effect of pressure is small, however—only a factor of 4 at 1300°C. The negative activation volume, -17.6 cm^3 at 1300°, was attributed to volume changes occurring in the melts during diffusion. Baker (1990) suggested that these negative activation volumes were the result of transient increases in the coordination (from 4 to 5, or possibly 6) of Si or Al cations during diffusion. This increase in coordination formed very short-lived (nanoseconds?) activated complexes whose creation and destruction could result in the exchange of Si and Al cations. Baker supported his conclusions with comparisons of the molar volume change during the reactions 1/2 nepheline + 1/2 albite = jadeite (-16.9 cm^3/mol) and coesite = stishovite (-12.5 cm^3/mol). Baker also found additional support for his mechanism in molecular dynamics simulations of silicate melts and in nuclear magnetic resonance studies of silicate melts and glasses.

Diffusion in a Temperature Gradient

A final category of chemical diffusion experiment whose importance is just beginning to be appreciated is that of diffusion in a temperature gradient. Referred to as "thermal" or "Soret" diffusion, this transport process involves chemical fractionation of an initially homogeneous melt reservoir that is subject to a static temperature gradient. Illumination of Soret effects in magmas has recently been accomplished by D. Walker and C. Lesher (Walker et al., 1981; Walker and Delong, 1982; Lesher and Walker, 1985), who have examined melts ranging in composition from lunar and mid-ocean ridge basalt to andesite, rhyolite, and simple 4-component liquids in the leucite-fayalite-silica system.

The numerous experiments performed by the Walker/Lesher group have some general characteristics in common. First, they are all similar in design: Each experiment employed a tubular capsule (molybdenum or graphite) of 1–2 mm I.D. and several mm length, into which was loaded the homogeneous glass powder of interest. The capsule was placed vertically in a standard 1/2-inch piston-cylinder assembly, with the upper end positioned at the hot spot of the graphite heater, subject to temperatures of 1500 to 1800°C. The lower end, well below the hot spot, was cooler by 200 to 300°C.

The Soret experiments are also similar in the general nature of the results: Invariably, once a steady state is achieved (which requires ~1 week at run conditions), the hot end of the originally-homogeneous melt is substantially enriched over the "cold" end in SiO_2, Na_2O, and K_2O and depleted in CaO, FeO, MgO, Cr_2O_3, MnO, and TiO_2. Only Al_2O_3 is ambivalent, showing in different melts both enrichment and depletion at the hot end, and sometimes no clear preference (See examples in Fig. 6).

Extraction of component diffusivities from experiments of the type just described is not straightforward, although Lesher and Walker (1985) obtained values for SiO_2 that make sense in the context of earlier results from more conventional experiments [e.g., those of Watson (1982) and of Powell et al. (1980)]. Like Watson, Lesher and Walker conclude that transport of most other oxide components is linked to and controlled by diffusion of SiO_2.

The Soret diffusion experiments are intriguing in their own right, and may have significant implications regarding element fractionation under natural magmatic conditions (See Walker et al. 1981; Walker and DeLong, 1982). However, the greatest (and certainly the unique) significance of the experiments lies in their ability to provide basic thermodynamic information on molten silicates.

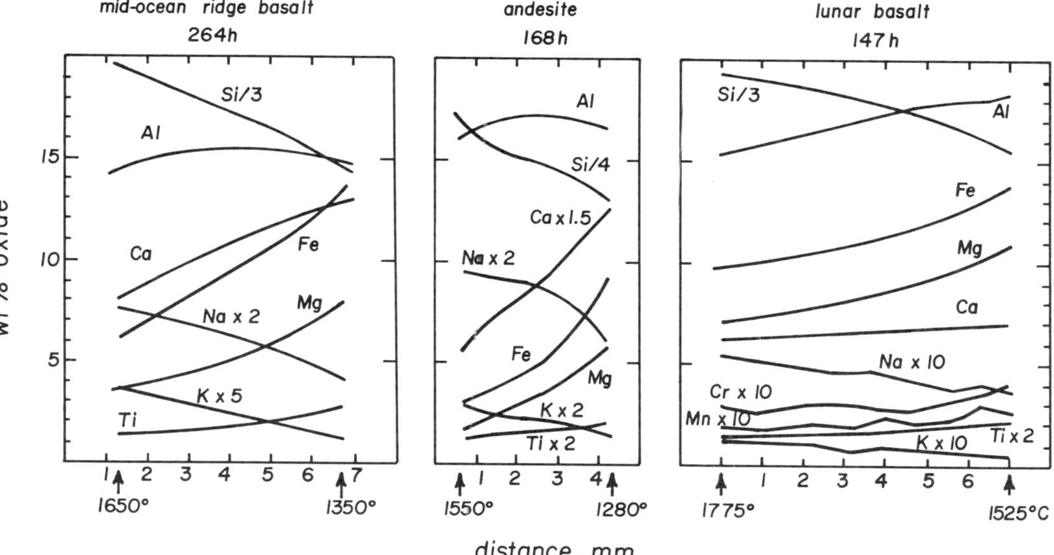

Fig. 6. Stable composition profiles that result from subjecting capsules containing natural magmatic liquids to temperature gradients of 250 to 300°C for a week or more. Note the temperatures at each end of the 4–7 mm capsules. The mid-ocean ridge basalt and andesite data are from Lesher and Walker (1985); the lunar basalt profiles were generated by Walker et al. (1981). See text for discussion.

Calculation of Diffusivities

Ability to calculate chemical diffusivities from first principles would avoid the necessity of making diffusion measurements on every composition of geological interest. Transition state theory (TST) was demonstrated by Baker (1990) to show some promise of directly achieving this goal for interdiffusion of silicate melts. The relative insensitivity of alkali diffusion to melt composition (Smith, 1974) makes these elements particularly good candidates for application of TST to compute diffusivities. For non-alkalies, detailed structural and thermodynamic data for the melts are required in order to use TST to compute diffusivities accurately. Unfortunately, such data are generally unavailable at this time. However, TST can also yield diffusivities indirectly from melt viscosities through the Eyring equation:

$$D = \frac{k_B T}{\lambda \eta} \tag{3}$$

where k_B is Boltzmann's constant, T is temperature in Kelvin, λ is the jump distance for the diffusing species during each diffusive step, and η is the melt viscosity. This equation was shown to hold for oxygen diffusion over a range of pressures for a variety of melt compositions (Shimizu and Kushiro, 1984). Additionally, Baker (1990) demonstrated good inverse relationships between viscosity and Si diffusivity in melts containing 65 wt% and 70 wt% SiO_2 during the interdiffusion of dacitic and rhyolitic melts at 1 atm. The inverse relationship between high-pressure diffusivities and viscosities was not as good, however, probably owing to the little-known effects of pressure on melt viscosities for dacitic to rhyolitic compositions (cf. Scarfe et al., 1987).

Figure 7 shows Si diffusivities measured during interdiffusion experiments on different compositions together with diffusivities calculated from the Eyring equation. These diffusivities include 1-atm measurements on anhydrous dacite-rhyolite (Baker, 1990), measurements on hydrous versions of the same compositions at 10 kbar and on hydrous rhyolites at 2 kbar (preliminary data by Baker), and a value from Watson's (1982) felsic melt/basalt interdiffusion study. The viscosities necessary for diffusivity calculations were computed using the method of Shaw (1972) for all melts except the high-pressure basalt, whose viscosity was estimated by comparison with experimental measurements on another basalt (Scarfe et al., 1987).

The relationship between measured and calculated diffusivities is remarkably good. Over the range of 10^{-11} to 10^{-7} cm^2/sec the Eyring equation can predict to within a factor of 3 the measured diffusivities of non-alkalies during interdiffusion. (As discussed above these diffusivities are interpreted to be the interdiffusivity of the network formers, Si and Al.)

Thus, the combination of Shaw's (1972) method for the calculation of melt viscosities and the Eyring equation provide a powerful tool for assessing the rates of diffusive interaction between juxtaposed melts of differing composition.

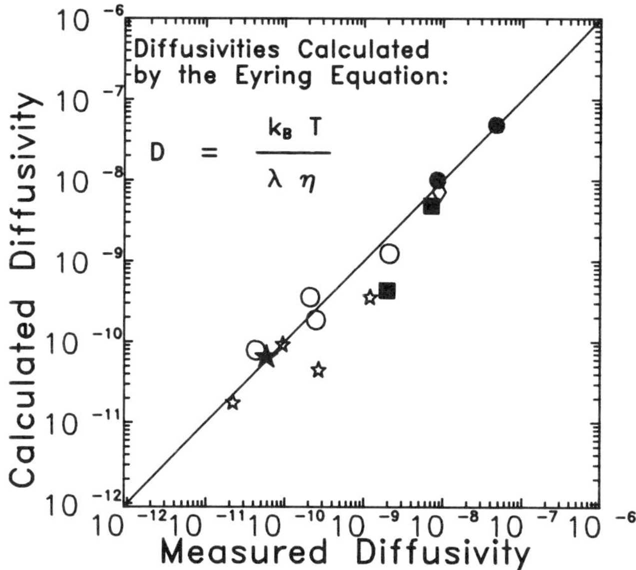

Fig. 7. Diffusivity calculated by the Eyring equation plotted against measured diffusivities from various interdiffusion studies. Open symbols are measurements on anhydrous compositions: *Diamonds*—Si diffusivity at 60% SiO_2 (Watson, 1982); *open circles*—Si diffusivity at 65% SiO_2 (Baker, 1990); *open stars*—Si diffusivity at 70% SiO_2 (Baker, 1990); *filled stars*—Al diffusivity during interdiffusion of two rhyolites (D.R. Baker, in prep.); *filled circles*—Si diffusivities at 65 and 70% SiO_2 during interdiffusion of dacitic and rhyolitic melts containing 6% H_2O at 10 kbar (D.R. Baker, in prep.); and *filled squares*—like filled circles but containing only 3% H_2O.

Major Areas of Geological Application

Although the nature of the experimental studies described in the preceding section gives some clues as to the intended applications of the results, some further elaboration seems appropriate. Generally speaking, igneous petrologists and geochemists wish to model or interpret the chemical dynamics of magmas; experimental data on diffusion constitute invaluable input for such models. Naturally, it must always be borne in mind that magmas are often physically dynamic systems, so that under some circumstances mechanical transport effects may predominate over diffusion. Nevertheless, there are many instances in which diffusion effects are clearly important, even if they rarely operate alone.

Magma Mixing and Contamination

Field and petrographic observations have documented numerous instances of interaction between two separately-generated magmas or between magmas and contacting solid materials. For summaries of references on this subject, the reader is referred to Reid et al. (1983), Gerlach and Grove (1982), Watson (1982), Dungan and Rhodes (1978), and Yoder (1973).

Some of the situations described by these authors involve interaction between two basic, low-viscosity magmas; under these circumstances, it is likely that diffusive processes are relatively unimportant, being overwhelmed by mechanical stirring effects. [A notable exception here may be instances in which a system of double-diffusive convection is established. See Turner (1973) for discussion.] The situation of contacting silicic and basic melts is, however, quite different—not only are there marked chemical contrasts to "drive" chemical diffusion, but also substantial differences in viscosity that may serve to restrict mechanical mixing. It is here that interdiffusion processes may be most significant.

In applying interdiffusion data acquired in the laboratory it must be recognized that the extent to which the composition of a magma is modified by contact with other molten materials will be highly dependent upon the physical circumstances or geometry of "exposure," simply because the latter determines the surface area across which diffusive transport can occur. For this reason, it seems unlikely that any specific case of magma mixing or contamination will be understandable in terms of a unique model of diffusional interaction. What is possible, however, is to show, by illustrative calculation, what mixing by diffusion can and cannot do.

Watson (1982) used interdiffusion data from his basalt/felsic melt diffusion couples to examine hypothetical contamination of basaltic magma by partially-molten felsic crustal xenoliths. The rate of bulk assimilation was assumed to be governed by the relatively slow diffusion of SiO_2. The experiments had clearly revealed, however, that much more rapid, *selective* transfer (i.e., non-linear diffusion) of alkalies was to be expected under some conditions, and the efficacy of this process could be assessed in relation to the bulk mixing rate. Watson concluded that contamination of basaltic magma by interaction with crustal materials is more likely for potassium than for any other element (Fig. 8). Less pronounced selective effects should be anticipated for other elements and isotopic systems. It must be recognized, of course, that if the interdiffusion process goes to completion with respect to all components, the end result will be no different from that achieved by any other mixing process.

Dissolution Rates of Crystals under Natural Melting Conditions

Chemical diffusion data acquired in the type of study done by Harrison and Watson (1983; 1984), Chekhmir (1984), Rapp and Watson (1985), Kuo and Kirkpatrick (1985), and Brearley and Scarfe (1984) are useful for many purposes, but the most direct application is to problems involving dissolution rates of natural crystals in contacting magmatic liquid. It is implicitly assumed in this application that the dissolution process is rate-limited by diffusion in the melt, not by crystal/liquid interface kinetics. This assumption has been justified by Harrison and Watson, at least over the range of experimental conditions, on the

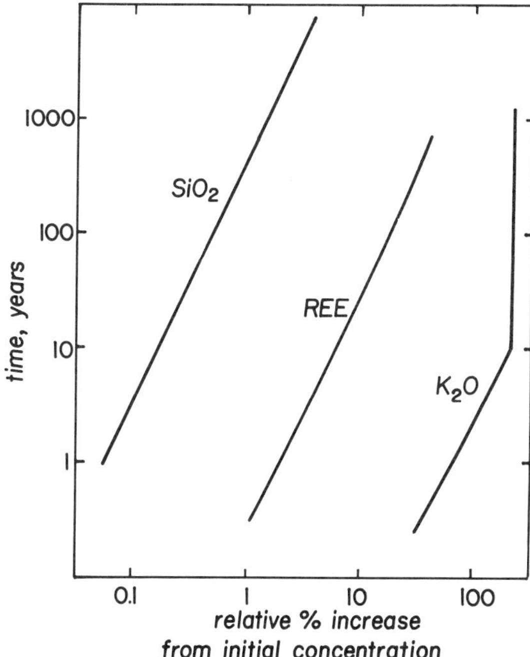

Fig. 8. Model contamination effects for K_2O, rare earths, and SiO_2 resulting from diffusive interaction of basaltic magma (initially 50 wt% SiO_2 and 0.1% K_2O) with partially-molten felsic xenoliths. For the model calculations, the mass ratio of basalt to xenoliths was assumed to be 9:1 and the xenoliths were assigned an initial radius of 50 cm. The main point of the diagram is to illustrate the kind of selective contamination possibly resulting from incomplete diffusive mixing. Diagram is simplified from Watson (1982).

grounds that diffusion gradients against dissolving accessory minerals usually conform very well to ideal diffusion models—a situation that could not generally obtain if interface kinetics were regulating the "release" of diffusant at the crystal surface.

Harrison and Watson (1983; 1984) applied their data to the general problem of inherited radiogenic isotopes in accessory phases of crustal granitoids. The documented presence of inherited radiogenic Pb in some zircons (e.g., Williams, 1978; Compston and Williams, 1982), for example, raises the question "Do zircons survive crustal fusion because they are insoluble in the melt even at peak melting conditions, or do they survive because their dissolution kinetics are very slow?" Based on their experimental diffusion data, Harrison and Watson concluded that unless an anatectic melt is quite poor in dissolved H_2O (<2 wt%) and diffusivities in the melt consequently low, the dissolution times for zircons (and for apatites) in crustal melting conditions are effectively instantaneous in a geological time frame. (This conclusion applies to strictly diffusion-controlled dissolution; mechanical stirring of the system would result in even shorter dissolution times). Thus, it can be concluded that accessory minerals survive crustal fusion simply because the host melts are saturated in them. This conclusion is important because it means that laboratory measurements of accessory mineral solubilities can be applied to the problem of residual accessory phases without much concern over major-element disequilibrium effects.

Although the experimental systems are very different, the motivation for the dissolution-rate studies at the University of Alberta (e.g., Brearley and Scarfe, 1984) is similar to that of Harrison and Watson. These workers sought to characterize the rates of olivine, pyroxene, spinel, and garnet dissolution during the ascent of deep-seated mafic magmas. Results to date indicate somewhat higher dissolution rates for olivine than for pyroxene, although both are sufficiently fast that unstable, mm-sized crystals would easily dissolve during ascent from the upper mantle. (30 and 60 minutes are required for complete dissolution of 1-mm olivines and pyroxenes, respectively).

Disequilibrium Effects Associated with Growth of Crystals

Trace elements. Albarede and Bottinga (1972) brought to the attention of geochemists the fact that diffusion in the melt can be a limiting factor in the attainment of crystal/liquid equilibrium with respect to trace elements. Simply stated, any element that is highly incompatible in a growing crystal and has a low diffusivity in the melt will tend to "pile up" ahead of a rapidly-advancing crystal surface. Because the equilibrium partition coefficient is maintained at the crystal/liquid interface, this "snow-plow" effect in the melt will cause the concentration of the incompatible element to rise in the crystal as well. The net result, then, could be a bulk crystal/liquid distribution of the trace element that bears little resemblance to an equilibrium partition coefficient. Although confirmation of the predicted trace-element enrichment against rapidly grown crystals is inherently difficult, Lindstrom et al. (1979) succeeded in showing indirectly that rare earth elements REE "pile up" against olivines grown at cooling rates of $\sim$400°C/hour.

The mathematics of this sort of disequilibrium process were addressed some time ago (e.g., Smith et al., 1955), so it is possible, in principle, to assess the likelihood of disequilibrium for any hypothetical situation, the necessary input being the growth rate of the crystal and the diffusivity (in the melt) of the trace element of interest. (The equilibrium crystal/liquid partition coefficient is needed as well, but the calculations are not sensitive to changes in this parameter as long as it has a very low value). At present, the real limitation to routine analysis of this type lies not in the availability of diffusion data but in the lack of information on crystal growth rates in natural systems.

Major elements. The considerations discussed above in reference to trace elements also apply to major components of magmas. If, for example, major component X is strongly concentrated in a particular mineral and diffuses slowly in the melt, then rapid growth of the mineral may cause depletion of X from the immediately-surrounding melt. Unlike the trace-element example, however, this situation is characterized by an automatic feedback mechanism: Depletion of component X in the melt due to rapid crystal growth is likely to slow or halt growth of the crystal, or to cause the crystal to change composition. This kind of behavior is thought to be responsible for oscillatory zoning in plagioclase feldspar (e.g., Bottinga et al., 1966; Sibley et al., 1976; Allègre et al., 1981; Loomis, 1982).

Major-element disequilibrium resulting from the interplay between crystal growth and diffusion in the melt has also been invoked to explain much larger-scale magmatic heterogeneities. McBirney and Noyes (1979), for example, attributed certain layering phenomena in basic intrusions to a crystallization/diffusion process like that noted above in reference to plagioclase feldspars. Several authors (McBirney, 1980; Chen and Turner, 1980; Turner and Gustafson, 1981) have made the interesting suggestion that compositional boundary layers developing ahead of solidification fronts in magma chambers may become gravitationally unstable due to depletion of heavy components (see Spera et al., 1984). Consequent detachment and buoyant rise of these liquids may be responsible for large-scale vertical stratification of magma chambers, possibly like that documented so thoroughly by Hildreth (1979).

Local saturation phenomena. An additional consequence of the simultaneous occurrence of rapid crystal growth and slow diffusion in the melt may be local

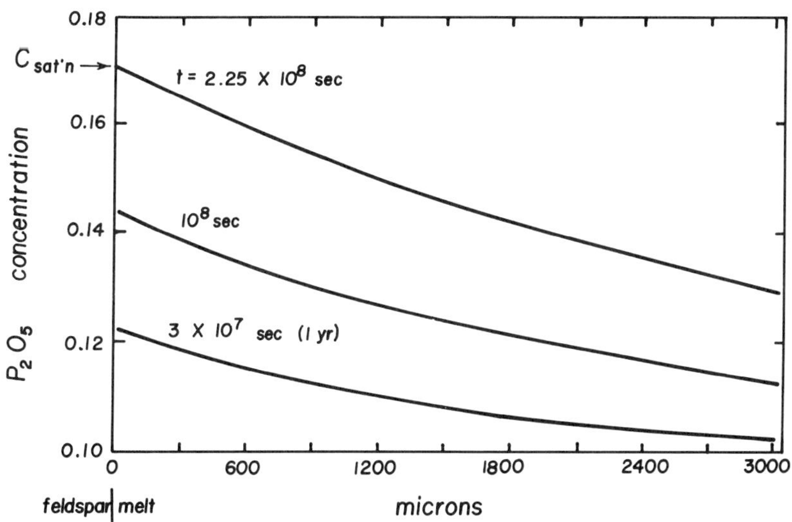

Fig. 9. Results of model calculations on 'pileup' of phosphorus in advance of a feldspar crystal surface (at left) growing at 10^{-9} cm/sec. The initial P_2O_5 concentration in the melt was assumed to be 0.1 wt% and the temperature 1000°C; the appropriate phosphorus diffusivity was taken from Harrison and Watson (1984). The level of P_2O_5 required for apatite saturation (0.17%) is reached after ~7 years of sustained growth; at this point, local precipitation of small apatite crystals will occur at the interface, and these will be included in the growing feldspar. See text and Harrison and Watson (1984) for details.

precipitation, against the advancing crystal surface, of a mineral whose stability-limiting component piles up due to sluggish transport away from the growing phase. This phenomenon was first proposed by Green and Watson (1982) to explain the common presence of tiny apatite inclusions in plagioclase phenocrysts of magmas clearly undersaturated in apatite. Since that time, Harrison and Watson (1984) obtained the phosphorus diffusion data necessary to rigorously model the local saturation process using assumed host-crystal (e.g., feldspar) growth rates. Because phosphorus diffusion is very slow in silicic melts, it is not necessary to call upon extremely rapid feldspar growth to cause sufficient phosphorus enrichment at the feldspar/melt interface for local precipitation of apatite (Fig. 9).

Diffusivities of most components, especially those diffusing as highly-charged ions, are slower in intermediate-to-silicic melts than in basic ones, so it seems likely that occurrence of local saturation phenomena will be largely restricted to more-evolved magma types. A possible exception to this generalization may be the inclusion of small chromite crystals in olivines of basic volcanics. D.M. Francis (personal communication, 1984) has noted instances in which chromite appears as inclusions in magnesian olivine, apparently having been incorporated at conditions inappropriate for bulk saturation of the magma in chromite.

Magma/Volatile Interactions

The diffusivity information pertaining specifically to the principal magmatic volatiles, H_2O and CO_2, has important applications in the area of magma dynamics. Shaw (1974b) used his diffusion data for H_2O to constrain some general models of convection and wall-rock interaction in granitic magma chambers. Sparks (1978), on the other hand, used the same data as input for calculations of the growth rates of H_2O vapor bubbles in ascending and erupting magmas. The same application was discussed less thoroughly by Watson et al. (1982) with regard to their diffusivity information on dissolved carbonate. In addition, these latter authors noted that because dissolved H_2O and CO_2 diffusivities are quite different, the possibility of diffusional fractionation of the two gases during bubble growth (and during other diffusion-controlled processes) must be recognized.

Thermodynamics of Molten Silicates

Lesher and Walker (1985) have recently shown that one of the principal benefits of studies specifically of Soret diffusion is basic thermodynamic data for silicate solutions. Although the idea of extracting information pertaining to homogeneous equilibrium from experiments involving kinetic processes may seem odd at first glance, it should be borne in mind that the "end result" of a Soret diffusion experiment is an array of composition gradients that is stable in the imposed temperature gradient. The existence of these gradients in fact reflects the relative changes in component free energies across the non-uniformly heated sample. Lesher and Walker were able to compute from their experiments not only heats of transport for SiO_2 (i.e., enthalpy changes associated with moving SiO_2 from one 'environment' to another) but also the interaction (Margules) parameters for binary asymmetric regular solution models that give good fits to their data. The computed interaction parameters compare favorably with those calculable from calorimetric and phase equilibrium studies. Thus it seems clear that thermal diffusion experiments hold considerable promise to shed light not only upon kinetic processes in magmas, but also upon silicate melt thermodynamics.

Future Directions of Research

The existing experimental data base on diffusion is generally good enough to provide, if not directly then certainly by extrapolation or interpolation, order-of-magnitude estimates of diffusivities for most magma components under most conditions realizable in nature. The effects of such magmatic variables as pressure and dissolved water content are now sufficiently well studied that an evaluation of their probable importance to diffusion can be made for any given situation. Needless to say, however, there remain significant gaps in our knowledge. The specific areas in need of additional data are to some extent a matter of opinion, but the following questions immediately come to mind:

1. What are the effects of dissolved volatiles other than H_2O (e.g., CO_2, F, Cl) on diffusion of other magma components?
2. Can the process of interdiffusion of two liquids be generally viewed in the simplistic terms proposed by Watson (1982)?
3. How prevalent is the kind of complexing implied by the coincident diffusivities of P and REE during monazite dissolution?
4. Is diffusion always the rate-limiting step in crystal dissolution?

These and related specific questions will probably be addressed soon enough by experimentalists. Even when the answers have been found, however, the topic of chemical diffusion in magmas will by no means be exhausted. The diversity and heterogeneity of solidified magmas attest very well to the need for applying diffusion data to problems of igneous petrogenesis, with the ultimate goal of understanding the relative contributions of chemical, mechanical, and thermal transport processes in magmatic systems.

Acknowledgments

An early draft of this paper was improved by comments from T.M. Harrison, S.R. Jurewicz, A.J.G. Jurewicz, and R.P. Rapp. A.W. Hofmann provided an insightful official review. Work on chemical diffusion at R.P.I. has been supported by the National Science Foundation through Grants EAR80-25887, EAR82-12453 and DMR85-10617.

References

Albarede, F. and Bottinga, Y. (1972) Kinetic disequilibrium in trace-element partitioning between phenocrysts and host lava. Geochim. Cosmochim. Acta **36**: 141–156.

Allègre, C.J., Provost, A. and Jaupart, C. (1981) Oscillatory zoning in plagioclase: Pathological case of crystal growth. Nature **294**: 223–228.

Arzi, A.A. (1978) Fusion kinetics, water pressure, water diffusion and electrical conductivity, interrelated. J. Petrol. **19**: 153–169.

Baker, D.R., and Watson, E.B. (1988) Diffusion of major and mace elements in compositionally complex Cl- and F-bearing silicate melts. J. Non-Cryst Solids **102**: 62–70.

Baker, D.R. (1990) Chemical interdiffusion of dacite and rhyolite: Anhydrous measurements at 1 atm and 10 kbar, application of Transition State Theory, and diffusion in zoned magma chambers. Contrib. Min. Pet. **104**: 407–423.

Bottinga, Y., Kudo, A. and Weill, D. (1966) Some observations on oscillatory zoning and crystallization of magmatic plagioclase. Am. Mineral. **50**: 792–806.

Bowen, N.L. (1921) Diffusion in silicate melts. J. Geol. **29**: 295–317.

Brearley, M. and Scarfe, C. (1984) Dissolution of upper mantle minerals in alkali basalt melt at 30 kbar: Implications for ultramafic xenolith survival. Geol. Soc. Am. Abstr. Progr. **16**: 454.

Chekhmir, A.S. (1984) Experimental study of diffusion processes in magmatic melts. Ph.D. thesis, Vernadskii Institute of Geochemistry and Analytical Chemistry, Moscow, USSR.

Chen, C.F. and Turner, J.S. (1981) Crystallization in a double-diffusive system. J. Geophys. Res. **85**: 2573–2593.

Compston, W. and Williams, I.S. (1982) Protolith ages from inherited zircon cores measured by high mass-resolution ion microprobe. Abstrs. Fifth Int. Conf. Geochron, Cosmochron. Isot. Geol.: 63–64.

Cooper, A.R. (1965) Model for multi-component diffusion. Phys. Chem. Glasses **6**: 55–61.

Crank, J. (1975) The Mathematics of Diffusion. Second Edition. Oxford University Press, 414 pp.

Delaney, J.R. and Karsten, J.L. (1981) Ion microprobe studies of water in silicate melts: Concentration-dependent diffusion in obsidian. Earth Planet. Sci. Lett. **52**: 191–202.

Dingwell, D.B. and Scarfe, C.M. (1984) Chemical diffusion of fluorine in jadeite melt at high pressure. Geochim. Cosmochim. Acta **48**: 2517–2525.

Dingwell, D.B. and Scarfe, C.M. (1985) Chemical diffusion of fluorine in melts in the system Na_2O–Al_2O_3–SiO_2. Earth Planet. Sci. Lett. **73**: 377–384.

Dowty, E. and Berkebile C.A. (1982) Differentiation and diffusion in laboratory charges of basaltic composition during melting experiments. Am. Mineral. **67**: 900–906.

Dungan, M.A. and Rhodes, J.M. (1978) Residual glasses and melt inclusions in basalts from DSDP legs 45 and 46: Evidence for magma mixing. Contrib. Mineral. Petrol. **67**: 417–431.

Dunn, T. (1982) Oxygen diffusion in three silicate melts along the join diopside-anorthite. Geochim. Cosmochim. Acta **46**: 2293–2299.

Dunn, T. (1983) Oxygen chemical diffusion in three basaltic liquids at elevated temperatures and pressures. Geochim. Cosmochim. Acta **47**: 1923–1930.

Freer, R. (1981) Diffusion in silicate minerals and glasses: A data digest and guide to the literature. Contrib. Mineral. Petrol. **76**: 440–454.

Fujii, T. (1981) Ca–Sr chemical diffusion in melt of albite at high temperature and pressure. EOS Trans. Am. Geophys. Union **62**: 428.

Gerlach, D.C. and Grove T.L. (198Z) Petrology of Medicine Lake highland volcanics: Characterization of endmembers of magma mixing. Contrib. Mineral. Petrol. **80**: 147–159.

Green, T.H. and Watson, E.B. (1982) Crystallization of apatite in natural magmas under high-pressure, hydrous conditions, with particular reference to 'orogenic' rock series. Contrib. Mineral. Petrol. **79**: 96–105.

Grove, T.L. and Raudsepp, M. (1978) Effects of kinetics on the crystallization of quartz normative basalt 15597: an experimental study. Proc. Ninth Lunar Planet. Sci. Conf.: 585–599.

Harrison, T.M. and Watson, E.B. (1983) Kinetics of zircon dissolution and zirconium diffusion in granitic melts of variable water content. Contrib. Mineral. Petrol. **84**: 66–72.

Harrison, T.M. and Watson, E.B. (1984) The behavior of apatite during crustal anatexis: Equilibrium and kinetic considerations. Geochim. Cosmochim. Acta **48**: 1467–1477.

Hildreth, W. (1979) The Bishop tuff: Evidence for the origin of compositional zonation in silicic magma chambers. Geol. Soc. Am. Spec. Paper **180**: 43–75.

Hofmann, A.W. (1980) Diffusion in natural silicate melts: A critical review. In: Hargraves, R.B. (ed) Physics of Magmatic Processes. Princeton Univ. Press, Princeton, New Jersey, pp 385–417.

Holmes, A. (1936) Transfusion of quartz xenoliths in alkali basic and ultra basic lavas, south-west Uganda. Mineral. Mag. **24**: 408–421.

Jambon, A. (1979) Diffusion of water in granitic melt. Carnegie Inst. Wash. Yearbook **78**: 352–355.

Karsten, J.L., Holloway, J.R. and Delaney, J.R. (1982) Ion microprobe studies of water in silicate melts: Temperature-dependent water diffusion in obsidian. Earth Planet. Sci. Lett. **59**: 420–428.

Kuo, L-C and Kirkpatrick, R.J. (1985) Kinetics of crystal dissolution in the system diopside-forsterite-silica. Am. J. Sci. **285**: 51–90.

Kushiro, I. (1983) Effect of pressure on the diffusivity of network-forming cations in melts of jadeitic compositions. Geochim. Cosmochim. Acta **47**: 1415–1422.

Lesher, C.E. and Walker, D. (1985) Solution properties of silicate liquids from thermal diffusion experiments. Geochim. Cosmochim. Acta **50**: 1397–1411.

Lindstrom, D.J., Lofgren, G.E. and Haskin, L.A. (1979) Experimental studies of kinetic effects on trace element partitioning. EOS Trans. Am. Geophys. Union **60**: 402.

Loomis, T.P. (1982) Numerical simulations of crystallization processes of plagioclase in complex melts: The origin of major and oscillatory zoning in plagioclase. Contrib. Mineral. Petrol. **81**: 219–229.

McBirney, A.R. and Noyes, R.M. (1979) Crystallization and layering of the Skaergaard intrusion. J. Petrol. **20**: 487–454.

McBirney, A.R. (1980) Mixing and unmixing of magmas. J. Volcanol. Geotherm. Res. **7**: 357–371.

Medford, G.A. (1973) Calcium diffusion in a mugearite melt. Can. J. Earth Sci. **10**: 394–402.

Muncill, G.E. and Lasagna, A.C. (1984) Chemical diffusion in plagioclase melts and petrologic implications. Geol. Soc. Am. Abstr. Progr. **16**: 603.

Oishi, Y., Nanba, M. and Pask, J. (1982) Analysis of liquid-state interdiffusion in the system $CaO-Al_2O_3-SiO_2$ using multiatomic ion models. J. Am. Ceram. Soc. **65**: 247–253.

Powell, M.A., Walker, D. and Hays J.F. (1980) Controlled cooling and crystallization of a eucrite: Microprobe studies. Proc. 11th Lunar Planet. Sci. Conf.: 1153–1168.

Rapp, R.P. and Watson, E.B. (1985) Kinetics of monazite dissolution and diffusion of rare earth elements in granitic melts of variable water content. Contrib. Mineral. Petrol. **94**: 304–316.

Reid, J.B., Evans, O.C. and Fates, D.G. (1983) Magma mixing in granitic rocks of the central Sierra Nevada, California. Earth Planet. Sci. Lett. **66**: 243–261.

Riebling, E.F. (1966) Structure of sodium alumino-silicate melts containing at least 50 mole % SiO_2 at 1500°C. J. Chem. Phys. **44**: 2857–2865.

Ross, A. (1982) The temperature and pressure dependence of silicon diffusion in a sodium alumino-silicate melt. Lunar and Planetary Science XIII: 659–660.

Ryerson, F.J. and Hess, P.C. (1978) Implications of liquid-liquid distribution coefficients to mineral-liquid partitioning. Geochim. Cosmochim. Acta **42**: 921–932.

Sato, H. (1975) Diffusion coronas around quartz xenocrysts in andesite and basalt from Tertiary volcanic region in northeastern Shikoku, Japan. Contrib. Mineral. Petrol. **50**: 49–64.

Scarfe, C.M., Mysen, B.O. and Virgo, D. (1987) Pressure dependence of viscosity of silicate melts. In B.O. Mysen (ed) Magmatic processes: Physicochemical principles, Geochemical Society Special Publication No. 1 University Park, Pennsylvania, pp 59–68.

Shaw, H.R. (1963) Obsidian-H_2O viscosities at 1000 and 2000 bars in the temperature range 700° to 900°C. J. Geophys. Res. **68**: 6337–6343.

Shaw, H.R. (1972) Viscosities of magmatic silicate liquids: An empirical method of prediction. Am. J. Sci. **272**: 870–893.

Shaw, H.R. (1974a) Diffusion of H_2O in granitic liquids. Part I. Experimental data. In: Hofmann, A.W., Giletti, B.J., Yoder, H.S. and Yund, R.A. (eds) Geochemical Transport and Kinetics. Carnegie Inst. Wash. Publ. **634**: 139–154.

Shaw, H.R. (1974b) Diffusion of H_2O in granitic liquids. Part II. Mass transfer in magma chambers. In: Hofmann, A.W., Giletti, B.J., Yoder, H.S. and Yund R.A. (eds) Geochemical Transport and Kinetics. Carnegie Inst. Wash. Publ. **634**: 155–170.

Shimizu, N. and Kushiro, I. (1984) Diffusivity of oxygen in jadeite and diopside melt at high pressures. Geochim Cosmochim. Acta **48**: 1295–1303.

Sibley, D.F., Vogel, T.A., Walker, B.M. and Byerly, G. (1976) The origin of oscillatory zoning in plagioclase: A diffusion and growth-controlled model. Am. J. Sci. **376**: 275–284.

Smith, H.D. (1974) An experimental study of the diffusion of Na, K, and Rb in magmatic silicate liquids. Ph.D. dissertation, Univ. Oregon.

Smith, V.G., Tiller, W.A. and Rutter, J.W. (1955) A mathematical analysis of solute redistribution during solidification. Can. J. Phys. **33**: 723–744.

Sparks, R.S.J. (1978) The dynamics of bubble formation and growth in magmas: A review and analysis. J. Volcanol. Geotherm. Res. **3**: 1–37.

Spera, F.J., Yuen, D.A. and Kemp, D.V. (1984) Mass transfer along vertical walls in magma chambers and marginal upwelling. Nature **310**: 764–767.

Stolper, E. (1982) Water in silicate glasses: An infrared spectroscopic study. Contrib. Mineral Petrol. **81**: 1–17.

Turner, J.S. (1973) *Buoyancy Effects in Fluids.* Cambridge University Press.

Turner, J.S. and Gustafson, L.B. (1981) Fluid motions and composition gradients produced by crystallization or melting at vertical boundaries. J. Volcanol. Geotherm. Res. **11**: 93–125.

Walker, D., Lesher, C.E. and Hays, J.F. (1981) Soret separation of lunar liquid. Proc. Lunar Planet. Sci. **12B**: 991–999.

Walker, D. and DeLong S.E. (1982) Soret separation of mid-ocean ridge basalt magma. Contrib. Mineral. Petrol. **79**: 231–240.

Watson, E.B. (1976) Two-liquid partition coefficients: Experimental data and geochemical implications. Contrib. Mineral. Petrol. **56**: 119–134.

Watson, E.B. (1979) Calcium diffusion in a simple silicate melt to 30 kbar. Geochim. Cosmochim. Acta **43**: 313–322.

Watson, E.B. (1981) Diffusion in magmas at depth in the earth: The effects of pressure and dissolved H_2O. Earth Planet. Sci. Lett. **52**: 291–301.

Watson, E.B. (1982) Basalt contamination by continental crust: Some experiments and models. Contrib. Mineral. Petrol. **80**: 73–87.

Watson, E.B. and Bender J.F. (1980) Diffusion of cesium, samarium, strontium, and chlorine in molten silicate at high temperatures and pressures. Geol. Soc. Am. Abstr. Progr. **12**: 545.

Watson, E.B. and Jurewicz, S.R. (1984) Behavior of alkalies during diffusive interaction of granitic xenoliths with basaltic magma. J. Geol. **92**: 121–131.

Watson, E.B., Sneeringer, M.A. and Ross, A. (1982) Diffusion of dissolved carbonate in magmas: Experimental results and applications. Earth Planet. Sci. Lett. **61**: 346–358.

Wendlandt, R.F. (1980) Oxygen diffusion in basalt and andesite melts. EOS Trans. Am. Geophys. Union **61**: 1142.

Williams, I.S. (1978) U-Pb evidence for the pre-emplacement history of granitic magmas, Berridale batholith, southeastern Australia. U.S. Geol. Surv. Open-file Rep. 78 701: 455–457.

Yoder, H.S. (1973) Contemporaneous basaltic and rhyolitic magmas. Am. Mineral. **58**: 153–171.

Zhang, Y., Stolper, E.M. and Wasserburg, G.J. (1989) The mechanism of water diffusion in silicate melts. EOS, Trans. Am. Geophys. Union **70**: 501.

Chapter 5
The Role of High Field Strength Cations in Silicate Melts

Paul C. Hess

Introduction

Early models of melt structure focused almost exclusively on the role of the polymerized aluminosilicate tetrahedral framework. Solubility relations, trace element partitioning patterns, redox equilibria and the occurrence of liquid immiscibility among other properties were explained in the following way. A highly polymerized melt is a relatively inflexible structure with few non-bridging oxygen. High field strength cations cannot substitute into the tetrahedrally coordinated structure (TO), and have difficulty in achieving coordination polyhedra of oxygen within the network. This difficulty limits the solubility of the cations whose addition leads eventually to crystallization or to silicate liquid immiscibility.

This simple model has had some success in rationalizing the myriad of phase, thermodynamic and physical properties of silicate melts. Nevertheless, the model is too narrowly focused and may lead to conclusions that are misleading or even incorrect. For example, some workers have argued that highly polymerized silicate melts contain few "sites" to accommodate non-tetrahedrally coordinated cations. This view, however, is incorrect. There are no "sites" of fixed geometry in silicate melts. There are no "free" oxygens available to provide coordination polyhedra to added cations. When a metal oxide is added to a silicate melt, the cation contributes new oxygen and must compete with the other cations for the oxygen already in the melt. These added cations must create their own sites in the melt and, in the process, must compete with other cations for the oxygen (or other anions). All oxygens are apportioned in such a way as to minimize the free energy of the melt while simultaneously satisfying the bonding requirements of all cations. Nevertheless, the structure of the silicate melt does strongly influence the solution properties of non-tetrahedrally coordinated species. The control, however, cannot be viewed as crystallographic where positions of fixed size and geometry act as filters for large and small cations. The control involves both steric and energetic effects.

This perspective is best illustrated with a simple example. A pure SiO_2 melt has no non-bridging bonds and no "sites" for non-framework cations. Yet, a complete liquid solution exists within the system $BaSiO_3$–SiO_2 at 1700°C (Levin et al., 1964). Obviously, "sites" for Ba were created in the SiOSi framework. In contrast, a large field of liquid immiscibility occurs in the $CaSiO_3$–SiO_2 system at the same temperature (Tewhey and Hess, 1979). In fact, the liquid becomes unstable with the addition of only 2% CaO to liquid silica. The same base liquid accepts an unlimited amount of $BaSiO_3$, yet is inhospitable to even small quantities of $CaSiO_3$. Arguing that "sites" were available for Ba^{+2} but not for Ca^2 is clearly nonsense. It is the energetics of the "site"-creating reaction that determines the solution behavior of a cation in a silicate melt. It is the competition between cations for oxygen (or other anions) that determine both the long- and short-range structures of silicate melts.

It is convenient, as well as dictated by necessity, to limit the discussion of silicate melt structure to questions of short-range order. Limitation is done most simply by organizing cation-oxygen bonds into three classes. The backbone of most silicate melts is the bond formed between cations tetrahedrally coordinated by oxygen. These TOT linkages are the most abundant and are some of the strongest bonds in silicate melts. The most common T cations include Si^{+4}, Al^{+3} and Fe^{+3} which, along with their associated charge balancing cations for the AlO_4 and $Fe^{+3}O_4$ tetrahedra, comprise a highly polymerized TOT structure that largely determines many of the physical and thermodynamic properties of silicate melts. For example, two thirds or more of the oxygen atoms in anhydrous melts of basalt to rhyolite composition coordinate with two T cations.

The remaining cations in the silicate melt are divided into two groups. Cations that disrupt and significantly weaken the TO structure are network-modifying cations and the bonds formed are identified by the shorthand MOT if M is either mono- or divalent and by FOT if the charge of the network-modifying cation is +3 or more. These oxygen are called non-bridging. The most common network-modifying cations are mono- and divalent cations, and more highly charged cations typically coordinated to 6 or more oxygen. The coordination numbers of network-modifying cations are ill-defined, however. The small amount of data that exist indicate that average coordination numbers for monovalent and divalent cations range in simple silicate glass from 4 to 6 (Navrotsky et al., 1985). However, a cation in tetrahedral coordination does not necessarily become a T-type cation. For example, Yin et al. (1983) conclude that Mg in $MgSiO_3$ glass has approximately 4.1 oxygen nearest neighbors at 2.08Å (with perhaps 2 more distant oxygen at 2.50Å also contributing to the coordination). Such large tetrahedra are unlikely to copolymerize efficiently with SiO_4 (Si–O = 1.62Å) or AlO_4 (Al–O = 1.72Å) tetrahedra. The large two-liquid field in the MgO–SiO_2 system (e.g., Hess, 1977) is clear testimony to the difficulty of mixing MgO_4 and SiO_4 tetrahedra. Moreover, the long Mg–O bonds coupled with their low Pauling bond strengths definitely distinguish MgO_4 species from TO_4 tetrahedra.

The substitution of Mg or other similarly coordinated network-modifying cations for T-type cations acts to depolymerize the melt and radically alter its physico-chemical properties. Thus, a network-modifying cation is identified by its relatively weak MO bonds and not necessarily by its coordination. A cation with a Pauling bond strength of less than $\frac{3}{4}$ should not behave as a T-type cation, even if its coordination is tetrahedral.

The principal network-modifying cations are monovalent and divalent cations not already utilized as charge-balancing cations for $T^{3+}O_4$ tetrahedra or those that coordinate oxygen with other highly charged cations (see later). In natural anhydrous silicate melts, the great majority of all MOT bonds are formed with Mg^{+2}, Fe^{+2} and Ca^{2+}, as most of the monovalent cations are tied up with $T^{3+}O_4$ tetrahedra (e.g., see Hess and Wood, 1982). The number of non-bridging bonds with M^+ cations in peraluminous and metaluminous melts are few and exist through the requirements of statistical thermodynamics. Only peralkaline melts ($K + Na > Al$) contain appreciable numbers of $M^{+1}OT$ bonds.

Cations of high charge and field strength have additional coordination possibilities. These cations have a strong affinity for oxygen and may outcompete T-cations for the available oxygen atoms. This competition between T-cations and the high field strength cations, here designated as F-cations, is given by two homogeneous equilibria:

$$TOT + FOF = 2FOT \qquad (1)$$

$$2MOT + FOF = 2FOM + TOT \qquad (2)$$

(A third reaction, MOT + FOF = FOT + FOM, is not a new equilibrium, but is just a linear combination of (1) and (2)). The solution characteristics of F-cations are rich in complexity. Relatively "weak" F cations become network-modifying species and form FOT bonds much like monovalent and divalent M cations (see equation (1)). These cations tend to coordinate with six or more oxygens in crystals and some of these oxygen may be bridging. As the F cation enters the TO structure, it will perturb the bridging oxygen and generally weaken the TOT framework. Thus, F-type cations are not as stable in highly polymerized liquids as are the low field strength cations. Many F cations, however, form complexes outside the TO structure. Three generalized F-type complexes occur in silicate melts. (1) F cations may polymerize to form FOF structures that coexist in homogeneous equilibrium with the TOT structure. An example of this occurs in the TiO_2–SiO_2 system, where macroscopic as well as microscopic segregation has occurred in the form of a large two-liquid field between SiO_2-rich and TiO_2-rich liquids (e.g., see Hess, 1977). (2) F cations may also copolymerize (or complex) with other F cations. For example, Ryerson and Hess (1978) showed that PO (REE) (REE = rare earth element; P = phosphorus) were stabilized in ferrobasaltic magma. (3) F cations may strip network-modifying cations from the TO structure (see equation (2)) and form FOM complexes. Complexes such as ZrOK, TiOK, and $Sn^{4+}OK$ are very stable in peralkaline melts (Watson, 1979;

Dickinson and Hess, 1985; Naski and Hess, 1985; Gwinn and Hess, 1989). Note that the existence of one type of F complex in a silicate melt doesn't preclude the existence of other types of complexes. In fact, thermodynamic laws demand that all possible F-type bonds coexist in homogeneous equilibrium. The abundance of some of these species, however, may be very small and virtually undetectable by spectroscopic means.

This paper is primarily concerned with the solution properties of cations of charge +3 or greater. After discussing the role of Si^{4+}, the rest of the paper is organized according to the Periodic Table, so that cations of similar valence states can be compared. The sections include a summary of the properties of Group 3, Group 4, Group 5 and Group 6 elements. Because of space constraints and the paucity of data, only brief mention will be made of some elements, and others are not mentioned at all.

Silicon

The basic structural unit of silicate melts is the SiO_4 tetrahedron*, which, by sharing oxygen with neighboring TO_4 tetrahedra, gives rise to the great diversity and complexity of TO liquid structure. The real question concerns the nature of Si environments. Is the liquid structure that of a "random network" or more like that of a crystal? Again, the question is addressed only by considering the immediate local environments around Si. There are five possible local environments, or Q^4, Q^3, Q^2, Q^1 and Q^0 where Q^n identifies a SiO_4 tetrahedron with n SiOSi bridging oxygen (DeJong et al., 1981). Crystalline silicates tend to be monodisperse, that is, they usually contain only one Q species. For instance, a tectosilicate contains exclusively Q^4 groups, sheet silicates Q^3, and pyroxene Q^2 groups. Amphiboles, however, contain Q^3 and Q^2 groups. The Q population in silicate melts is likely to be more diverse.

Although the ratio of bridging to non-bridging oxygen is fixed by the composition of most glasses (assuming that all cations are network-modifiers or network-formers), the distribution of non-bridging oxygen on SiO_4 tetrahedra will vary according to the nature of the network-modifying cation. Randomly distributed network-modifying cations will produce a broad Q population, although the Q values are peaked around some most probable distribution (Lacy, 1965). A more crystal-like liquid structure will be characterized by a narrow Q population that is determined by the stochiometry of the glass, e.g., a glass of $MSiO_3$ contains

* ^{29}Si NMR spectra of $Na_2O \cdot 2SiO_2$ glasses containing more than 38 mol % P_2O_5 show a major peak at −213 ppm, very similar to the peak observed for six coordinated silicon in SiP_2O_7 (Dupree et al., 1988). The above is the only occurrence of six coordinated Si in glasses obtained at one atmosphere. Room temperature glasses of SiO_2 and $CaMgSi_2O_6$ composition that were compressed to pressures from 200 to 400 kb are believed to contain four and six coordinated Si (Williams and Jeanloz, 1988).

primarily Q^2 groups, whereas a glass between $MSiO_3$ and $M_2Si_2O_5$ compositions contains mainly Q^2 and Q^3 groups.

Raman, infrared, NMR, and X-ray emission spectroscopy can be applied to discriminate between the two extreme models. Although such information is invaluable, the description of silicate glass structure in terms of a Q population is less determinant than needed to describe the more-extended silicate structure including the identification of chain, ring, sheet and network polymers. The latter information is not attainable at present. A brief review of the limited understanding of the Q structure of silicate glass is given below.

Although there is a striking similarity between crystal and glass Raman spectra, the bands in the glass spectra are broader, some are shifted in frequency and some are of peculiar shape (Brawer and White, 1975). Most importantly, there are peaks or shoulders in the glass spectra that have no parallel in the crystal spectra (Brawer and White, 1975; Virgo et al., 1979; Matson et al., 1983; McMillan, 1984). The extra peaks and shoulders probably correspond to the Si–O stretching vibrations from different Q groups in the liquid.

The glasses definitely are not crystal-like in their Q distribution. It is very difficult, however, to calculate Q abundances from the intensities of the Raman bands. Besides problems of deconvolution, the intensities of Raman bands depend on the molecular polarizabilities of the vibrating species, which are difficult to calibrate and quantify (McMillan and Hoffmeister, 1988).

With the advent of ^{29}Si magic-angle NMR spectroscopy, it may be possible to determine the types and concentrations of the various Q groups in simple silicate glasses (Oldfield and Kirkpatrick, 1985). The results to date hold great promise, but unfortunately have a number of uncertainties that need to be resolved. For example, Dupree et al. (1984) concluded that $Na_2O \cdot 2SiO_2$ and $Na_2O \cdot SiO_2$ glasses contain only Q^3 and Q^2 species respectively, whereas glasses of intermediate composition contain both Q^2 and Q^3 species. In contrast, Schramm et al. (1984) used five Gaussians, one for each Q species, to deconvolute the MAS NMR spectra of Li_2O–SiO_2 glass. They found that a distribution of Q^4, Q^3 and Q^2 species characterized glasses with 17% to 40% Li_2O. For example, the Q population for glass with 33% Li_2O was estimated at 22% Q^2, 57% Q^3 and 15% Q^4 with uncertainties of about 30%. A crystal-like glass would contain 100% Q^3, whereas a random distribution calculated after Lacy (1965) would contain about 40% Q^3.

The most definitive work to date utilizes static NMR spectra, relying on differences in the chemical shift anisotropy rather than on the isotropic chemical shift (Stebbins, 1988; Brandriss and Stebbins, 1988; Schneider et al., 1987). The ^{29}Si spectra of $Li_2Si_2O_5$ and $Na_2Si_2O_5$ glasses have well-resolved Q^4 and Q^3 peaks, and from mass balance considerations, a Q^2 (and, perhaps Q^1 and Q^0) peak is inferred. The Q^4 abundance for the $Li_2Si_2O_5$ glass (11.5%) compares reasonably well with the estimates of Schramm et al. (1984).

These results support the conclusions obtained from vibrational spectroscopy that glasses have a broader Q distribution than is indicated by the stoichiometry of the glass. The distributions are not as broad as calculated for a random distribution of non-bridging oxygen, however. In general, the higher the field strength of the network-modifying monovalent or divalent cation, the broader the distribution. It is incorrect, however, to associate a broader Q distribution with increased randomness. High field strength cations are locally concentrated rather than dispersed within the silicate melts. This effect leads to segregation of network-modifying cations from network-forming cations and eventually to the onset of liquid-liquid phase separation (Hess, 1971). DeJong and Brown (1980) and DeJong et al. (1981) predicted that the tendency towards a bimodal Q distribution was greatest in systems with network-modifying cations located at the beginning of the series Mg > Ca > Sr > Ba > Li > Na > K > Rb > Cs. The limited data so far gathered seem to support these ideas. Melts in the beginning of the series show the greatest deviation from ideality and the greatest tendency to undergo silicate liquid immiscibility (Hess, 1977). The onset of silicate liquid immiscibility is the most obvious expression of ordering within a silicate melt, and thus is an expression of preference for specific Q distributions.

The Q distribution determined by spectroscopy typically reflects the quenched-in structures of silicate liquids at the glass transition. Since the glass transition temperature is roughly $\frac{2}{3}$ to $\frac{3}{4}$ of the melting temperature (see Richet and Bottinga, 1986), extrapolations of several hundreds of degrees are required to determine the structure of the melt at the liquidus temperatures. Brandriss and Stebbins (1988) have shown that the homogeneous equilibrium

$$2Q^3 \leftrightarrows Q^2 + Q^4 \tag{3}$$

of a $Na_2Si_2O_5$ glass gets displaced slightly to the right if the glass transition temperatures are increased by as little as 80°C. Given this observation, caution is obviously recommended when glass structures are used to rationalize liquid structures at liquidus temperatures. There are other distinctions, moreover, between glass and liquid states.

The NMR spectra of liquids at temperatures above the glass transition contain only a single narrow peak in contrast to the broader and more complex spectra of glasses (Stebbins, 1988). These effects are produced in fluids in which the atomic motions responsible for configurational changes occur at time scales shorter than the NMR timescale. If this hypothesis is correct, it means that the silicon-oxygen-network modifying cation bonds are continuously and rapidly broken and reformed leading to an average spectrum. The Q distribution in silicate liquids apparently is not a static one, but represents the time and space average of an equilibrium distribution. The lifetime of an Si–O bond is estimated to be very brief, in the micro-second range. The fact that the vibrational spectrum of $Na_2Si_2O_5$ liquid is similar to that of the glass (Sweet and White, 1969) suggests that the lifetime of the Si–O bond is longer than the vibrational time scale. It is this property that gives the liquid its distinguishing physical and thermodynamic character and it reaffirms the earlier argument that fixed sites do not occur in liquids.

Group III Elements

Boron

Boron exists in three-fold coordination with oxygen in B_2O_3–SiO_3 glasses (Bray, 1978). The addition of even small amounts of B_2O_3 causes a significant depression in the cristobalite liquidus (Levin et al., 1969), suggesting that B efficiently destroys the SiOSi framework and dramatically lowers the activity of SiO_2. This effect is consistent with BO_3 trigonal groups being distributed within the SiO_2 framework. The liquidus assumes a sigmoidal shape at high B_2O_3 contents, perhaps indicating the existence of metastable liquid immiscibility. Low-angle X-ray scatter experiments found no evidence for the existence of a metastable two-liquid field (Vasilevskaya et al., 1980), but did discover 10-15Å-sized regions enriched in SiO_2 or B_2O_3. Thus, ordering of boron occurs in boron-rich silicate melts and may presage the onset of silicate liquid immiscibility.

The addition of alkali or alkaline earth oxides to borosilicate glass causes BO_3 units to convert to BO_4 tetrahedra, the latter utilizing alkalies and alkaline earths as charge-balancing cations (Dell et al., 1983; Park et al., 1979). The conversion, however, is not complete at all compositions. In Na_2O–B_2O_3–SiO_2 glasses with Na_2O/B_2O_3 (moles) < 0.5, ^{11}B NMR studies show that all sodium atoms are associated with BO_4 tetrahedra (Fig. 1). As Na_2O/B_2O_3 is increased (for a given silica content) proportionally fewer BO_3 units are transformed to BO_4 groups until a maximum in BO_4/BO_3 is reached. Beyond this critical Na_2O/B_2O_3 ratio, the BO_4/BO_3 ratio steadily declines to very small values at high Na_2O/B_2O_3. Moreover, the BO_4/BO_3 ratio at a given Na_2/B_2O_3 increases with the SiO_2 content.

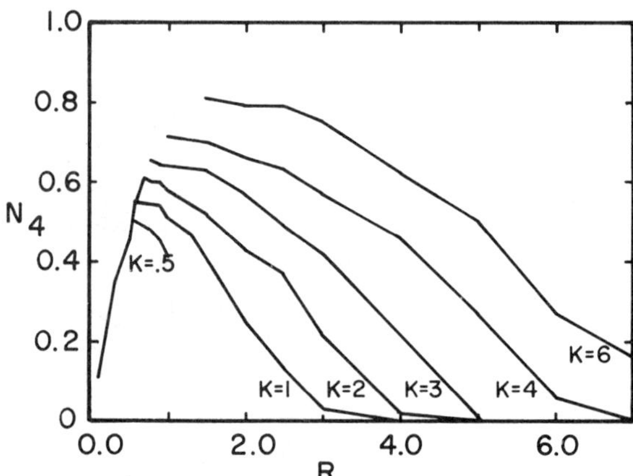

Fig. 1. Mole fraction of tetrahedrally coordinated BO_4 groups (N_4) in Na_2O–B_2O_3–SiO_2 glass plotted as functions of $R = Na_2O/B_2O_3$ (moles) and $K = SiO_2/B_2O_3$ (moles). Modified from Dell et al. (1983).

These data show that the BO_3 to BO_4 transformation is enhanced by increasing SiO_2 content, and that BO_4 tetrahedra copolymerize with, and are stabilized by, the SiOSi network. However, the solution of BO_4 tetrahedra into the SiOSi solvent is limited and saturation is reached at a boron to silicon tetrahedron ratio of about 1:4 (Dell et al., 1983). If a random distribution of BO_4 tetrahedra is assumed, each boron tetrahedron will be linked to 4 SiO_4 tetrahedra. These groups are expressed by the formula $NaBSi_4O_{10}$ and have a mineral analog in the borate phase, reedmergnerite ($NaBSi_3O_8$).

BO_4 tetrahedra are only partially soluble within the liquid silicate framework as is confirmed in several additional studies. For example, adding MgO to B_2O_3–SiO_2 liquids creates a large field of immiscibility between borosilicate and Mg-bearing borosilicate liquids with $MgO/B_2O_3 > 1$ (Levin et al., 1969). The MgO-rich borate glasses contain few BO_4 tetrahedra (Park et al., 1979) Thus, BO_4 bearing glasses with Mg (and probably other divalent cations) as charge-balancing cations are less stable than those with monovalent charge-balancing cations. This conclusion is supported by molecular orbital calculations which show that the bond angles subtended by BOSi bonds are very different from those subtended by SiOSi bonds (Navrotsky et al., 1985). The different angular requirements should limit the ability of BO_4 tetrahedra to substitute within the SiOSi tetrahedral framework. These results may also apply to Al-bearing melts.

The ^{11}B MAS NMR spectra of glasses along the join $NaAlSi_3O_8$ (Ab)—$NaBSi_3O_8$ (Rd) indicate that boron exists in both trigonal and tetrahedral coordination, with the percentage of the trigonal species increasing from 29% in Rd_{100} to 60% in Rd_{20} compositions (Geisinger et al., 1988). Most of the Si, Al and tetrahedrally coordinated B form a copolymerized network, but it is believed that $SiO_4{}^{4-}$ and $BO_3{}^{3-}$ are not nearest neighbors. Domains of Si, Al, B (tetrahedral) rich-glass apparently coexist with domains enriched in B (trigonal). Thus, if the albite-rich glass is an adequate model for anhydrous rhyolite melts, it follows that B is predominantly in trigonal coordination in metaluminous-peraluminous melts, whereas $BO_4{}^{-5}$ species predominate in peralkaline melts (Dell et al., 1983). However, the role of H_2O and other volatiles and the concentration of other highly charged cations (e.g., P^{+5}) may significantly modify these predictions (see later).

Nevertheless, even small amounts of boron have a dramatic effect on the solidus temperatures of highly polymerized liquids. Chorlton and Martin (1978) obtained solidus temperatures for tourmaline-bearing granite that are 60–90°C lower than the minimum melting temperatures in the dry granite system. The addition of B also lowers the liquidus temperatures of H_2O-undersaturated and H_2O-saturated alkali aluminosilicate melts (Pichavant, 1981, 1987).

Boron increases the solubility of H_2O in silicate melts (London, 1987; Pichavant, 1987), and is preferentially partitioned into the vapor phase (London et al., 1988). The hygroscopic nature of boron in silicate melts is clearly indicated. Caution must therefore be exercised when trying to apply the spectroscopic results obtained on anhydrous systems to H_2O-bearing siliceous systems that better approximate natural granite or pegmatite magmas.

Aluminum

It is widely accepted that aluminum occurs in tetrahedral coordination with oxygen in silicate melts whenever appropriate charge balancing cations exist in sufficient abundance. The charge balancing cations are mainly of low charge. Calorimetric studies of glasses in systems of SiO_2–$M^{n+}{}_{1/n}AlO_2$ show that the sequence $Cs \simeq Rb > K > Na > Li \simeq Ba > Pb > Sr > Ca > Mg$ is arranged according to the charge balancing ability of the cations (Roy and Navrotsky, 1984). This means that charge balancing cations of low field strength maximize the thermodynamic stability of aluminosilicate melt structures. These observations are rationalized as follows. Charge balancing cations are located within cages made up of SiOSi and AlOSi coordinated oxygens much as observed in framework aluminosilicate minerals (Taylor and Brown, 1979). Low field strength cations do not strongly polarize the bridging oxygen and only weakly perturb the aluminosilicate structure (Navrotsky et al., 1985; Hess and Wood, 1982). In contrast, high field strength cations form relatively strong bonds with the bridging oxygen. The strong bonds perturb the aluminosilicate structure by narrowing the TOT bond angle and lengthening the TOT bonds (the AlO bond is lengthened more than the SiO bond), causing a decreasing positive TO overlap population and a significant weakening of the TO bonds (Geisinger et al., 1985; Navrotsky et al., 1985). The overall structure is destabilized and such liquids are prone to crystallization or silicate liquid immiscibility.

These predictions are confirmed in several studies. (1) CaO, MgO aluminosilicate liquids have significantly lower viscosities than comparable Na_2O, K_2O aluminosilicate melts (e.g., see Mysen, 1988). (2) Temperatures of metastable two-liquid fields in aluminosilicate systems are much greater in liquids containing divalent charge-balancing cations than in those with monovalent charge-balancing cations (Varshal, 1975). (3) In coexisting immiscible silicate melts, $(Na, K)AlO_2$ tetrahedral species are concentrated in the highly polymerized SiO_2-rich melt, whereas $(Ca, Mg)_{0.5}AlO_2$ species prefer the low SiO_2 melt (Hess and Wood, 1982), because the high field strength charge-balancing cations perturb and destabilize the highly polymerized TOT liquid.

The hierarchy established among the charge-balancing cations also confirms the notion that the solution properties of cations in an aluminosilicate melt are not very sensitive to cation size (Ellison and Hess, 1989c). The largest monovalent cation, Cs, heads the series for the monovalent cations, and the largest of the divalent cations, Ba, leads the divalent cations. Yet crystalline cesium feldspar $CsAlSi_3O_8$ has not been synthesized, presumably because the large Cs atom cannot fit within the voids of the aluminosilicate framework (Liebau, 1985). The Cs is accommodated, however, within the more open and flexible crystalline $CsAlSi_2O_6$ and $CsAlSiO_4$ structures. While the more flexible properties of the melt are certainly understandable, the question remains why Cs is more stable than the smaller monovalent cations in liquid aluminosilicate. It is not because there are more large voids than smaller ones in silicate melts. This point is demonstrated quite nicely by the solution properties of the noble gases in silicate melts.

Lux (1987) determined that Henry's Law solubility coefficients in silicate melts were in the order of He > Ne > Ar > Kr > Xe; an order inversely related to the size of the noble gas atom. In this instance, the small atoms have greater stability than the large atoms, and there is an energy cost in trying to dissolve the larger noble gas atoms in the silicate melts. In contrast to the noble gases, the stability of the charge balancing cation (of fixed charge) is proportional to the size of the cation. The ionic radius of Cs in twelve-fold coordination with oxygen is similar to that of Kr, whereas that of Na is similar to He. If it is assumed that the free energy of forming a void in the silicate melt is related to the size of the void formed, then it must be concluded that strong bonds are formed between the large Cs ion and the bridging oxygen coordinated to aluminum. The energy gained through bonding with these oxygen more than offsets the energy required to form (or deform) the site for the large charge-balancing cation.

The coordination of Al in peraluminous melts where Al is in excess of the charge-balancing cations is a complex function of composition. The ^{27}Al MAS NMR spectra of roller quenched, phase separated Al_2O_3–SiO_2 glasses clearly show peaks at about 60, 30 and 0 ppm (Risbud et al., 1987). The peaks at 60 and 0 ppm are easily assigned to aluminum in tetrahedral and octahedral coordination respectively. The assignment of the peak at 30 ppm is less certain but Al in five-fold coordination in andalusite and in other phases resonates at this value. Accepting this assignment, then Al(4) appears to coexist with Al(5) in SiO_2-rich glasses (Al_2O_3 < 15 wt%), whereas Al(6) becomes more prominent in the SiO_2-poor glass. SiO_2-poor glasses, therefore, contain Al in 4-, 5- and 6-fold coordination, although the exact mechanism of charge compensation of oxygen is not known.

The coordination of Al in more complex peraluminous melts is known with less confidence. The ^{27}Al MAS NMR spectra of some peraluminous glasses in the CaO–Al_2O_3–SiO_2 system exhibit sharp peaks with chemical shifts at +48 to +56 ppm and −3 to +11 ppm, chemical shifts which typically are associated with AlO_4 and AlO_6 coordinations, respectively (Engelhardt et al., 1985). It is noteworthy, however, that only the most peraluminous glasses with Al_2O_3/CaO > 1.5 (moles) have octahedrally coordinated Al. Less peraluminous glasses and all metaluminous glasses have no detectable Al(6). All the Al in these glasses occurs as AlO_4 tetrahedra. This is true even for compositions with Al_2O_3/SiO_2 > 1 (moles) wherein AlOAl bonding is required by stoichiometry. Tetrahedral coordination of Al appears to be favored in silicate melts and glasses at low pressures.

This conclusion is supported by Al K-XANES and Al K-EXAFS studies of peraluminous Na_2O–Al_2O_3–SiO_2 glasses, where Al remains in tetrahedral coordination in all glasses with Al/Na ≤ 1.6 (moles) (McKeown et al., 1985). Similarly, Raman spectra of peraluminous glasses indicate that the SiO_4 tetrahedra become increasingly polymerized with the substitution of Al for K (Dickinson et al., 1984). This substitution is explained by the formation of aluminum triclusters (Lacy, 1963) in which a single oxygen is coordinated by three AlO_4 tetrahedra (or perhaps some more complex combination of AlO_4, SiO_4 and even AlO_5 polyhedra). The formation of the tricluster and the concomitant ordering of Al around oxygen induces SiO to polymerize, e.g.

$$3Al_2SiO_5 = 3Al_2O_3 \cdot SiO_2 + 2SiO_2 \quad (4)$$

The equilibrium above is the melt analogue of the mineral reaction

$$3 \text{ sillimanite} = \text{mullite} + \text{cristobalite} \quad (5)$$

wherein mullite contains the tetrahedral triclusters. The enthalpy change of the mineral reaction is positive, showing that mullite owes its stability to a greater entropy (Robie et al., 1978). Whether these energy considerations hold true in the melt is debatable. It is noteworthy, however, that mullite has a large liquidus field in the SiO_2-rich, peraluminous portions of the M_2O–MO–Al_2O_3–SiO_2 systems (Levin et al., 1964). The mullite liquidus, in fact, extends nearly to the charge-balanced compositions. It is logical to conclude, absent evidence to the contrary, that tetrahedrally coordinated Al resides primarily in some form of tricluster in peraluminous melts (but see McKeown et al., 1985). Only the most peraluminous melts with $(M^+ + 1/2M^{2+})/Al > 1.6$ contain significant concentrations of AlO_6.

It is interesting that the O1S binding energy of oxygen in peraluminous glasses is identical to that of the bridging oxygen in peralkaline glasses (Bruckner et al., 1978). This similarity means that the differences in the electron density of bridging oxygen located within SiOSi, SiOAl and Al triclusters are either below detection limits or are completely averaged out throughout the network. The data suggest that the aluminosilicate structure in peraluminous glasses is not strongly perturbed by the formation of Al triclusters. Temperature-composition coordinates of the metastable two liquid field in the MgO–Al_2O_3–SiO_2 system support this thesis. Figure 2 contains a T°C versus $MgO/Al_2O_3 + MgO$ section through the solvus at 75% SiO_2 (Galachov et al., 1976a). Note that aluminosilicates with excess Mg are less stable than comparable peraluminous liquids. Melts with Al triclusters are therefore more stable with respect to unmixing than melts containing divalent network-modifying cations in excess of Al. This is not true for melts containing aluminum and monovalent charge-balancing cations.

There are probably few occurrences of Al-triclusters in natural silicate melts. Peraluminous melts contain $Al > K + Na + Ca$ but usually there are enough other divalent cations such as Mg and Fe^{+2} to act as charge-balancing cations. Exceptions include high SiO_2 rhyolites and granites, and highly peraluminous pegmatites (London et al., 1989). Thus, the existence of Al triclusters or even AlO_6 octahedra cannot totally be dismissed. For example, ^{27}Al NMR studies of CaO–Al_2O_3–P_2O_5 glasses with $CaO/Al_2O_3 > 1$ (moles) have both AlO_4 and AlO_6 polyhedra (and perhaps even AlO_5 species if the chemical shifts at +37, +4 and −21 ppm correspond to ^{27}Al shifts in Al_2O_2–SiO_2 glasses of +60 (Al, 4), +30, (Al, 5) and 0 (Al, 6) ppm) (Müller et al., 1983). Apparently, the strong

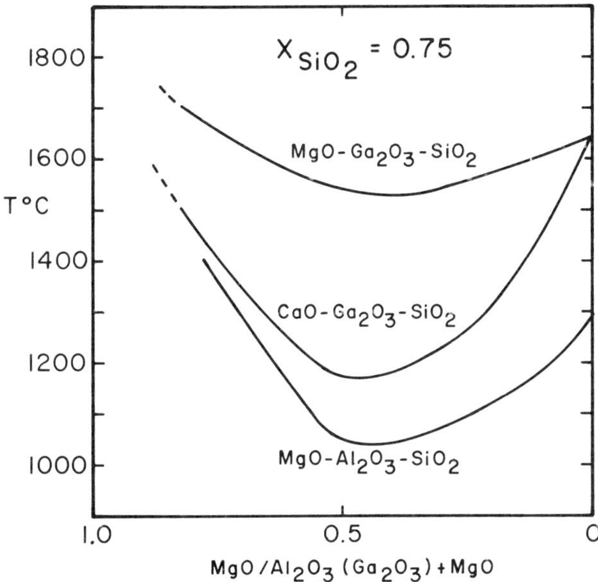

Fig. 2. Approximate temperature-composition section at $xSiO_2 = 0.75$ (moles) through the metastable two liquid field in the system $MgO-Ga_2O_3-SiO_2$, $CaO-Ga_2O_3-SiO_2$ and $MgO-Al_2O_3-SiO_2$. Note that the critical temperatures in the peraluminous compositions are significantly less than in Al_2O_3-poor compositions. Also, immiscibility temperatures in the Ga system are much greater than in the equivalent Al system. (Data from Galachov et al., 1983, 1979a.

association of Ca and P^{+5} (see later) require that aluminum seek coordinations other than tetrahedral. Thus, P and other highly charged cations may introduce interesting but, typically, minor perturbations to the solution behavior of Al (recall that Si is also forced into six fold coordination in P_2O_5-rich glasses).

Gallium

The basic features found in the Raman spectra of sodium aluminosilicate glass are present also in the spectra of corresponding Ga substituted glass (Matson and Sharma, 1985; Fleet et al., 1984). These data plus the well-known Al–Ga diadochy in crystalline feldspars and other silicate minerals (Kroll et al., 1978; Gazzoni, 1973), are convincing evidence that the structural roles of Ga and Al in silicate melts are nearly identical. Thus, Ga should occur in tetrahedral coordination with oxygen whenever appropriate charge-balancing cations are available. The nature of the charge-balancing cations are probably identical to those that were found suitable for Al.

This model predicts that the phase relations of Ga-bearing melts should mimic those of Al-bearing melts. In fact, the metastable two liquid solvi in the MgO(CaO)–Ga_2O_3–SiO_2 and MgO(CaO)–Al_2O_3–SiO_2 are very similar (Galachov et al., 1976a, 1983). The solvi are saddle shaped and reach a minimum temperature at compositions near MgO/Al_2O_3 and $MgO/Ga_2O_3 = 1$ (moles) (Fig. 2). Melts of charge-balanced Ga^{+3} and Al^{+3} have the optimum stability. Thus, both Ca and Mg are appropriate charge-balancing cations for GaO_4 tetrahedra. Moreover, the temperatures of the saddle in the MgO–Ga_2O_3–SiO_2 are greater than those of the CaO–Ga_2O_3–SiO_2 system as also observed in comparable Al_2O_3-bearing liquids.

However, the critical temperatures of the gallium solvi are significantly greater than those of the aluminum solvi; the maximum temperature of the saddle point is about 1500°C in the MgO–Ga_2O_3–SiO_2 solvus, whereas it is less than 1100°C in the MgO–Al_2O_3–SiO_2 system (Fig. 2). Charge balanced GaO_4 tetrahedra are less stable within the TOT framework than are AlO_4 tetrahedra. Similarly, the stability of Al in peraluminous glasses appears to be greater than Ga in pergallic glasses. Galachov et al. (1976b) have determined that the critical temperature rises smoothly from 1300°C in the Al_2O_3–SiO_2 system to 1640°C in the Ga_2O_3–SiO_2 system. The difference in stability of the single liquid is consistent with the greater Ga–O bond lengths and the larger Ga–O polyhedra as compared to Al–O species.

The GaO_4 tetrahedra are more stable in silicate melts than are $Fe^{3+}O_4$ tetrahedra. Dickenson and Hess (1986) determined that $Fe^{3+}O_4$ tetrahedra are stable in K_2O–Ga_2O_3–SiO_2 melts as long as the K content exceeds that of Ga^{3+} and Fe^{3+} (moles) (Fig. 3). The same pattern is observed in K_2O–Al_2O_3–SiO_2

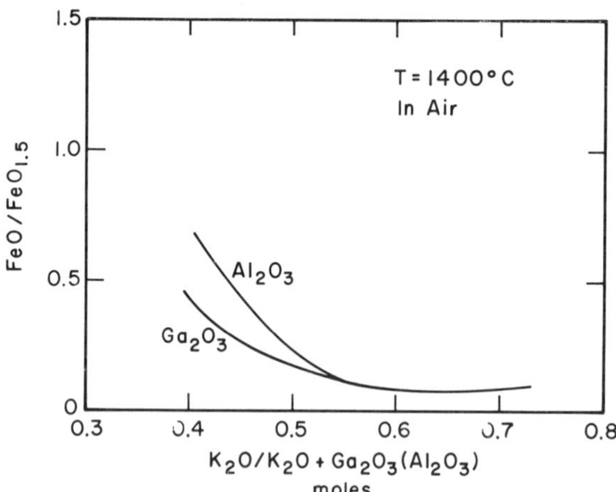

Fig. 3. Redox ratio of Fe^{2+}/Fe^{3+} plotted against $K_2O/K_2O + (Al_2O_3, Ga_2O_3)$ for quenched liquids in the system K_2O–(Al_2O_3, Ga_2O_3)–SiO_2 with 2% "Fe_2O_3". All data for compositions of 78 mol% SiO_2, in air and at 1400°C. Redox ratios in peralkaline liquids are low and approximately constant. Redox ratios rise sharply in peraluminous glasses where there are insufficient K atoms to charge balance AlO_4, GaO_4, and $Fe^{+3}O_4$ tetrahedra. From Dickenson and Hess (1981, 1986).

melts (Dickenson and Hess, 1981). The Fe^{3+} is reduced to Fe^{2+} in melts with $K < Ga + Fe^{3+}$, because there are not enough potassium atoms to charge balance the $Fe^{3+}O_4$ tetrahedra as well as the GaO_4 tetrahedra. Taken together these data show that the network-forming tendency of these trivalent cations is in the order Al > Ga > Fe^{+3}.

Sc, Y, and REE

The remaining group III elements considered here, Sc, Y, and the trivalent rare earth elements (REE), commonly occur in 6- to 9-fold coordination in crystalline oxide compounds (Felsche, 1973). Furthermore, their large octahedral radii (0.75Å to 1.03Å (Shannon, 1976)) makes substitution into the TOT framework highly unlikely. These cations must coordinate with non-bridging oxygen (FOT) and/or oxygen shared by network-modifying (FOM) or other F-type (FOF) cations. The main question is which of these solution complexes are most abundant in silicate melts and what factors control their stability.

Phase equilibria provide some useful guides. Binary silicate compounds of R_2SiO_5 to $R_2Si_2O_7$ compositions where R = Sc, Y, and REE, are near liquidus phases in all the binary systems (Levin et al., 1969). R–O–Si bonds are indeed stable and should exist in silicate melts. Note, however, that the silicate compounds are very depolymerized: the $R_2Si_2O_7$ compound contains only $\frac{1}{2}$ bridging oxygen per SiO_4 tetrahedron, while the R_2SiO_5 compound contains only non-bridging oxygen and "free oxygen" (ROR) (Felshe, 1973). Moreover, SiO_2-rich silicate melts with R/Si > 1 are unstable at high temperatures; a large two-liquid field exists to temperatures in excess of 2000°C (Levin et al., 1969). It is implied that these cations have a limited stability in highly polymerized melts (see Introduction). If true, their distribution within a melt is not random, but they should concentrate in liquid structures of relatively low TOT polymerization.

Parts of this model find confirmation in the Raman spectra of a series of K_2O–$5SiO_2$ glasses where La is substituted for K on an equal oxygen basis (Ellison and Hess, 1987; 1990). This substitution is particularly instructive because it preserves the nominal non-bridging oxygen content of the glass. The spectra are relatively easy to interpret because the bands associated with the vibration of KOSi are polarized, whereas the LaOSi bands are relatively depolarized (Fig. 4). The spectra show clearly that La and K ions occupy more or less separate coordination environments in the glass. The main evidence is the growth of depolarized bands at 1030 cm^{-1} and 940 cm^{-1} with the substitution of La for K while the polarized 1100 cm^{-1} band remains in a fixed position. The 1030 cm^{-1} and 940 cm^{-1} bands are assigned to the LaOSi vibrations on Q^3 and Q^2 silicate tetrahedra, whereas the 1100 cm^{-1} is assigned to the Q^3 vibration of K. The La for K substitution therefore leads to a redistribution of Q environments from a glass that contained mostly Q^3 tetrahedra (0% La) to one that contains two different Q^3 species and Q^2 species. This effect can be expressed as the homogeneous equilibrium

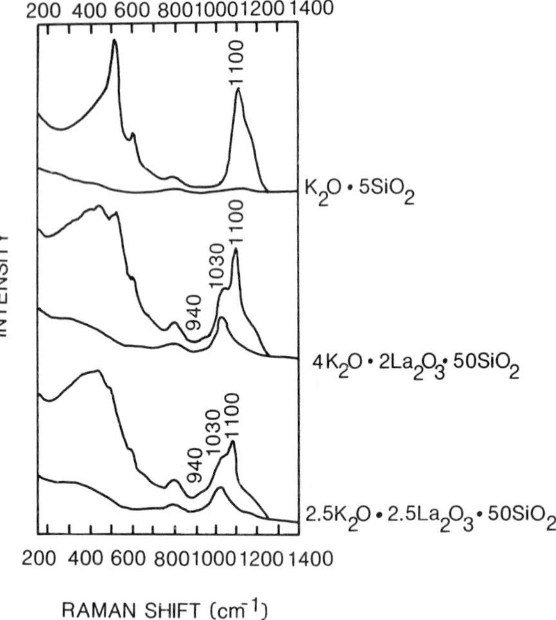

Fig. 4. Parallel (upper curves) and cross-polarized (lower curves) Raman spectra of $K_2O \cdot 5SiO_2$ and $(10 - x)K_2O$–$(x/3)La_2O_3$–$5O\ SiO_2$ glasses. Note the growth of the bands at 1030 cm^{-1} and 450 cm^{-1} as La_2O is substituted for K_2O on an equal oxygen basis (Ellison and Hess, 1987).

$$2Q^3 = Q^4 + Q^2 \tag{6}$$

where the La for K substitution drives the reaction to the right. Note that the growth of a new band at 480 cm^{-1} is also consistent with the development of regions rich in SiOSi bonds and Q^4 species.

The spectra generally support the concept that highly charged cations have limited stability in highly polymerized structures (i.e., higher Q environments). But the Q species must be much more polymerized than anticipated if crystal structures are to be used to approximate liquid structures. As discussed above, binary crystalline La-silicates occur only as pyro- (Q^1) and ortho-silicates (Q^0), yet La is located on Q^2 and Q^3 species in the K_2O–SiO_2 glass. It is possible that these more polymerized La species are stabilized by K. This explanation is not viable, however. Note that the spectrum of the La/K = 1 glass contains the Q^3 band for SiOK. Whenever lanthanides and potassium coordinate the same non-bridging oxygen in chain silicates, the K/La ratio typically is ≥ 1 (Liebau, 1985). If this ratio applies to the melt, the discrete band at 1100 cm^{-1} band should disappear, but this doesn't happen. It follows that few of the K and La ions coordinate the same non-bridging oxygen, and that LaOSi bonds on Q^2 and Q^3 species exist. This effect demonstrates again that a silicate melt is a more hospitable and flexible environment than the corresponding crystals. Crystal structures should therefore be used as guides but not constraints on liquid structures.

It is logical to extend the La model to other group III trivalent cations since the phase equilibria are similar. Indeed, the Raman spectra of Gd and Yb in K_2O–$4SiO_2$ substituted glasses are virtually identical to those of the La glasses (Ellison and Hess, 1987b, 1991) as are the spectra of Sc, Y and La Na_2O–SiO_2 bearing glasses (Krol and Smets, 1984; Nelson et al., 1984). The coordinations, Q population and liquid structures of these R^{3+} network-modifying cations are virtually identical, with only minor deviations from the expected norm.

Group IV Elements

Carbon

Under suitably oxidizing conditions, carbon dissolves in silicate melts as molecules of CO_2 and as variously distorted carbonate complexes (Fine and Stolper, 1985). The solubility of CO_2 is the sum of the concentrations of the various forms of the dissolved CO_2. The solubility of CO_2 and its speciation is sensitively dependent on the bulk composition of the melt as well as on the ambient pressure and temperature. Fine and Stolper (1985), for example, determined that the ratio of molecular CO_2 to CO_3^{2-} increased significantly with increasing SiO_2 contents of melts in the $NaAlO_2$–SiO_2 join. A homogeneous equilibrium of the sort

$$2NaAlO_2 + CO_2 = Na_2CO_3 + Al_2O_3 \tag{7}$$

might adequately describe the solubility relations. In this equilibrium, CO_2 strips the charge-balancing Na cation from AlO_4 tetrahedra to form Na_2CO_3 complexes. The Al is forced to assume a new coordination, possibly the tricluster configuration that is believed to be characteristic of peraluminous melts. The net effect is to weaken the aluminosilicate liquid and thereby reduce its viscosity (Brearley and Montana, 1989). An alternative model proposed by Fine and Stolper (1985) is that the melts are not fully polymerized and contain sufficient non-bridging oxygen to allow the following equilibrium to occur:

$$2NaOSi + CO_2 = Na_2CO_3 + SiOSi \tag{8}$$

Equilibria such as these are believed to control the solubility of CO_2 in less-polymerized melts (Eggler and Rosenhauer, 1978) where network-modifying cations such as Ca^{2+}, Mg^{2+} and Fe^{2+} are available to stabilize the CO_3^{2-} groups. The lower viscosity of CO_2–$NaAlSi_2O_3$ melts compared to CO_2-free melts, however, supports the former mechanism.

The CO_2 equilibria deduced from experiments on simple systems have direct application to natural melts. Stolper and Holloway (1988) found that CO_2 exists only as CO_3^{2-} ions in basalts below 2 kb. The CO_2/CO_3^{2-} ratio increases in the sequence from andesite, rhyodacite to rhyolite melts, with only molecular CO_2 existing in rhyolite (Fogel and Rutherford, 1990).

Increasing pressure increases the solubility of CO_2 and decreases the CO_2/CO_3^{2-} ratio in albite melts (Stolper et al., 1987). Similar relationships probably obtain in less polymerized melts. The temperature of the equilibrium

$$\text{Diopside} + \text{Forsterite} + CO_2 = \text{Enstatite} + \text{Melt} \quad (9)$$

is only slightly depressed by increased PCO_2 at low pressures. At pressures above 25 kb, however, the temperature of this reaction is depressed by about three hundred degrees and the solubility of CO_2 increases by more than an order of magnitude (Eggler, 1975). This remarkable phase change is believed to reflect a dramatic increase in the stability of CO_3^{2-}. Albite melts, in contrast, show no such large increase in the solubility of CO_2 up to 30 kb (Stolper et al., 1987), reflecting the difficulty of stripping Na from its charge-balancing position. Pressures over 60 kb are probably required to increase dramatically the CO_2 solubility in aluminosilicate melts, because ^{27}Al NMR studies of $NaAlSi_2O_6$ melts suggest that AlO_4 tetrahedra gradually transform to AlO_6 octahedra at these pressures (Ohtani et al., 1985), thus freeing Na from its charge-balancing role to combine with the CO_3^{2-} ions.

Germanium

The ionic radius of Ge^{4+} in four-fold coordination with oxygen is identical with that of Al in four-fold coordination. It is not surprising that Ge, like Al, may readily substitute for Si in crystals and glasses. Raman spectra of GeO_2, $KAlGe_3O_8$, $KAlGe_2O_6$ and $KAlGeO_4$ glasses are very similar to those of equivalent Si-bearing glasses (Matson and Sharma, 1985). Moreover, despite the size differences between SiO_4 and GeO_4 tetrahedra, Si and Ge apparently mix completely to form a continuous network structure ranging from pure SiO_2 to pure GeO_2 composition (Sharma et al., 1984) and the enthalpy of mixing of $NaAlGeO_3$–$NaAlSiO_3$ glasses is ideal (Capabianco and Navrotsky, 1982). Still, the similarity between Ge and Si is not complete. For example, germanate glasses of low alkali contents contain GeO_6 octahedra, whereas only SiO_4 tetrahedra exist in the corresponding alkali-SiO_2 glasses (Furukawa and White, 1980). Nevertheless, the data strongly suggest that Ge occurs as TO_4 network-forming tetrahedra in naturally occurring silicate melts.

Titanium

The coordination of titanium in silicate melts and glasses appears to be less than simple. Polyhedra containing 4, 5 and 6 oxygens around Ti, or mixtures of these coordinations, have been proposed. Support for the existence of tetrahedrally coordinated Ti comes from studies by Raman spectroscopy (Furukawa and White, 1979; Mysen et al., 1980). A distinctive feature of the Raman spectra of TiO_2-bearing silicate glasses is a very intense band in the 850–940 cm^{-1} region. This band was assigned to the Ti–O–M stretching mode on Ti tetrahedra even though the only compound examined with tetrahedral Ti (Ba_2TiO_4) has the most intense band at 745 cm^{-1}. The higher Ti–O–M stretching frequencies in the glass spectra were attributed to the effects of polymerization of TiO_4 tetrahedra. Indeed the polymerization of SiO_4 tetrahedra causes the Si–O–M stretching frequencies to increase by several 100 cm^{-1}. Unfortunately, this hypothesis cannot be tested for TiO_4 tetrahedra because there are no crystalline compounds with polymerized TiO_4 units. In contrast, polymerized TiO_5 and TiO_6 units are common features of titanate crystalline compounds.

Raman spectra of crystalline powders of $K_2Ti_2O_5$ and $K_2Ti_3O_7$ have intense bands in the 850–900 cm^{-1} region, yet these phases contain only 5 or 6 co-ordinated Ti (Dickinson, 1984). Glassy $K_2Ti_2O_5$ also has a strong band near 900 cm^{-1}. Dickinson argued that the intense band stemmed from anomalously shortened Ti–O bands in a distorted TiO_6 octahedron. In support of this argument, a combined XAS and neutron scattering study of K_2O–TiO_2–$2SiO_2$ glasses determined that Ti was 5-coordinated, with four oxygens at 1.96Å and one at 1.65Å (Yarker et al., 1986). Other studies of multicomponent Ti-bearing glasses determined that the average Ti-O distance was 1.85Å (Dumas and Petiau, 1986) consistent with the distorted tetragonal pyramid model or perhaps to a mixture of 4 and 6 coordinated Ti.

Other spectral data confirm that the solution properties of Ti are complex. Honada and Soga (1980) conclude from x-ray emission spectra of Na_2O–TiO_2–SiO_2 glasses that TiO_6 octahedra are abundant at low concentrations $<10\%$ but convert to lower average coordinations at higher TiO_2 concentrations. EXAFS data for TiO_2–SiO_2 glasses were modelled as mixtures of TiO_6 and TiO_4 polyhedra (Greegor et al., 1983). Photoacoustic spectroscopic studies of binary titanate glasses indicate that 4 and 6 coordinated Ti do occur, the 4:6 ratio decreasing with increasing TiO_2 concentration and with increasing field strength of the coordinating cation (Yoshimaru et al., 1984).

It is safe to conclude that the spectral data allow for the existence of at least two types of TiO bonds. Whether these bond types are linked to TiO_4 and TiO_6 or distorted TiO_5 remains to be seen. Nevertheless, the ability of TiO_2 to accommodate itself within a silicate melt has several interesting applications. For example, the activity coefficient of TiO_2 (at fixed T°C) increases by a factor of about 4 from basalt to metaluminous granite (Ryerson and Hess, 1978; Watson, 1976). This increase means that the crystallization of important minerals such as sphene, ilmenite, and titanomagnetite among others will occur at much lower concentrations of TiO_2 in SiO_2-rich melts than in SiO_2-poor melts at comparable temperatures. The study of Dickinson and Hess (1985) showed that the activity coefficient of TiO_2 decreased very sharply with the K_2O/Al_2O_3 ratio in peralkaline granitic melts but changed only slightly in peraluminous melts. The "peralkaline effect" is controlled by an equilibrium such as

$$2SiOK + TiOTi = 2TiOK + SiOSi \tag{10}$$

where TiOK complexes are produced at the expense of SiOK non-bridging species. The average stoichiometry of the titinate complex is estimated to be

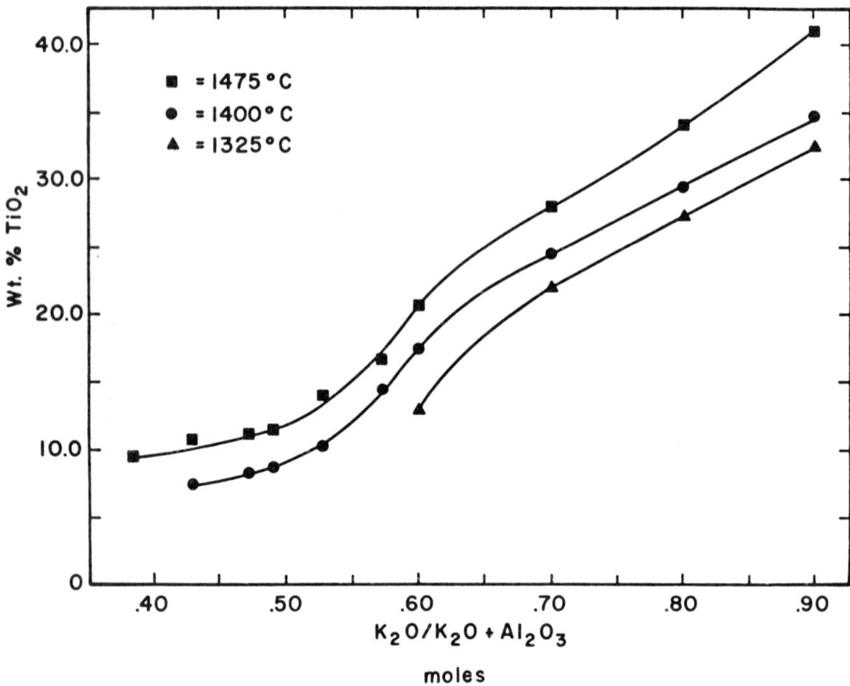

Fig. 5. Solubility of rutile (wt% TiO_2) in melts of K_2O–Al_2O_3–SiO_2 composition with initial $xSiO_2 = 80$ mol%. Note that the solubility is comparatively low in peraluminous melts ($K_2O/(K_2O + Al_2O_3) < 0.5$) but increases sharply in peralkaline melts. Since the activity of TiO_2 is fixed at constant temperature, the activity coefficient of TiO_2 varies inversely to the mole fraction of TiO_2. From Dickinson and Hess (1985).

between $K_2Ti_2O_5$ and $K_2Ti_3O_7$ (Dickinson and Hess, 1985; Gwinn and Hess, 1989). In peraluminous melts, K^{+1} is used to form $KAlO_4$ complexes and therefore is not accessible to Ti. The solution behavior in the peraluminous melts is therefore controlled by the stability of TiOTi complexes. It is not strongly influenced by the formation of AlOTi complexes, because the solubility of rutile decreases with the aluminum content (Fig. 5). An equilibrium like (10) also explains why the cotectic between forsterite-enstatite and cristobalite-enstatite shifts to less SiO_2-rich compositions upon addition of TiO_2 (Kushiro, 1975; Ryerson, 1985). The equilibrium

$$2SiOMg + TiOTi = 2TiOMg + SiOSi \quad (11)$$

is shifted to the right, thereby polymerizing the melt and increasing the activity coefficient of SiO_2.

A question often raised is whether Ti behaves as a network-forming (in the sense of TiO_4 and SiO_4 tetrahedra copolymerizing into an extended network) or a network-modifying cation. The alternatives offered in this question, however, are incomplete. Two liquid partitioning studies provide direct information about the network-forming abilities of a cation. The partitioning of Ti between immiscible silicate melts in simple systems (Watson, 1976; Visser and Koster van Gros, 1979b; Wood and Hess, 1980; Hess and Wood, 1982), in synthetic ferrobasaltic and granitic immiscible liquids (Ryerson and Hess, 1978; Hess et al., 1975, 1978; Rutherford et al., 1974) and in natural occurring immiscible melts in lunar and terrestrial igneous rocks (Roedder and Weiblen, 1972; Philpotts, 1982) is always in favor of the less polymerized liquid. Thus, even if Ti is four coordinated, it does not partition into the polymerized aluminosilicate network. The phase equilibria and spectroscopic data are most easily explained if Ti prefers to form polymerized titinate complexes with network-forming cations. These titinate complexes coexist with the aluminosilicate network, although each is soluble to varying degrees in the other. Titanium, therefore, has characteristics of a T and F-type cation, although F-type behavior is expected to dominate in most silicate melts and glasses.

Sn^{+4}, Zr^{+4}, Hf^{+4}

There is only a limited amount of data on the solution behavior of the Group IV cations Sn^{+4}, Zr^{+4}, and Hf^{+4}. What data exists, however, can be used in conjunction with the Ti data to develop a realistic representation of their solution characteristics. For example, cassiterite, SnO_2, has a rutile structure and malayaite, $CaSnO(SiO_4)$ is isomorphous with sphene $CaTiO(SiO_4)$. Mossbauer studies indicate that Sn^{+4} occur as SnO_6 octahedra in silicate glasses (Wong and Angell, 1976). Thus, it is expected that Sn^{+4} would follow Ti^{+4} in silicate melts. Experiments in the laboratory at Brown confirm these predictions. By application of cryoscopic techniques, Dickinson and Hess (1985) showed that the activity coefficient of TiO_2, decreases slightly with $K_2O/K_2O + Al_2O_3$ ratio in peraluminous melts but then decreases dramatically in peralkaline melts. Similar behavior is noted for the activity coefficient of SnO_2 in identical melts (Naski and Hess, 1985). The solubility of cassiterite is increased sharply by the availability of K_2O in excess of that needed to charge balance tetrahedrally coordinated aluminum. The homogeneous equilibrium

$$2SiOK + SnOSn = 2SnOK + SiOSi \tag{12}$$

shows that Sn^{+4} strips K from its non-bridging position and results in the formation of more polymerized SiOSi melt. This "peralkaline effect" is a common characteristic of high field strength cations.

Similar types of equilibria occur in Al_2O_3-free melts. Pyare and Nash (1982) determined that the redox ratio Sn^{+2}/Sn^{+4} decreases sharply with alkali oxide concentration in R_2O (R = K, Li, Na)-SiO_2 melts. This redox equilibrium is also explained by the formation of $Sn^{+4}OK$ bonds

$$2SiOK + Sn_{0.5}{}^{+2}OSi + 1/4O_2 = 2Sn_{0.25}{}^{+4}OK + 3/2SiOSi \quad (13)$$

Thus, decreasing the activity of SiOSi by the addition of R^+ causes Sn^{+4} to from at the expense of Sn^{+2}, as indeed is observed.

Zircon saturation experiments on dry or H_2O-bearing granitic melts (Watson and Harrison, 1983; Watson, 1979) and unpublished work at Brown have established that zircon solubility is increased sharply in peralkaline melts. A similar solution mechanism is implied, i.e.,

$$SiOR + ZrOSi = ROZr + SiOSi \quad (14)$$

Equation (14) predicts that the solubility of zircon should increase in peralkaline melts as observed. Moreover, this effect emphasizes that the stability of the ROZr complex is as important as the degree of SiOSi polymerization in determining the solubility of zircon.

There is only limited data on the solution characteristics of HfO_2. The solubility of hafnon, $HfSiO_4$, and zircon, $ZrSiO_4$, are virtually identical in melts along the (Na, K)AlO_2–SiO_2 join (Ellison and Hess, 1986). This similarity is not surprising since Hf and Zr behave coherently in natural geochemical systems. It is not hard to predict that HfO_2 will display the peralkaline effect in silicate melts.

Raman spectra of $(xK_4(1 - x)Zr)O_2 \cdot 8SiO_2$ glasses demonstrate that ZrOSi and SiOSi bonds are created at the expense of KOSi bonds as Zr is exchanged for K on an equal oxygen basis (Ellison and Hess, 1989b). The K_2O–$4SiO_2$ glass contains only Q^3 and Q^4 species. The Zr-bearing glasses contain Zr-coordinated Q^0 species as well as the K-coordinated Q^3 species, although the assignment of the Zr band is somewhat speculative. If the assignments are correct, then there is a redistribution of Q species such that Q_3 species are converted to Q_0 species, e.g.,

$$4Q^3 = Q^0 + 3Q^4 \quad (15)$$

The growth of the 450 cm^{-1} band with ZrO_2 content is consistent with the creation of SiOSi-rich regions and the production or Q^4 species via reaction (15).

The non-bridging oxygen belonging to the Q^0 species are likely to be coordinated simultaneously to both K and Zr. Such a coordination scheme would nicely explain the peralkaline effect observed in the study of zircon solubility. Watson (1979) found that one mole of ZrO_2 would be dissolved for every two moles of excess $K_2O + Na_2O$ in peralkaline melts. He proposed that alkali-zirconate complexes of $K_4Zr(SiO_4)_2$ stoichiometry exist in the melt. These melts were H_2O saturated at 2 kb and at 700 to 800°C. In dry melts at one atmosphere and several hundreds of degrees higher, one mole of ZrO_2 is dissolved for every one mole or excess K_2O, implying a complex of K_2ZrSiO_5 stoichiometry (unpublished). Local electrical neutrality on the non-bridging oxygen in the K_2ZrSiO_5 complex would be achieved exactly if K and Zr were six coordinated; each non-bridging oxygen would be coordinated to two K, one Zr, and one Si atom. Indeed, the coordination of Zr apparently is six in Na_2O–SiO_2 glasses (Waychunas and Brown, 1984).

Group V Elements

Phosphorus

Phosphorus forms PO_4 tetrahedra in both phosphate and silicate glasses and melts. In SiO_2 melts, PO_4 tetrahedra copolymerize with the slightly smaller SiO_4 tetrahedra. However, one oxygen atom is formally double bonded to each phosphorus atom, resulting in a significant distortion and weakening of the phosphosilicate framework (Mysen et al., 1981; Ryerson and Hess, 1980). This effect is expressed in a dramatic freezing-point depression of the cristobalite liquidus, and, of course, in a large drop in the activity coefficient of SiO_2. In fact, when calculated on a $PO_{2.5}$ basis, the freezing point depression is greater than that for any common network-modifying or network-forming species (Ryerson, 1985).

Phosphorus remains in PO_4 coordination in compositionally more complex melts. Even a small amount of P_2O_5 added to Na_2O–$3SiO_2$ and Na_2O–SiO_2 glasses produces significant changes in the Raman spectra (Nelson and Tallant, 1984). New spectral lines are observed that correspond to the vibration of NaOP bonds. However, the frequencies of the prominent SiONa and SiOSi vibrations are unchanged; only their intensities are diminished. The addition of phosphate to sodium silicate glass results in the formation of discrete Na-phosphate groups and minimal substitution of PO_4 tetrahedra in the SiOSi structures. This solution mechanism is expressed by the homogeneous equilibrium

$$2\mathrm{SiONa} + \mathrm{POP} = 2\mathrm{PONa} + \mathrm{SiOSi} \tag{16}$$

Note that phosphate reacts with the non-bridging species, SiONa, to form a discrete phosphate, PONa, and a more polymerized silicate structure (Ryerson and Hess, 1980; Mysen et al, 1981). The competition for non-bridging species is given by

$$\mathrm{SiONa} + \mathrm{POSi} = \mathrm{PONa} + \mathrm{SiOSi} \tag{17}$$

Phosphorus is thus distributed between network-forming, network-modifying, and discrete PONa species (Ryerson, 1985). Chromatographic studies of low SiO_2 melts show that both POSi and PONa species coexist in homogeneous equilibrium (Masson et al, 1974). At higher SiO_2 contents, the scavenging of Na or other network-modifying cations results in local ordering of melt into TOT- and MOP-rich regions, a preliminary step to the onset of macroscopic phase separation. In fact, the addition of P_2O_5 to liquids in the K_2O–FeO–Al_2O_3–SiO_2 system causes the two-liquid solvus to expand to much higher temperatures and pressures (Visser and Koster van Groos, 1979).

Phosphorus is most stable in silicate melts containing appropriate network-modifying cations that can be utilized to form POM complexes. This model explains several interesting experimental observations:

1. The addition of P_2O_5 to multisaturated liquids (enstatite-forsterite,

cristobalite-enstatite) shifts the boundary curves to lower silica contents thereby increasing the γSiO_2 in the cotectic liquids (Kushiro, 1975). This shift is consistent with an increase in the number of SiOSi bonds created by the equilibrium given in (16) (e.g., Hess, 1980).

2. P_2O_5 is partitioned into FeO, CaO-rich basaltic melts rather than in the coexisting immiscible granite melts (Watson, 1976; Ryerson and Hess, 1978). P_2O_5 is stabililized by the availability of the divalent network-modifying cations in the less-polymerized liquid.

3. The solubility of apatite is greatest in ultramafic melts rich in network-modifying cations and least in granitic melts under identical temperatures and pressures (Watson, 1979). Moreover, the solubility relations are similar whether done under anhydrous or water-saturated conditions. Hence, the extent of silicate polymerization is not as important as access to appropriate network-modifying cations.

4. The solubility of $LaPO_4$ ("monazite") exhibits a peralkaline and peraluminous effect in potassium aluminosilicate melts (Ellison and Hess, 1988). The solubility is at a minimum at $K_2O/Al_2O_3 = 1$ (moles) but increases with both increasing peralkalinity and increasing peraluminosity ($K/Al < 1$). The peralkaline effect, seen also in hydrous melts (Montel, 1986), is described by reaction Eq (16) with alkalis in excess of aluminum acting as the network-modifying cation. The peraluminous effect is described by the formation of $AlPO_4$ from the aluminum in excess of potassium or

$$AlOAl + POP = 2AlOP \qquad (18)$$

This model is supported by examination of the peraluminous glasses by Raman and both static and MAS NMR techniques (Gan and Hess, 1989; Mysen et al., 1981).

5. The observations that rare earth elements are strongly associated with phosphate complexes (Ryerson and Hess, 1978) suggest that phosphates have a strong affinity for R^{3+} cations. Indeed, the addition of phosphorus causes a decrease in the Fe^{2+}/Fe^{3+} ratio in peraluminous melts (Dickenson and Hess, 1983; Gwinn and Hess, in progress). It is likely that phosphorus will strongly influence the geochemistry of other trivalent trace elements such as the rare earths, B, Ga, and Sc among others (see also London, 1987).

As^{+5}, V^{+5}, Ta^{+5}, Nb^{+5}

The remaining Group V cations considered here include As^{5+}, V^{5+}, Ta^{5+}, and Nb^{5+}. A common feature of P^{5+} and the Group V cations is their ability to form a large number of compounds with alkali oxides and the alkaline earths (Levin et al., 1964, 1969). For example, five $Na_2O{-}nV_2O_5$ compounds are stable at high temperatures with $n = \frac{1}{3}, \frac{1}{2}, \frac{9}{16}$, 1 and 4. Moreover, F_2O_5-rich compounds (recall that F does not mean fluorine in this context) are most favored for F^{+5} cations with the largest ionic radii. For example, P^{5+} with $r = 0.35$Å (for octahedral

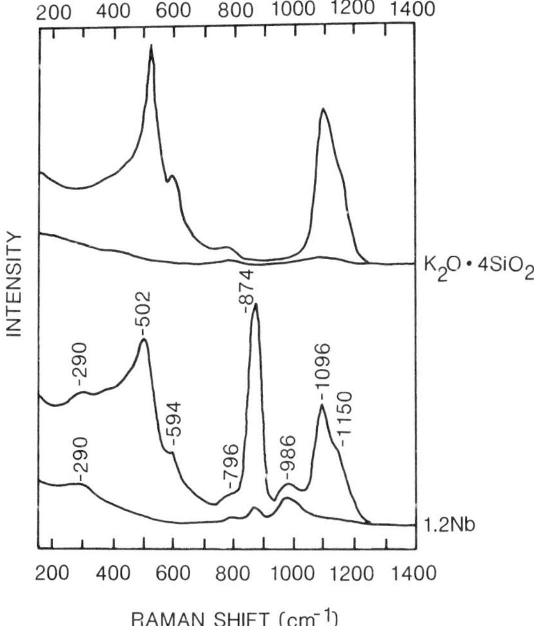

Fig. 6. Parallel (upper curves) and cross-polarized (lower curves) Raman spectra of $K_2O \cdot 4SiO_2$ and $7K_2O \cdot 0.6Nb_2O_5 \cdot 4OSiO_2$ glasses. (Ellison and Hess, 1989)

coordination with oxygen) has $n \leq 1$ whereas Nb^{5+} with $r = 0.64$Å has $n \leq 10$. These data hint that (1) Group V cations tend to complex with mono- and/or divalent cations in silicate melts and (2) large Group V cations form the most polymerized complexes; i.e., complexes with the greatest F_2O_5/R_2O ratios.

Raman spectra for Nb-bearing glasses tend to support these conclusions. Parallel and cross-polarized Raman spectra of $((1 - X)5K_2\ X\ Nb_2)O_5$–$20SiO_2$ glasses (Fig. 6) have a remarkably intense, polarized band at 874 cm^{-1} and a depolarized band at 986 cm^{-1}, both of which are absent in the $5K_2O.20SiO_2$ glass (Ellison and Hess, 1989b). The band at 874 cm^{-1} is similar to the strongest band in the spectra of MNb_2O_6 crystals where M is a divalent cation and NbO_6 octahedra have one NbOM and 5 NbONb bonds (Fukumi and Sakka, 1988). Increasing the number of NbOM bands decreases the wave number of this band slightly. The 874 cm^{-1} bond is therefore assigned to the NbOK band on NbO_6 octahedra. On the other hand, the depolarized band at 986 cm^{-1} is typical of the assymetric vibrations produced by oxygen coordinated by Si and a cation of high field strength (Ellison and Hess, 1989b). This band is assigned to an SiONb vibration, possibly of the Q_3 species. The structure of the Nb_2O_5 glass is therefore dominated by two equilibria

$$2KOSi + NbONb = 2KONb + SiOSi \tag{19}$$

$$SiOSi + NbONb = 2NbOSi \tag{20}$$

and possibly consists of niobate-rich and silicate-rich domains. The niobate domains contain NbOK and probably most of the NbOSi species. The SiO_2-rich domains contain few NbOSi species and most of the KOSi and SiOSi. This distribution is also implied by the existence of a large two-liquid field that spans the Nb_2O_5–SiO_2 system from about 20% SiO_2 to 95% SiO_2 at 1700°C (Levin et al., 1964). It is clear that more SiO_2, probably as SiONb, is dissolved in the Nb_2O_5-rich immiscible melt than Nb_2O_5 is dissolved in the SiO_2-rich melt.

A somewhat more complex picture emerges for the solution of arsenic in K_2O–SiO_2 glasses (Verweij, 1981). In metasilicate glass ($50K_2O$–$50SiO_2$–$0.5As_2O_5$), arsenic occurs primarily as K_3AsO_4 monomers. As the K_2O/SiO_2 content decreases, the arsenate ions polymerize to form dimers or perhaps trimers ($As_3O_{10}{}^{-5}$) and a small fraction of the arsenic is reduced and forms arsenite chains of $As_mO_{2m+1}{}^{-(m-2)}$. The As^{5+} ions strip K from Si and form isolated arsenate ions outside the network. The arsenite (As^{3+}) ions may form non-bridging oxygen with Si, or possibly, use K as a charge-balancing cation, and copolymerize with SiO_4 tetrahedra. This alternative is reasonable because As^{3+} and Fe^{3+} have similar ionic radii and Fe^{3+} also is reduced to Fe^{+2} in glasses with decreasing K_2O/SiO_2 (Paul and Douglas, 1965).

The homogeneous equilibria that determine the solution character of Nb and the other 5+ cations is

$$2SiOM + F\text{–}O\text{–}F = 2MOF + SiOSi \qquad (21)$$

where M is an alkali or 1/2 alkaline earth and F is As^{5+}, V^{5+}, Ta^{5+}, or Nb^{5+}. Hence F^{5+} outcompetes Si^{+4} for network-modifying cations. Moreover, it is possible that F^{5+} may even outcompete AlO_4 for their charge balancing species, perhaps forcing some Al cations into tricluster-type coordination.

Group VI Elements

Mo^{6+}, W^{6+}, Cr^{6+}

It is not difficult to argue that the solution behavior of F^{6+} cations will be similar to the F^{5+} group. The Raman spectra of $Na_2O \cdot 3SiO_2 \cdot 0.1Cr_2O_3$ have an intense narrow band at 850 cm^{-1} and a weaker, but still intense band at 900 cm^{-1} (Nelson et al., 1983). Bands at these wave numbers exist in aqueous solutions containing $CrO_4{}^{-2}$ (846 cm^{-1}) and $Cr_2O_7{}^{2-}$ (904 cm^{-1}) species (Michel and Cahay, 1986), and therefore are assigned to Na_2CrO_4 and $Na_2Cr_2O_7$ species respectively in the silicate glass. The Raman spectra of Mo-doped $K_2O \cdot 4SiO_2$ glasses (Fig. 7) contain an intense polarized band at 890 cm^{-1} that is absent from the $K_2O \cdot 4SiO_2$ glass (Ellison and Hess, 1989b). A strong band at 892 cm^{-1} exists in the spectra of crystalline Na_2MoO_4 and at 896 cm^{-1} for $MoO_4{}^{2-}$ species in aqueous solutions (Dean and Wilkinson, 1983). The 890 cm^{-1} band is assigned to the Mo–O–K vibration in the glass. MO appears to form orthomolybdenate species

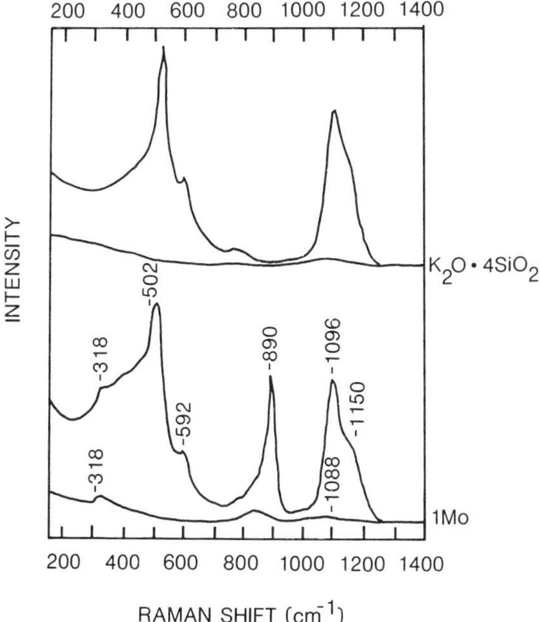

Fig. 7. Parallel (upper curves) and cross-polarized (lower curves) Raman spectra of K_2O–$4SiO_2$ and $7K_2O \cdot MoO_3 \cdot 40SiO_2$ glasses. (Ellison and Hess, 1989)

which are located entirely outside the silicate structure because there are no depolarized bands between 900–1100cm^{-1} as would be expected from the vibrations of Mo–O–Si bands.

The F^{6+} cations apparently are segregated into F-rich domains and few, if any, F–O–Si bonds are formed. The equilibrium constant for the reaction

$$2KOSi + F^{6+}OF^{6+} = 2F^{6+}OK + SiOSi \quad (22)$$

is expected to be the largest for any group of cations. The $^{+}6$ cations should be most strongly partitioned into the depolymerized regions of silicate melts.

Discussion

Theory

The stability of highly charged cations in a silicate melt depends on their ability to compete with network-forming cations for available oxygen and for real or "potential" network-modifying cations. Potential network-modifying cations are those that exist as charge-balancing cations for tetrahedrally coordinated Al, Fe^{3+}, B, and Ga. The key to understanding silicate melts is to establish a hierarchy of interactions of network-modifying cations, principally mono- and divalent-cations, with F cations, Si and the trivalent network-forming cations.

The hierarchy of interactions within each group, that is, in the M–F, M–Si and M–T^{3+} (where T^{3+} stands for network-forming cations of +3 formal charge) groups, are relatively well founded. Calorimetric studies have established the sequence Cs $\simeq$ Rb, K, Na, Li $\simeq$ Ba, Pb, Sr, Ca, Mg for M–Al interactions (Roy and Navrotsky, 1984). There is only a limited amount of information for other T^{3+} cations. The redox ratio of iron, Fe^{2+}/Fe^{3+}, increases in the sequence K, Na, Li in M_2O–SiO_2 melts (Douglas et al., 1965) and in the sequence Ba, Sr, Ca, Mg for MO–SiO_2 melts (Mysen et al., 1984). Moreover, the redox ratio is approximately the same in comparable Ba and Li melts if differences in experimental conditions are taken into account. Since Fe^{3+} occurs largely as $Fe^{3+}O_4$ tetrahedra in these glasses (e.g., Mysen, 1988), the redox ratio is an index of the charge-balancing capacity of these cations with $Fe^{3+}O_4$ tetrahedra, i.e., the lower the redox, the more favorable the M–$Fe^{+3}O_4$ interactions. If this theory is correct, then the charge-balancing hierarchy is K, Na, Li $\simeq$ Ba, Sr, Ca, Mg, a sequence already established for AlO_4 tetrahedra.

The hierarchy for M–O–Si interactions is obtained from phase equilibria (Hess, 1977), cryoscopic measurements (Ryerson, 1985; Hess, 1980), heats of formation of crystalline silicates (Hess, 1980) and free energies of mixing of silicate melts (Hess, 1980; Blander and Pelton, 1987). The hierarchy is Cs > Rb > K > Na > Li > Ba > Sr > Mn^{2+} > Mg > Fe^{2+} > Co^{2+}, a sequence that agrees with the more limited set obtained above.

The interactions between M and F cations cannot be established with great certainty. The data for phosphates suggest that the hierarchies will more or less parallel the trends established above. The heats of formation of glassy phosphates become more positive (less favorable) as Na, Li, Ca, Zn (Jeffes, 1975). The heats of formation of binary crystalline phosphates (Tardy and Vieillard, 1977) normalized to one mole of reactants are in the sequence Cs, Rb, K, Na, Ba, Li, Sr, Ca, Mg, Mn^{2+}, Pb, Ni, where the phosphates at the end of the series are less stable than those at the beginning. Subsets of the same series exist also for the enthalpies of formation of crystalline binary tungstates, sulfates, nitrates and carbonates (Navrotsky, 1985; Pankratz et al., 1984).

It is dangerous to assume without justification that the trends observed in the crystalline state are directly transferable to the molten state. Steric factors, the long range order, and the relatively few bonding options of the crystalline state introduce constraints that do not exist, or are muted in the liquid state. An increase of configurational entropy of the liquid, for example, can compensate for unfavorable contributions to the free energy from the heats of formation. Given these concerns, it is surprising that such regularities exist in the bonding of cations with oxygen. This statement underlies the assumptions that the interatomic potentials in condensed phases are dominated by short range forces (Gibbs, 1982). Thus, it may be possible to translate the interatomic potential for a given M–O–S interaction from minerals to melts.

The patterns deduced above are not inviolate. Neighbors in the series can be interchanged and local anomalies in the series no doubt exist in other systems. The sequence between Li, Ba and Sr is particularly prone to deviate from the "norm". The series does have a physical basis, however. The series from Cs to Mg is one of decreasing "basicity" of the network-modifying cation. As used here, basicity refers to the state of the oxygen and its ability to donate electrons to Si or to other highly charged F or T cations. A non-bridging oxygen in a K_2O–SiO_2 melt, for example, is coordinated to Si and K. Depending on its coordination, the Pauling bond strength of K will be $+\frac{1}{4}$ to $+\frac{1}{6}$, whereas that of Si is $+1$. The relatively weak K–O bond causes the Si–O (non-bridging) bond to shorten and strengthen. A "basic" network-modifying cation, therefore, is one that does not compete aggressively with more highly charged cations for the non-bridging oxygen. The polarization of the non-bridging oxygen (or "aspherical distortions") and the build-up of electrons on the Si–O bond reduces the free energy of the solution relative to free energy of a mechanical mixture of unpolarized oxygen in K_2O and SiO_2 phases. There is no one unambiguous basicity scale, but the ionic potential (formal charge divided by ionic radius) or Pauling bond strength are useful proxies. However, the coordinations of the cations in the melt must be determined in order to calculate these parameters.

It follows from the above arguments that the most stable melts are those which contain small cations of large charge (acid cations) and large cations of small charge (basic cations). The TiO_2–SiO_2 melts are unstable and have a wide field of immiscibility because Ti^{4+} and Si^{4+} compete about equally for oxygen. Adding alkali oxide to these melts creates polarized, non-bridging oxygen and eventually removes the two-liquid field from the liquidus. A more interesting question is how the non-bridging oxygen are apportioned to Ti^{4+} and Si^{4+}, and, in general, how Si, T^{3+} and F cations compete for the network-modifying cations.

It is a worthwhile exercise to extend these ideas to the general problem of distributing network-modifying cations among Si and other highly charged cations. The distributions should create the greatest number of highly polarized non-bridging oxygen, consistent with the requirement that the enthalpy and entropy minimize the free energy of the system. The enthalpic contributions are most favorable when the most basic network-modifying cation combines with the most acid cation (see also Dowty, 1987; Navrotsky, 1985). The entropic contribution (considered here to be solely due to configurational effects) will generally counter the tendency to localize the network-modifying cations on specific complexes. The peralkaline effect is one manifestation of these acid-base interactions. The peralkaline effect is described by the homogeneous equilibrium

$$\mathrm{SiO(K,Na) + FOF = FO(K,Na) + SiOSi} \quad (23)$$

where the polarization of oxygen in the FO(K, Na) complex is sufficient to shift non-bridging oxygen from a Si-network to a F-based network. Phase equilibria and spectroscopic studies indicate that Zr^{4+}, Ti^{4+}, Sn^{4+}, P^{5+}, Nb^{5+}, As^{5+} and Mo^{6+} are capable of stealing network-forming cations from the silicate network. In order to parameterize the acid-base equilibria, the stability of the FOF and SiOSi complexes and the free energy of creating non-bridging oxygen on different

Q species, where Q^n is now generalized to apply to F cations as well as Si, must be estimated. The simple rule of oxygen polarization therefore cannot be used in isolation and may very well give the wrong answer. This point is illustrated below.

The peralkaline effect for Zr^{4+} is nicely demonstrated in both natural and synthetic systems (Watson, 1978; Ellison and Hess, 1986). Yet, zirconium which is six coordinated in peralkaline melts (Waychunas and Brown, 1985) is less acidic than Si according to the ionic potentials, Pauling electronegativity or Pauling bond strength—SiOK should then be favored over ZrOK. Clearly, it is not so favored. This statement would be true even for Zr in four-coordinated complexes if the Zr–O bond distances scale to Si–O distances. The same analysis would not predict the peralkaline effect for Sn^{4+} or Ti^{4+} (but would correctly predict the peralkaline stability of P^{5+}, Nb^{5+}, As^{5+} and Mo^{6+}).

The exact reason for the above situation is not clear. One possible explanation is that the SiOSi bond is much more stable in SiO_2-rich melts than are TiOTi, SnOSn or ZrOZr bonds. In other words, there is a large energy penalty in creating polymerized FOF domains in SiO_2-rich melts. Moreover, the addition of F cations (as oxides) to silicate melts polymerizes the silicate network but creates F complexes that are less polymerized than the original melt. Thus, there is a net shift of network-modifying cations from high Q silicate domains to low Q F domains. This shift may contribute to the energy change of reaction Eq. (24). Finally, the absence of long range order in silicate melts may allow anomalously short and, therefore, stable bonds to form that shift the balance of the equilibria in the FOK direction. The postulate of a short TiOK bond in alkali silicate glasses is one example. This "explanation," however, begs the question, because it gives no insight as to why or when such bonds occur.

Summary and Conclusions

Some of the chemical parameters that strongly influence the geochemical behavior of highly charged cations are reasonably well known and are summarized below.

1. $(K_2O + Na_2O)/Al_2O_3$

F cations form stable complexes with monovalent cations. The highly charged cations are stabilized in peralkaline melts versus peraluminous or metaluminous melts because alkali atoms in excess of aluminum are free to complex with F cations. The discussion in the previous section and the evidence contained in the Raman spectra of Y, Zr, Nb, and Mo doped K_2O–SiO_2 glasses (Ellison and Hess, 1989b) indicate that the tendency to form FOK complexes is proportional to the charge of the F cation. Thus, the peralkaline effect should be stronger for Mo^{+6} than, say, Y^{3+}. This effect has significant implications for the behavior of trace elements in peralkaline melts. Crystal-liquid partition coefficients for these elements in peralkaline melts will be significantly lower than those of peraluminous/metaluminous melts under the same physical conditions—minor phases that concentrate these elements, e.g., zircon, apatite, monazite, ilmenite, niobates, etc., are relatively late crystallizing in peralkaline melts. Such elements behave incompatibly over a much greater range of crystallization in peralkaline melts than in peraluminous/metaluminous liquids. Finally, the redox state of multivalent cations, $F^n/F^{(n+m)}$, where n and m are positive integers, will decrease with increasing $(K_2O + Na_2O)/(Al_2O_3)$ in peralkaline liquids, e.g.

$$5SiOK + 3As^{3+}{}_{0.33}OSi + \tfrac{1}{2}O_2 \rightarrow 5As^{5+}{}_{0.20}OK + 4SiOSi$$

$$2SiOK + Sn^{2+}{}_{0.5}OSi + \tfrac{1}{4}O_2 \rightarrow 2Sn^{4+}{}_{0.25}OK + 1.5SiOSi \tag{24}$$

(e.g., Pyare and Nath, 1982).

A knowledge of the agpaitic index (i.e., $K_2O + Na_2O/Al_2O_3$) is required to predict the solubility of F cations. But the question is, how dominant is this index? Also, what elements can loosen the strong bonds between alkalis and tetrahedrally coordinated AlO_4? It is clear that Fe^{3+}, Ga^{3+}, Ti^{4+}, Zr^{4+}, Sn^{4+} and P^{5+} do not compete successfully for these alkali atoms. No data is available for other highly charged cations, but it is possible that the +5 and +6 cations would have more success in stealing the alkalis from Al. A few elements, in fact, out-compete Al for alkalies, including B, F and to a lesser degree, C. Addition of B_2O_3 under H_2O-saturated conditions to the system $NaAlSi_3O_8$–$KAlSi_3O_8$–H_2O leads to a reduction in the temperatures of the liquidus of feldspars but not of quartz (Pichavant, 1987). Moreover, the albite liquidus contracts relative to the more potassic feldspars. The lowering of the activity coefficient of $NaAlSi_3O_8$ is explained by an equilibrium such as

$$2NaAlSi_3O_8 + B_2O_3 = 2NaBO_2 + Al_2Si_6O_{15} \tag{25}$$

where B_2O_3 and $NaBO_2$ represent trigonally and tetrahedrally-coordinated boron, and the $Al_2Si_6O_{15}$ is a schematic representation of an Al complex without charge-balancing cations. Thus, B has extracted some alkali cations, apparently mostly Na, from the aluminosilicate network (Burnham and Nekvasil, 1986). The reaction is not quantitative, however. Konijnendijk (1975) has shown that the addition of Al_2O_3 to Na_2O–B_2O_3–SiO_2 melts has induced the equilibrium

$$2NaBO_2 + Al_2O_3 = B_2O_3 + 2NaAlO_2 \tag{26}$$

to proceed to the right. It may be concluded that there is an active competition for alkalies between B and Al (see also Oestrike, et al., 1986).

The addition of F to the H_2O-saturated granite system depresses the liquidus and contracts the feldspar field relative to the quartz field, also the $NaAlSi_3O_8$-rich field relative to the $KAlSi_3O_8$-rich field (Manning, 1981). Manning (1981) argues that F interacts with aluminosilicate ions to form aluminum-fluoride complexes. This interaction may be explained by the homogeneous equilibrium

$$NaAlSi_3O_8 + 4F = AlF_3 + NaF + Si_3O_8 \tag{27}$$

which rationalizes the decrease in the $NaAlSi_3O_8$ activity coefficient and the increase in the SiO_2 activity coefficient.

The solubility of CO_2 in totally polymerized melts is very small, consequently, the liquidus is only modestly depressed even at high pressures (Stolper et al., 1987). At constant temperature and pressure, the proportion of CO_2 as $CO_3{}^{2-}$ increases with increasing $NaAlO_2$ content along the SiO_2–$NaAlO_2$ join (Fine and Stolper, 1985). As was discussed earlier, this change may be ascribed to the formation of Na_2CO_3 in the melt, where the Na is stripped from $AlO_4{}^{3-}$ units. This effect is not strong, however, as less than 0.8 wt% CO_2 as $CO_3{}^{2-}$ is incorporated in albite melts at 30 kb; this is about 50% of the molecular CO_2.

2. *Phosphorus*

Trivalent cations including B, the rare earths and others, are likely to form stable $F^{3+}OP$ associations with phosphorus (and perhaps other group V cations). Crystal-liquid partition coefficients for these elements should be sensitive to the P_2O_5 content of melts (Ryerson and Hess, 1978). Conversely, the activity of P_2O_5 in silicate melts will depend on the availability of B and excess Al. The peraluminous effect on monazite saturation is one manifestation of the stabilization of phosphorus by trivalent cations (Ellison and Hess, 1988). The high P_2O_5 contents of certain pegmatites relative to high silica rhyolites (Pichavant et al., 1987; London, 1987) might be a reflection of high B_2O_3 contents and the peraluminous composition of the pegmatite.

The addition of P_2O_5 to a silicate melt will affect the redox state of certain multivalent cations and, in doing so, influence the activity of the given metal oxide. For example, increasing the P_2O_5 content of Fe-bearing peraluminous granitic melts increases the Fe^{3+}/Fe^{2+} ratio perhaps through the formation of $Fe^{3+}PO_4$ associations (Gwinn and Hess, in preparation). In contrast, this effect does not occur in peralkaline melts presumably because $KFe^{3+}O_2$ bonds are relatively more stable than $Fe^{3+}OP$ bonds.

3. *Titanium*

Titanium is known to form crystalline compounds with Group IV and Group V cations. For example, Zr, Hf, and Nb are strongly partitioned into lunar ilmenite and armalcolite (McKay and Wagstaff, 1985; Quick et al., 1981). By analogy, it is likely that associations of (Zr, HF, Nb)OTi, or more generally, $F^{5+}OTi$ species, are stable in silicate melts. Since $Ti/F^{5+} \gg 1$ in most natural melts, the solution behavior of these cations will be a function of the Ti concentration and the nature of the MOTi associations. For example, in peraluminous melts, the associations of F^{5+} with Ti would surely be important. However, in peralkaline melts there must exist a competition between MOTi and $F^{5+}OTi$ species that cannot be evaluated at this time.

The competion between KOTi and $KFe^{3+}O_2$ has been determined, however (Gwinn and Hess, 1989). By determining the saturation surface of pseudobrookite, it was established that the content of TiO_2 and Fe_2O_3 in peralkaline melts was controlled by the amount of excess K_2O (K_2O in excess of that needed to charge balance Al_2O_3). Thus, Fe^{3+} and Ti^{4+} compete about equally for the network-modifying alkalies.

4. Other F-type cations

The homogeneous equilibria of F-type trace elements will be affected by the concentrations and associations of other F-type cations. For example, Schreiber and Balazs (1981) showed that the uranium redox equilibria were strongly altered by addition of Cr to the silicate melt. The prevailing redox state of an element in a magma directly controls how it is partitioned among crystalline phases, so it is very important to establish an "electromotive force series" that will order the interactions of multivalent cations (Schreiber, 1985). It is equally important to derive a series that measures the relative strengths of FOF and MOT associations, and orders the interactions between F, T, and M species.

5. H_2O

Water dissolves in silicate melts both in molecular form and as hydroxyl ions produced during the reaction

$$H_2O + O^{2-} = 2OH$$

(Stolper, 1982). The formation of hydroxyls depolymerizes the melt according to

$$SiOSi + HOH = 2SiOH$$

and should sharply alter the thermodynamic properties of both network-forming and network-modifying species. However, the effect on high field strength cations, particularly those that are incorporated as FOF and FOM species might be more modest if the OH species do not directly complex with the F cations. Green and Watson (1982) found that the solubility of apatite was more or less the same in H_2O-bearing and anhydrous granitic melts at high pressure. This finding suggests that the chemical potential of CaOP was not sensitive to the depolymerization of the SiOSi structure by hydroxyl formation. Moreover, the peralkaline effect for Zr and P is clearly demonstrated in both anhydrous and H_2O-saturated melts.

6. CO_2

Carbon dioxide dissolves in silicate melts in both molecular form and as CO_3 ions. Those trace elements that participate in carbonate formation will be stabilized in CO_2-bearing silicate melts. Rare earth elements and other elements that are strongly partitioned into carbonate magma (Wendlandt and Harrison, 1979) are likely to be stabilized in CO_2-bearing melts relative to CO_2-free melts.

Acknowledgements

Thanks are due to the many colleagues who have contributed significantly to the ideas in this paper. The paper also benefited from the reviews by B.G. Varshal and Alexandra Navrotsky. The work was supported by grants from the U.S. National Sciences Foundation (EAR 81-15996; EAR 84-16769; EAR 87-19358).

References

Blander, M., and Pelton, A.D. (1987) Thermodynamic analysis of binary liquid silicates and prediction of ternary solution properties by modified quasi-chemical equations. *Geochim. Cosmochim. Acta*, **51**, 85–96.

Brandriss, M.E., and Stebbins, J.F. (1988) Effects of temperature on the structure of silicate liquids: ^{29}Si NMR results. *Geochim. Cosmochim. Acta*, **52**, 2659–2670.

Brawer, S.A., and White, W.B. (1975) Raman spectroscopic investigation of the structure of silicate glasses I. The binary alkali silicates. *J. Chem. Phys*; **63**, 2421–2432.

Brawer, S.A., and White, W.B. (1977) Raman spectroscopic study of hexavalent chromium in some silicate and borate glasses. *Mat. Res. Bull.*, **12**, 281–288.

Bray, B.J. (1978) NMR studies of borates. In: *Borate Glasses: Structure, Properties, Applications*, edited by L.D. Pye, V.D. Frechette, and N.F. Kreidl Plenum Press, New York.

Brearly, M., and Montana, A. (1989) The effect of CO_2 on the viscosity of silicate liquids at high pressure. *Geochim. Cosmochim Acta.*, **53**, 2609–2616

Bruckner, R., Chun, H.U., and Goretzki, H. (1978) Photoelectron spectroscopy (ESCA) on alkali silicate—and soda aluminosilicate glasses. *Glastechn. Ber.*, **51**, 1–7.

Burnham, C.W., and Nekvasil, H. (1986) Equilibrium properties of granite pegmatite melts. *Am. Mineral Jahns. Mem.* **71**, 239–263.

Capabianco, C., and Navrotsky, A. (1982) Calorimetric evidence for the ideal mixing of silicon and germanium in glasses and crystals of sodium feldspar composition. *Amer. Mineral.*, **67**, 718–724.

Chorlton, L.B., and Martin, R.F. (1978) The effect of boron on the granite solidus. *Can. Mineral.*, **16**, 239–244.

Dean, K.J., and Wilkinson, G.R. (1983) Precision Raman investigations of the ν_1 mode of vibration of SO_4^{-2}, WO_4^{-2} and MoO_4^{-2} in aqueous solutions of different concentrations. *J. Raman Spectr.*, **14**, 130–134.

DeJong, B.H.W.S., and Brown, G.E. (1980) Polymerization of silicate and aluminate tetrahedra in glasses, melts and aqueous solutions II. The network modifying effects of Mg^{+2}, K^+, Na^+, Li^+, OH^-, F^-, Cl^-, H_2O, CO_2 and H_3O+ on silicate polymers. *Geochim. Cosmochim. Acta*, **44**, 1627–1642.

DeJong, B.H.W.S., Keefer, K.D., Brown, G.E., and Taylor, C.M. (1981) Polymerization of silicate and aluminate tetrahedra in glasses, melts and aqueous solutions—III. Local silicon environments and internal nucleation in silicate glasses. *Geochim. Cosmochim. Acta*, **45**, 1291–1308.

Dell, W.J., Bray, P.J., and Xiao, S.Z. (1983) ^{11}B NMR studies and structural modeling of Na_2O–B_2O_5–SiO_2 glasses of high soda content. *J. Noncrystall. Solids*, **58**, 1–16.

Dickenson, M.P., and Hess, P.C. (1981) Redox equilibria and the structural role of iron in aluminosilicate melts. *Contrib. Mineral. Petrol.*, **78**, 352–357.

Dickenson, M.P., and Hess, P.C. (1986) The structural role of Fe^{+3}, Ga^{+3}, and Al^{+3} and homogeneous iron redox equilibria in K_2O–Al_2O_3–Ga_2O_3–SiO_2–Fe_2O_3–FeO melts. *J. Noncrystall. Solids*, **86**, 303–310.

Dickinson, J.E. (1984) Raman spectra of potassium titanate crystals and glass. Implications for the structural role of titanium in silicate melts. *Geol. Soc. Amer. Abstr.*, **16**, 486.

Dickinson, J.E., and Hess, P.C. (1981) Zircon saturation in lunar basalts and granite. *Earth Planet. Sci. Lett.*, **57**, 336–344.

Dickinson, J.E., Hess, P.C., Dickenson, M.P., and Danckwerth, P.A. (1984) Aluminum distribution in alkali aluminosilicate glasses. *Geol. Soc. Amer. Abstr.*, **16**, 488.

Dickinson, J.E., and Hess, P.C. (1985) Rutile solubility and Ti coordination in silicate melts. *Geochim. Cosmochim. Acta*, **49**, 2289–2296.

Douglas, R.W., Nath, P., and Paul, A. (1965) Oxygen ion activity and its influence on the redox equilibrium in glasses. *Phys. Chem. Glasses*, **6**, 216–223.

Dowty, E. (1987) Vibrational interactions of tetrahedra in silicate glasses and crystals: II Calculations on melilites, pyroxenes, silica polymorphs and feldspars. *Phys. Chem. Mineral.*, **14**, 122–138.

Dumas, T., and Petiau, J. (1986) EXAFS study of titanium and zinc environments during nucleation in a cordierite glass. *J. Noncrystall. Solids*, **81**, 201–220.

Dupree, R., Holland, D., McMillan, P.W., and Pettifer, R.F. (1984) The structure of soda-silicate glasses: A MAS NMR study. *J. Noncrystall. Solids*, **68**, 399–410.

Dupree, R., Holland, D., Mortuza, M.G., Collins, J.A., and Lockyer, M.W.G. (1988) An MAS NMR study of network-cation coordination in phosphosilicate glasses. *J. Noncrystall. Solids*, **106**, 403–407.

Eggler, D.H., and Rosenhauer, M. (1978) Carbon dioxide in silicate melts: II Solubilities of CO_2 and H_2O in $CaMgSi_2O_6$ (diopside) liquids and vapors at pressure to 40 kb. *Am. J. Sci.*, **278**, 64–94.

Ellison, A.J., and Hess, P.C. (1986) Solution behavior of +4 cations in high silica melts: Petrologic and geochemical implications. *Contrib. Mineral. Petrol.*, **94**, 343–351.

Ellison, A.J., and Hess, P.C. (1987) Raman spectroscopic studies of quenched glasses of the system 10 K_2O – X R_2O_3 – 50 SiO_2 (R = La, Gd, Yb; X = 0.2, 5, 10). *Mat. Res. Soc.*, p. 302.

Ellison, A.J., and Hess, P.C. (1988) Peraluminous and peralkaline effects upon monazite solubility in high silica liquids. *EOS*, **69**, 498.

Ellison, A.J., and Hess, P.C. (1989b) Solution mechanisms of Period V cations in potassium silicate glasses: Inferences from Raman spectra. *EOS*, **70**, 487.

Ellison, A.J., and Hess, P.C. (1989c) Solution properties of rare earth elements in silicate melts: Inferences from immiscible liquids. *Geochim. Cosmochim. Acta*, **53**,

Ellison, A.E., and Hess, P.C. (1990) Lanthanides in silicate glasses: A vibrational spectroscopic study. *J. Geophys. Res.*, **95**, 15,717–15,726.

Ellison, A.J., and Hess, P.C. (1991) Vibrational spectra of high–silica glasses of the system KO–SiO–LaO. *J. Noncrystall. Solids*, **127**, 247–259.

Englehardt, G., Nofz, M., Forkel, K., Wihsmann, F.G., Magi, M., Samoson, A., and Lippmaa, E. (1985) Structural studies of calcium aluminosilicate glasses by high resolution solid state ^{29}Si and ^{27}Al magic angle spinning nuclear magnetic resonance. *Phys. Chem. Glasses*, **26**, 157–165.

Felsche, J. (1973) The crystal chemistry of the rare earth silicates. In *Structure and Bonding, Vol. 13*, edited by J.D. Donitz, et al., pp. 100–197, Springer-Verlag, New York.

Fine, G., and Stolper, E. (1985) The speciation of carbon dioxide in sodium aluminosilicate glasses. *Contrib. Minerol. Petrol.*, **91**, 105–112.

Fleet, M.E., Herzberg, C.T., Henderson, G.S., Crozier, E.P., Osborne, M.D., and Scarfe, C.M. (1984) Coordination of FeGa and Ge in high pressure glasses by Mössbauer, Raman and X-ray spectroscopy and geological implications. *Geochim. Cosmochim. Acta*, **48**, 1455–1466.

Fogel, R.A., and Rutherford, M.J. (1990) The solubility of carbon dioxide in rhyolitic melts: A quantitative FTIR study. *Amer. Mineral.*, **75**, 1311–1326.

Fukumi, K., and Sakka, S. (1988) Coordination state of Nb^{+5} ions in silicate and gallate glasses as studied by Raman spectroscopy. *J. Mat. Sci.*, **23**, 2819–2823.

Furukawa, T., and White, W.B. (1980) Raman spectroscopic investigation of the structure and crystallization of binary alkali germanate glasses. *J. Mat. Sci.*, **15**, 1648–1662.

Furukawa, T., and White, W.B. (1979) Structure and crystallization of glasses in $Li_2Si_2O_5$–TiO_2 system determined by Raman spectroscopy. *Phys. Chem. Glasses*, **20**, 69–80.
Galachov, F.Ya., Aver'yanov, V.I., Vavilonova, V.T., Areshev, M.P., and Makeeva, N.M. (1983) Metastable liquid phase separation in MgO (CaO)–Ga_2O_3–SiO_2 systems. *Soviet J. Glass Phys. Chem.*, **8**, 104–110.
Galachov, F.Ya., Aver'yanov, V.I., Vavilonova, V.T., and Areshev, M.P. (1976a) The metastable liquid phase separation region in the MgO–Al_2O_3–SiO_2 system. *Soviet J. Glass Phys. Chem.*, **2**, 405–408.
Galachov, F.Ya., Aver'yanov, V.I., Vailonova, V.T., and Areshev, M.P. (1976b) The metastable two liquid region in the Ga_2O_3–Al_2O_3–SiO_2 and Al_2O_3–SiO_2 system. *Soviet J. Glass Phys. Chem.*, **2**, 127–132.
Gan, H., and Hess, P.C. (1989) Phosphorus effects upon the structure of potassium aluminosilicate glass: Inference from Raman and NMR, *EOS*, **70**, 1375
Gazzoni, G. (1973) Al–Ga and Si–Ge diadochy in synthetic $BaAl_2Si_2O_8$ and $SrAl_2Si_2O_8$. *Z. Krist.*, **137**, 24–34.
Geisinger, K.L., Gibbs, C.V., and Navrotsky, A. (1985) A molecular orbital study of bond length and angle variations in framework structures. *Phys. Chem. Mineral.*, **11**, 266–283.
Geisinger, K.L., Oestrike, R., Navrotsky, A., Turner, G.L., and Kirkpatrick, R.J. (1988) Thermochemistry and structure of glasses along the join $NaAlSi_3O_8$–$NaBSi_3O_8$. *Geochim. Cosmochim. Acta*, **52**, 2405–2414.
Ghiorso, M.S., Charmichael, I.S.E., Rivers, M.L., and Sack, R.O. (1983) The Gibbs free energy of mixing of natural silicate liquids; an expanded regular solution approximation for the calculation of magmatic intensive variables. *Contrib. Mineral. Petrol.*, **84**, 107–145.
Gibbs, G.V. (1982) Molecules as models for bonding in silicates. *Amer. Mineral.*, **67**, 421–450.
Greegor, R.B., Lytie, F.W., Sandstrom, D.R., Wong, J., and Schultz, P. (1983) Investigation of TiO_2–SiO_2 glasses by x-ray absorption spectroscopy. *J. Noncrystall. Solids*, **55**, 27–43.
Green, T.H., and Watson, E.B. (1982) Crystallization of apatite in natural magmas under high pressure, hydrous conditions, with particular reference to "orogenic" rock series. *Contrib. Mineral. Petrol.*, **79**, 96–105.
Gwinn, R. and Hess, P.C. (1989) Iron and titanium solution properties in peraluminous and peralkaline rhyolitic liquids. *Contrib. Mineral. Petrol.*, **101**, 326–338.
Hallas, E., Haubenreissen, U., Haeriert, M., and Mueller, D. (1983) NMR-Untersuchungen an Na_2O–Al_2O_3–SiO_2 Glaesern mit Hilfe der Chemischen Verschiebung von ^{27}Al-Kerner. *Glastechn. Ber.*, **56**, 63–70.
Hart, S.R., and Davis, K.E. (1978) Nickel partitioning between olivine and silicate liquids. *Earth Planet. Sci. Lett.*, **40**, 203–220.
Hess, P.C. (1971) Polymer model of silicate melts. *Geochim. Cosmochim. Acta*, **35**, 289–306.
Hess, P.C. (1977) Structure of silicate melts. *Can. Mineral.*, **15**, 162–178.
Hess, P.C. (1980) Polymerization model for silicate melts. In: *Physics of Magmatic Processes* edited by R.B. Hargraves, pp. 1–48.
Hess, P.C., Rutherford, M.J., Guillemette, R.N., Ryerson, F.J., and Tuchfield, H.A. (1975) Residual products of fractional crystallization of lunar magmas: an experimental study. *Proc. 6th Lunar Sci. Conf.*, **1**, 895–909.
Hess, P.C., and Wood, M.I. (1982) Aluminum coordination in metaluminous and peralkaline silicate melts. *Contrib. Mineral. Petrol.*, **81**, 103–112.
Honada, T. and Soga, N. (1980) Coordination of titanium in silicate glasses. *J. Noncrystall. Solids*, **38**, 105–110.

Jeffes, J.H.E. (1975) The thermodynamics of polymeric melts and slags. *Silicates Ind.*, **40**, 325–340.
Johnston, W.D. (1965) Oxidation–reduction equilibria in molten Na_2O–$2SiO_2$ glass. *J. Am. Ceram. Soc.*, **48**, 184–190.
Konijnendijk, W.L. (1975) The structure of borosilicate glasses. *Philips. Res. Repts., Suppl.* **1**, Eindhoven.
Krol, D.M., and Smets, B.M.J. (1984) Group III ions in sodium silicate glass. Part 2: Raman study. *Phys. Chem. Glasses*, **25**, 119–125.
Kroll, H., Phillips, M.W., and Pentinghaus, H., (1978) The structure of the ordered synthetic feldspars $SrGa_2Si_2O_8$, $BaGa_2Si_2O_8$ and $BaGa_2Ge_2O_8$. *Acta Crystall.*, **B34**, 354–365.
Kushiro, I. (1975) On the nature of silicate melts and its significance in magma genesis: regularities in the shift of liquid boundaries involving olivine, pyroxene, and silica minerals. *Amer. J. Sci.*, **275**, 411–431.
Lacy, E.D. (1963) Aluminum in glasses and in melts. *Phys. Chem. Glasses*, **4**, 234–238.
Lacy, E.D. (1965) A statistical model of polymerization/depolymerization relationships in silicate melts and glasses. *Phys. Chem. of Glasses*, **6**, 171–180.
Levin, E.M., Robbins, C.R., and McMurdie, H.F. (1964) Phase diagrams for ceramists. *Amer. Ceram. Soc.*
Levin, E.M., Robbins, C.R., and McMurdie, H.F. (1969) Phase diagrams for ceramists, *Amer. Ceram. Soc.* (Suppl.)
Liebau, F. (1985) *Structural Chemistry of silicates.* Springer-Verlag, New York.
London, D. (1987) Internal differentiation of rare element pegmatites: Effects of boron, phosphorus and fluorine. *Geochim. Cosmochim. Acta*, **51**, 403–420.
London, D., Hervig, R.H., and Morgan, G.B. VI (1988) Melt-vapor solubilities and element partitioning in peraluminous granite-pegmatite systems: experimental results with Macusani glass at 200 MPa. *Contrib. Mineral. Petrol.*, **99**, 360–373
London, D., Morgan, G.B. VI, and Hervig, R.L. (1989) Vapor-undersaturated experiments with Macusani glass $+H_2O$ at 200 MPa and the internal differentiation of granitic pegmatites. *Contrib. Mineral. Petrol.*, **102**, 1–17.
Lux, G. (1987) The behavior of noble gases in silicate liquids: Solution, diffusion, bubbles and surface effects, with applications to natural samples. *Geochim. Cosmochim. Acta*, **51**, 1549–1560.
Manning, D.A.C. (1981) The effect of fluorine on liquidus phase relationships in the system Qz–Ab–Or with excess water at 1 Kb. *Contrib. Mineral Petrol.*, **76**, 206–215.
Manning, D.A.C., and Pichavant, M. (1983) The role of fluorine and boron in the generation of granitic melts. In: *Migmatites, Melting and Metamorphism*, edited by M.P. Atherton and C.D. Gribble. Shiva Publishing, Lmt.
Masson, C.R., Jamieson, W.D., and Mason, F. (1974) Ionic constitution of metallurgical slags. In: *The Richardson Conference on Physical Chemistry of Process Metallurgy.* IMM, London.
Matson, D.W., Sharma, S.K., and Philpotts, J.A. (1983) The structure of high-silica alkali-silicate glasses—a Raman spectroscopic study. *J. Noncrystall. Solids*, **58**, 323–352.
Matson, D.W., and Sharma, S.K. (1985) Structures of the sodium alumino—and gallo-silicate glasses and their germanium analogs. *Geochim. Cosmochim. Acta*, **49**, 1913–1924.
McKay, G., and Wagstaff, J. (1985) Ilmenite partitioning revisited: Confirmation of zirconium results for high-Ti mare basalt. *Lunar Planet. Sci.*, **XVI**, 542–543.
McKeown, D.A., Waychunas, G.A., and Brown, G.E., Jr. (1985) EXAFS study of the

coordination environment of aluminum in series of silica-rich glasses and selected minerals with the Na_2O–Al_2O_3–SiO_2 system. *J. Noncrystall. Solids*, **74**, 349–371.
McMillan, P.F. (1984) Structural studies of silicate glasses and melts—applications and limitations of Raman spectroscopy. *Amer. Mineral.*, **69**, 622–644.
McMillan, P.F., and Hoffmeister, A.M. (1988) Infrared and Raman spectroscopy. In: *Spectroscopic Method in Mineralogy and Geology*, edited by F.C. Hawthorne *Reviews in Mineral.*, **18**, 99–160.
Michel, G., and Cahay, R. (1986) Raman spectroscopic investigations on the chromium VI equilibria: Part 2—Species present, influence of ionic strength and $Cr_2O_4^{-2}$–$Cr_2O_7^{-2}$ equilibrium constant. *J. Raman Spectrosc.*, **17**, 79–82.
Montel, J.-C. (1986) Experimental determination of the solubility of Ce-monazite in SiO_2–Al_2O_3–K_2O–Na_2O melts at 800°C, 2 kbar, under H_2O-saturated conditions. *Geology*, **14**, 659–662.
Müller, D., Berger, G., Grunze, I., Ludwig, G., Hallas, E., and Haubenreisser, U. (1983) Solid state high-resolution ^{27}Al nuclear magnetic resonance studies of the structure of CaO–Al_2O_3–P_2O_5 glasses. *Phys. Chem. Glasses*, **24**, 37–42.
Mysen, B.O., and Virgo, D. (1980) Trace element partitioning and melt structure: an experimental study at 1 atm pressure. *Geochim. Cosmochim. Acta*, **44**, 1917–1930.
Mysen, B.O., Ryerson, F.J., and Virgo, D. (1980) The influence of TiO_2 on the structure and derivative properties of silicate melts. *Amer. Mineral.*, **65**, 1150–1165.
Mysen, B.O., Ryerson, F.J., and Virgo, D. (1981) The structural role of phosphorus in silicate melts. *Amer. Mineral.*, **66**, 106–117.
Mysen, B.O., Virgo, D., and Seifert, F.A. (1984) Redox equilibria of iron in alkaline earth silicate melts: Relationships between melt structure, oxygen, fugacity, temperature, and properties of iron-bearing silicalc liquids. *Amer. Mineral.*, **69**, 834–848.
Mysen, B.O. (1988) *Structure and Properties of Silicate Melts*. Elsevier, New York.
Naski, G.C., and Hess, P.C. (1985) SnO_2 solubility: Experimental results in peraluminous and peralkaline high silica glasses. *EOS* (*ABS*), **66**, 412.
Navrotsky, A., Geisinger, K.L., McMillan, P., and Gibbs, G.V. (1985) The tetrahedral framework in glasses and melts—inferences from molecular orbital calculations and implications for structure, thermodynamics, and physical properties. *Phys. Chem. Mineral.*, **11**, 284–298.
Nelson, C., Furukawa, T., and White, W.B. (1983) Transition metal ions in glasses: Network modifiers or quasi-molecular complexes. *Mat. Res. Bull.*, **18**, 959–966.
Nelson, C., and Tallant, D.R. (1984) Raman studies of sodium silicate glasses with low phosphate contents. *Phys. Chem. Glasses*, **25**, 31–38.
Nelson, C., Tallant, D.R., and Schelnutt, J.A. (1984) Raman spectroscopic study of scandium in sodium silicate glasses. *J. Noncrystall. Solids*, **68**, 87–98.
Ohtani, E., Taulelle, F., and Angell, C.A. (1985) Al^{+3} coordination changes in liquid silicates under pressure. *Nature*, **314**, 78–81.
Oldfield, E., and Kirkpatrick (1985) High-resolution nuclear magnetic resonance of inorganic solids. *Science*, **227**, 1537–1544.
Pankratz, L.B., Stuve, J.M., and Gokcen, N.A. (1984) Thermodynamic Data for Mineral Technology. Bureau of Mines, Bull. **677**, U.S. Dept. of the Interior.
Park, M.J., Kim, K.S., and Bray, P.J. (1979) The determination of the structures of compounds and glasses in the system MgO–B_2O_3 using ^{11}B NMR. *Phys. Chem. Glasses*, **20**, 31–34.
Paul, A., and Douglas, R.W. (1965) Ferrous-ferric equilibrium in binary alkali silicate glasses. *Phys. Chem. Glasses*, **6**, 207–211.

Philpotts, A.R. (1982) Composition of immiscible liquids in volcanic rocks. *Contrib. Mineral. Petrol.*, **80**, 201–218.
Pichavant, M. (1981) An experimental study of the effect of boron on water-saturated haplogranite at 1 kbar pressure. Geological Applications. *Contrib. Mineral. Petrol.*, **76**, 430–439.
Pichavant, M. (1987) Effects of B and H_2O on liquidus phase relations in the haplogranite system at 1 kbar. *Amer. Mineral.* **72**, 1056–1070.
Pyare, R., and Nash, P. (1982) Stannous-stannic equilibrium in molten binary alkali silicate and ternary silicate glasses. *J. Amer. Ceram. Soc.*, **65**, 549–554.
Quick, J.E., James, O.B., and Albee, A.L. (1981) Petrology and petrogenesis of lunar breccia 12013. *Proc. Lunar Planet. Sci.*, **12**, 117–172.
Richet, R., and Bottinga, Y. (1986) Thermochemical properties of silicate glasses and liquids: A review. *Rev. Geophysics*, **24**, 1–25.
Risbud, S.H., Kirkpatrick, R.J., Taglialavore, A.P. and Montez, B. (1987) Solid state NMR evidence for 4-, 5-, and 6-fold aluminum sites in roller-quenched SiO_2–Al_2O_3 glasses. *J. Amer. Cer. Soc.*, **70**, C-10-C-12.
Robie, R.A., Heminway, B.S. and Fisher, J.R. (1978) Thermodynamic Properties of Minerals and Related Substances at 298.15 K and 1 bar (10^5 Pascals) Pressure and at Higher Pressure. U.S. Geological Survey Bull. 1452.
Roedder, E., and Weiblen, P.W. (1972) Petrology of melt inclusions, Apollo 11 and Apollo 12 and terrestrial equivalents. *Proc. 2nd Lunar Sci. Conf.* **1**, 507–528.
Roy, B.N., and Navrotsky, A. (1984) Thermochemistry of charge coupled substitutions in silicate glasses: The systems $M^{n+}{}_{1/n}$ AlO_2–SiO_2 (M = Li, Na, K, Rb, Cs, Mg, Ca, Sr, Ba, Pb). *J. Amer. Ceram. Soc.*, **67**, 606–610.
Rutherford, M.J., Hess, P.C., and Daniel, G.H. (1974) Experimental liquid line of descent and liquid immiscibility for basalt 70017. *Proc. 5th Lunar Sci. Conf.*, **1**, 569–583.
Ryerson, F.J. (1985) Oxide solution mechanisms in silicate melts: Systematic variations in the activity coefficient of SiO_2. *Geochim. Cosmochim. Acta*, **49**, 637–649.
Ryerson, F.J., and Hess, P.C. (1978) Implications of liquid-liquid distribution coefficients to mineral-liquid partitioning. *Geochim. Cosmochim. Acta*, **42**, 921–932.
Ryerson, F.J. and Hess, P.C. (1980) The role of P_2O_5 in silicate melts. *Geochim. Cosmochim., Acta*, **44**, 611–624.
Schneider, E., Stebbins, J.F., and Pines, A. (1987) Speciation and local structure in alkali and alkaline earth silicate glasses: constraints from ^{29}Si NMR spectroscopy. *J. Noncrystall. Solids*, **89**, 371–383.
Schramm, C.M., DeJong, B.H.W.S., and Parziale, V.E. (1984) ^{29}Si magic angle spinning NMR study on local silicon environments in amorphous and crystalline lithium silicates. *J. Amer. Chem. Soc.*, **106**, 4396–4403.
Schreiber, H.D. (1987) An electromotive series of redox couples in silicate melts: A review and application to geochemistry. *J. Geophys. Res.*, **92**, 9225–9232.
Schreiber, H.D., and Balazs, G.B. (1981) Uranium redox equilibria in interactions with chromium in molten silicates. *Lunar Planet. Sci.* **XII**, 940–942.
Schreiber, H.D., Merkel, R.C., Schreiber, V.L., and Balazs, G.B. (1987) Mutual interactions of redox couples via electron exchange in silicate melts: Models for geochemical melt systems. *J. Geophys. Res.*, **92**, 9233–9245.
Shannon, R.D. (1976) Revised effective ionic radii and systematic studies of interatomic distances in halides and chalcogenides. *Acta. Crystall.*, **A32**, 751–767.
Sharma, S.K., Matson, D.W., Philpotts, J.A. and Roush, T.L. (1984) Raman study of the structure of glasses along the join SiO_2–GeO_2. *J. Non. Crystal. Solids.*, **68**, 99–114.

Simpson, D.R. (1977) Aluminum phosphate variants of feldspar. *Amer. Mineral.*, **62**, 351–355.

Smets, B.M.J., and Krol, D.M. (1984) Group III ions in sodium silicate glass. Part 1. X-ray photoelectron spectroscopy study. *Phys. Chem. Glasses*, **25**, 113–118.

Stebbins, J.F. (1988) Effects of temperature and composition on silicate glass structure and dynamics: ^{29}Si NMR results. *J. Noncrystall. Solids*, **106**, 359–369.

Stolper, E. (1982) Water in silicate glasses: An infrared spectroscopic study. *Contrib. Mineral. Petrol.*, **81**, 1–17.

Stolper, E., Fine, G., Johnson, T. and Newman, S., (1987) Solubility of carbon dioxide in albitic melt. *Amer. Mineral.*, **72**, 1071–1085.

Stolper, E., and Holloway, J.R. (1988) Experimental determination of the solubility of carbon dioxide in molten basalt at low pressure. *Earth Planet. Sci. Lett.*, **87**, 397–408.

Sweet, J.R., and White, W.B. (1969) Study of sodium silicate glasses and liquids infrared spectroscopy. *Phys. Chem. Glasses*, **10**, 246–251.

Tardy, Y., and Vieillard, P. (1977) Relationships among Gibbs free energies of formation of phosphates, oxides and aqueous ions. *Contrib. Mineral. Petrol.*, **63**, 75–88.

Taylor, M., and Brown, G.E. Jr. (1979) Structure of mineral glasses: I-The feldspar glasses $NaAlSi_3O_8$, $KAlSi_3O_8$ and $CaAl_2Si_2O_8$. *Geochim. Cosmochim. Acta*, **43**, 61–75.

Tewhey, J.D., and Hess, P.C. (1979) Silicate immiscibility and thermodynamic mixing properties of liquids in the $CaO–SiO_2$ system. *Phys. Chem. Glasses*, **20**, 41–53.

Varshal, B.G. (1975) Liquation phenomena and the structure of glasses in three component aluminosilicate systems. *Soviet J. Glass Phys. Chem.*, **1**, 40–43.

Verweij, H. (1981) Raman study of arsenic-containing potassium silicate glasses. *J. Amer. Ceram. Soc.*, **64**, 493–498.

Virgo, D., Mysen, B.O., and Kushiro, I. (1979) Anionic constitution of silicate melts quenched at 1 atm from Raman spectroscopy. *Science*, **208**, 1371–1373.

Visser, W., and Koster van Groos, A.F. (1979a) Effect of pressure on liquid immiscibility in the system $K_2O–FeO–Al_2O_3–SiO_2–P_2O_5$. *Amer. J. Sci.*, **279**, 1160–1175.

Visser, W., and Koster van Groos, A.F. (1979b) Effects of P_2O_5 and TiO_2 on the liquid-liquid equilibria in the system $K_2O–FeO–Al_2O_3–SiO_2$. *Amer. J. Sci.*, **279**, 970–989.

Watson, B.E. (1976) Two liquid partition coefficients: Experimental data and geochemical implications. *Contrib. Mineral. Petrol.*, **56**, 119–134.

Watson, E.B. (1977a) Partitioning of manganese between forsterite and silicate liquid. *Geochim. Cosmochim. Acta*, **41**, 1363–1374.

Watson, E.B. (1979) Zircon saturation in felsic liquids: experimental data and applications to trace element geochemistry. *Contrib. Mineral. Petrol.*, **70**, 407–419.

Watson, E.B., and Harrison, T.M. (1983) Zircon saturation revisited: temperature and composition effects on a variety of crystal magma types. *Earth. Planet. Sci. Len.*, **64**, 295–304.

Waychunas, G.A., and Brown, G.E., Jr. (1984) Application of EXAFS and XANES spectrosocpy to problems in mineralogy and geochemistry. In: *EXAFS and Near Edge Structure III*, edited by K.O. Hodgson, et al. *Springer, Proceedings of Physics* **2**, Springer-Verlag, New York.

Wendlandt, R.F., and Harrison, W.J. (1979) Rare earth partitioning between immiscible carbonate and silicate liquids and CO_2 vapor: Results and implications for the formation of light rare-earth enriched rocks. *Contrib. Mineral. Petrol.*, **69**, 409–419.

Williams, Q., and Jeanloz, R. (1988) Spectroscopic evidence for pressure induced coordination changes in silicate glasses and melts. *Science*, **239**, 902–905.

Wong, J., and Angell, C.A. (1976) *Glass Structure By Spectroscopy.* Marcel Dekker, Inc., 864 pp.

Wood, M.I., and Hess, P.C. (1980) The structural role of Al_2O_3 and TiO_2 immiscible silicate liquids in the system SiO_2–MgO–CaO–FeO–TiO_2–Al_2O_3. *Contrib. Mineral. Petrol.*, **72**, 319–328.
Wyllie, P.J. (1980) The origin of kimberlite. *J. Geophys. Res.*, **85**, 6902–6910.
Yarker, C.A., Johnson, P.A.V., Wright, A.C., Wong, J., Greegor, R.B., Lytle, F.W., and Sinclare, R.N. (1986) Neutron diffraction and EXAFS evidence for TiO_5 units in vitreous K_2O–TiO_2–$2SiO_2$. *J. Noncrystall. Solids*, **79**, 117–136.
Yin, C.D., Okuno, M., Morikawa, H., and Maruma, F. (1983) Structure analysis of $MgSiO_3$ glass. *J. Noncrystall. Solids*, **55**, 131–141.
Yoshimaru, K., Veda, Y., Morinaga, K. and Yanagase, Y. (1984) Glass forming region and Ti^{+4} coordination number in R_2O–TiO_2 (Rb, K, Na) and BaO–TiO_2 binary glasses. *J. Ceram. Soc. Jap.*, **92**, 17–22.

Chapter 6
The Mobility of Mg, Ca, and Si in Diopside–Jadeite Liquids at High Pressures

Nobumichi Shimizu and Ikuo Kushiro

Introduction

Mobilities of elements in magmas have an important bearing on quantitative understanding of geochemical processes. Mixing of magmas, assimilation of xenoliths, contamination of magmas, and interaction between magma and wall-rocks are all related to diffusive transport of elements in magmas. Diffusion also plays an important role in processes of crystal growth from magma. Thus, the determination of element mobilities in magmas has an immediate geologic interest and its significance is to constrain time and/or spatial scale lengths of these processes.

Mobilities of elements in silicate liquids provoke more fundamental interest as well. How are mobilities of network-forming elements (e.g., Si and O) affected by network polymerization? Does network polymerization affect mobilities of network-modifying cations? What are diffusion species? Do diffusion species change as a function of network polymerization? How does pressure affect mobilities of network-forming and network-modifying cations? These are questions of fundamental importance since they relate element mobilities to the structure of silicate liquids.

Intrinsic mobilities (i.e., self-diffusion) of elements were experimentally determined in this study and an attempt is made to answer some of the questions raised above.

Interpretations of mobility data in relation to other "structural" observations are not necessarily straightforward. For instance, the viscosity of a silicate liquid can be related directly to the mobility of oxygen in terms of the Eyring equation. Similarities in activation energies of oxygen self-diffusion and viscous flow (e.g., Oishi et al., 1975, Dunn, 1982; Shimizu and Kushiro, 1984) lend support to this contention. However, it is well known that the Eyring equation fails to relate viscosity and cation mobilities (e.g., Watson, 1979), indicating that cation motion could be decoupled from the bulk transport phenomena.

However, the self-diffusion data obtained here are not directly applicable to geochemical processes listed above because of fundamentally different diffusion mechanisms. For instance, in diffusive mixing of magmas, the diffusion process occurs in the presence of chemical potential gradients (i.e., chemical diffusion), in which the flux of one component is generally dependent on concentration gradients of all the others, and thus mobilities of individual components are closely related to one another. If the volume of the system is maintained constant during mixing, a flow of components in one direction must be compensated by a counter-flow of other components in the opposite direction.

Even in a simple ternary oxide system such as $CaO–Al_2O_3–SiO_2$, the compositional diffusion path between two end-member compositions within the system is not a straight line, but a curve of S- or Z-shape (see for instance, Cooper, 1974; Oishi, 1965). On the other hand, self-diffusion occurs in the absence of chemical potential gradient through random walk-type processes, representing the intrinsic mobility of the element. The self-diffusion data are directly applicable in geochemical processes involving isotopic exchange and homogenization without bulk flow of material.

The diopside-jadeite system was chosen for the present study because it is a simple structural analog for natural magmas with varying chemistry. For instance, viscosity in the diopside liquid at 1 atm, 1400°C is approximately 10 poise, whereas the jadeite liquid at 1 atm, 1350°C has 6.3×10^4 poise (Scarfe et al., 1979). The range of viscosity covered in this system thus corresponds to that produced by variable SiO_2 contents in magmas ranging from olivine tholeiite to rhyolite. On the basis of the Raman spectroscopic data it is suggested that the diopside liquid has chain-like units of $(Si_2O_6)_n^{4n-}$ as its major structural construction (Mysen et al., 1982). Taylor and Brown (1979) concluded that the jadeite liquid has a three-dimensional network structure similar to stuffed tridymite or cristobalite with network-modifying Na ions occupying "channel" or "cavity" positions. SiO_2-rich magmas are considered to have a similar structure. It is thus suggested that the structural variation of liquids (i.e., polymerization of network tetrahedra) in this system corresponds to that of magmas with varying SiO_2 contents from basaltic to rhyolitic. Variations of intrinsic mobilities of network formers such as Si, Al, and O in this system could have direct relevance to one of the questions raised earlier.

Similarly, variations of intrinsic mobilities of network modifiers (e.g., Ca) can be studied as a function of network polymerization as well.

The diopside-jadeite system is interesting also in view of the pressure-induced viscosity change of liquids in relation to the diffusivity change of ions. The jadeite liquid shows a negative pressure dependence of viscosity (viscosity decreases with increasing pressure) at constant temperature (Kushiro, 1976), whereas the diopside liquid shows a positive pressure dependence (Scarfe et al., 1979). It was demonstrated by the ion dynamics simulations that the diffusivities of network-forming ions in the jadeite liquid (i.e., Si^{4+}, Al^{3+}, O^{2-}) increase with increasing pressure to about 200 kbar at 6000K (Angell et al., 1982). This increase was confirmed by direct measurements of diffusion of oxygen in the jadeite liquid at 5–20 kbar (Shimizu and Kushiro, 1984) and interdiffusion of Si^{4+}–Ge^{4+}, and Al^{3+}–Ga^{3+} in $NaAl(Si, Ge)_2O_6$ and $Na(Al, Ga)Si_2O_6$ liquids at 6–20 kbar (Kushiro, 1983). On the other hand, the diffusivity of oxygen in the diopside liquid was shown to decrease with increasing pressure by direct measurements at 10–17 kbar (Shimizu and Kushiro, 1984). The ion dynamics simulations also showed a decrease of diffusivity of oxygen as well as silicon with increasing pressure at 4000 and 6000 K (Angell et al., 1987). It is evident from these studies that the viscosity and the diffusivities of network-forming ions are inversely related and that the jadeite and diopside liquids have opposite signs in the pressure-induced changes of viscosity and diffusivities of network-forming ions. It is thus quite interesting and worthwhile to determine where in the system the change of the sign from one to the other takes place. Also, the diffusivities of silicon as well as other cations in both the jadeite and diopside liquids have not been determined by direct measurements. The ion dynamics simulations were carried out at excessively high temperatures (i.e., 6000 and 4000 K) and the calculated diffusivities may include considerable uncertainties. It is, therefore, desirable to determine the diffusivities of cations in these liquids by direct measurements at temperatures closer to natural magmas.

Experimental

The experimental procedures used in this study are similar to those of Shimizu and Kushiro (1984). For a given chemical composition, a diffusion couple was made of powdered glass of identical chemistry with contrasting isotopic compositions. The "heavy" glass was made using mixtures of oxides (MgO and SiO_2) and carbonate ($CaCO_3$) with enriched isotopes. The ^{25}Mg (with abundance of 97.87 atomic %), ^{30}Si (83.95%) and ^{44}Ca (98.68%) were mixed with those of normal isotopic compositions to obtain $^{25}Mg/^{24}Mg$, $^{30}Si/^{28}Si$ and $^{44}Ca/^{40}Ca$ ratios significantly different from the normal values: $^{25}Mg/^{24}Mg = 0.380$ (as opposed to 0.127 for the natural abundance ratio), $^{30}Si/^{28}Si = 0.245$ (0.033), and $^{44}Ca/^{40}Ca = 0.280$ (0.022). The heavy glass powder was first packed tightly into the lower-half portion of a Pt capsule and the "normal" glass powder, made from oxides and carbonates with normal isotopic compositions, was then loaded. The sealed Pt capsule was then placed in a graphite heater with tapered edges, with AlSiMag powder as a filler. After a diffusion run, the capsule was sliced in half longitudinally, mounted in epoxy and polished to 0.3-micron with alpha alumina abrasives.

A Cameca IMS 3f ion-microprobe was used to obtain isotope ratios as a function of position along longitudinal transects. A beam of negatively charged oxygen ions with a current of approximately 0.2 nA was focussed to a spot of approximately 10 microns in diameter. Positively charged secondary ions of Mg, Si and Ca were detected by a 17-stage Allen type electron multiplier and counted by a TTL counting circuit. A moderate offset (−40v) was applied to the secondary ion acceleration potential to suppress interference of $^{24}Mg^{16}O$ on ^{40}Ca (energy filtering; Shimizu and Hart, 1982a) and to reduce the effect of isotopic fractionation (Shimizu and Hart, 1982b). Typically, the isotope ratios were measured with precisions ranging from 0.3 to 0.5% (2 sigma).

Based on the concentrations of the tracer isotopes (i.e., ^{25}Mg, ^{30}Si, and ^{40}Ca) calculated from the measured isotope ratios as a function of position, self-diffusion coefficients for individual elements were calculated using a diffusion model with a composite semi-infinite media (Crank, 1975). The solution to the diffusion equation is:

$$C = \frac{C_1 + C_2}{2} + \frac{C_1 - C_2}{2} \operatorname{erf}\left(\frac{|x|}{2\sqrt{Dt}}\right) \tag{1}$$

where C denotes the concentration of the tracer isotope at position x (measured from the initial interface), C_1 and C_2 are the initial concentrations in the right ($x > 0$) and the left ($x < 0$) halves respectively, D is the diffusion coefficient, and t the time in seconds. The diffusion coefficient D was calculated from the slope of a straight line given by a linearized version of Eq. (1),

$$y \equiv \operatorname{erf}^{-1}\left[\frac{2C - (C_1 + C_2)}{C_1 - C_2}\right] = \frac{|x|}{2\sqrt{Dt}} \tag{2}$$

Because x is a distance measured from the original interface and the original interface was optically invisible, its position was estimated as the position corresponding to $C = (C_1 + C_2)/2$. The validity of this approach was borne out by the fact that the locations of the initial interface independently estimated for Mg, Ca, and Si agreed with each other within the spacing of spots (less than 50 microns) in all samples studied (see also Fig. 1). The least squares method of York (1969) was used to calucate the diffusion coefficients using Eq. (2) with a conservative blanket error of 2% on both x and y. As shown in Table 1, the diffusion coefficients were determined with a standard error ranging from 2.1% to 16% for Mg, 1.2% to 16% for Ca and 3.3% to 14% for Si. Table 1 also contains results of time study, i.e., the diffusion coefficients were determined for a given composition at a given set of P and T with different run durations. For instance, it is demonstrated that 1/5 hours were sufficient for Di100 composition at 14 kbar 1650°C. Absence of convection was ascertained simply by the fact that three parallel longitudinal transects in a sample (Di100 run at 10 kbar, 1650°C) yielded identical diffusion profiles for all elements concerned.

Results and Comments

The self-diffusion coefficient was determined for Ca, Mg, and Si for diopside (Di100), jadeite (Di0), Di60Jd40, Di40Jd60 and Di20Jd80 compositions at pressures ranging from 5 to 20 kbar and temperatures from 1400 to 1650°C. The results are summarized in Table 1.

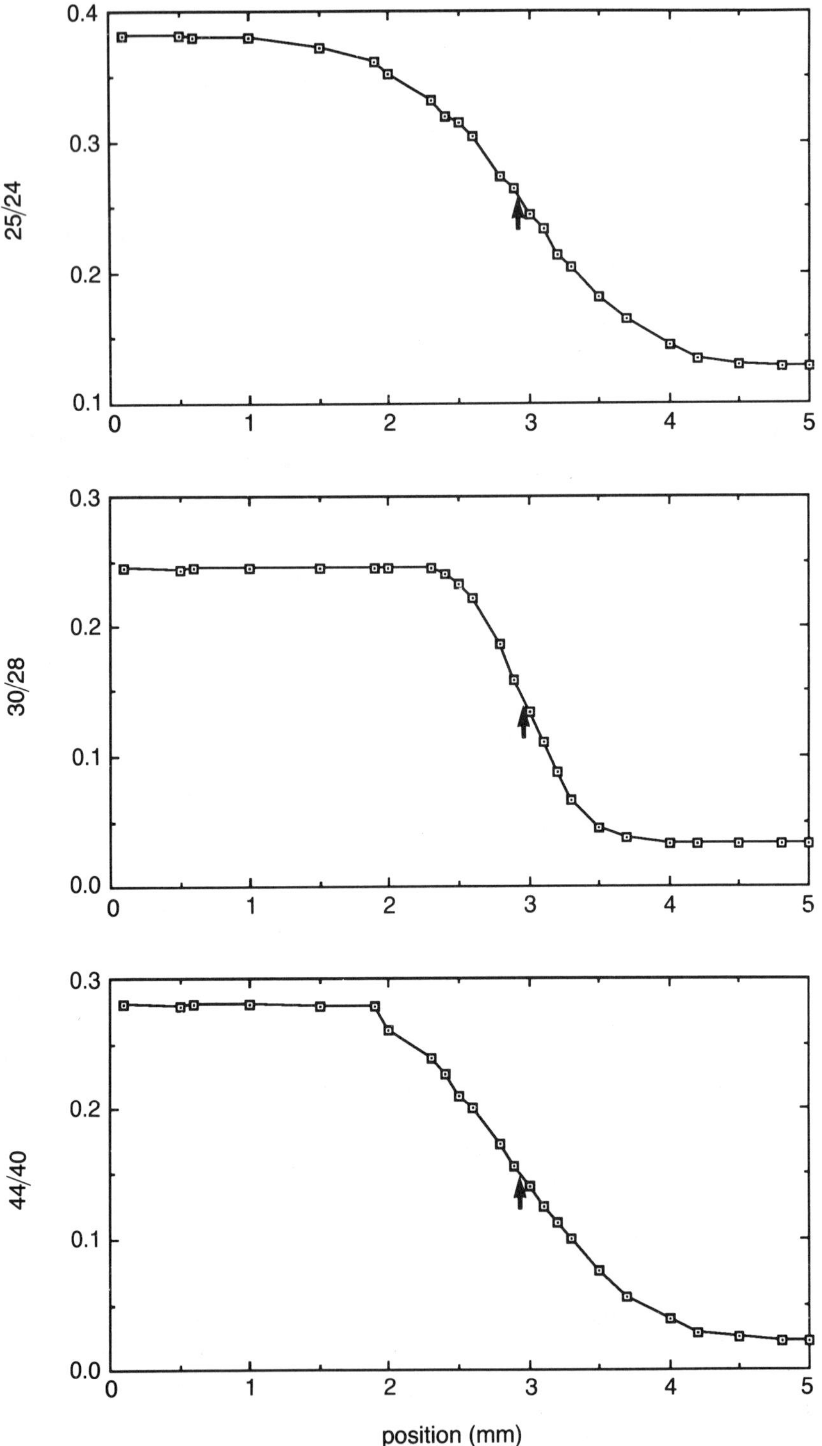
25/24
0.4
0.3
0.2
0.1
0
1
2
3
4
5
30/28
0.3
0.2
0.1
0.0
0
1
2
3
4
5
44/40
0.3
0.2
0.1
0.0
0
1
2
3
4
5
position (mm)

Table 1. Experimental results of diffusion coefficient measurements

Comp.	*P* (kb)	*T*(°C)	t (sec)	D(cm²/sec) Mg(+/−)	Ca(+/−)	Si(+/−)
Di100	5	1650	1200	4.60(.46/.36)E−6	4.83(.21/.19)E−6	7.76(.81/.62)E−7
	10	1650	720	4.46(.60/.43)E−6	3.79(.62/.42)E−6	8.24(.94/.70)E−7
	10	1650	1500	4.74(.59/.50)E−6	3.94(.19/.17)E−6	7.91(1.1/.92)E−7
	10	1600	1500	3.29(.07/.07)E−6	2.77(.36/.30)E−6	5.18(.45/.40)E−7
	10	1550	1200	2.96(.38/.28)E−6	2.37(.10/.09)E−6	4.49(.28/.23)E−7
	10	1550	3600	2.61(.41/.24)E−6	2.38(.26/.19)E−6	4.22(.44/.39)E−7
	14	1650	720	3.97(.12/.11)E−6	3.08(.28/.23)E−6	7.44(.34/.30)E−7
	14	1650	2700	3.99(.39/.33)E−6	3.48(.46/.39)E−6	7.74(.66/.49)E−7
	17	1650	1500	3.34(.39/.29)E−6	2.57(.10/.09)E−6	6.64(.38/.33)E−7
Di60	10	1550	3600	1.57(.07/.07)E−6	1.74(.11/.11)E−6	1.67(.05/.05)E−7
	10	1650	3600	3.30(.14/.14)E−6	3.14(.14/.13)E−6	3.37(.11/.12)E−7
	20	1550	3600	1.38(.12/.10)E−6	1.39(.09/.08)E−6	1.38(.09/.09)E−7
	20	1650	3600	2.63(.14/.13)E−6	2.25(.15/.15)E−6	3.31(.15/.14)E−7
Di40	10	1550	3600	8.87(.28/.28)E−7	1.30(.05/.05)E−6	8.92(.40/.37)E−8
	20	1550	3600	8.35(.45/.42)E−7	9.32(.50/.45)E−7	7.64(.33/.30)E−8
Di20	10	1400	7200	1.19(.10/.10)E−7	2.21(.34/.29)E−7	3.62(.38/.33)E−9
	10	1550	3600	4.12(.27/.24)E−7	9.49(.11/.11)E−7	2.06(.11/.11)E−8
	10	1650	3000	8.52(.60/.54)E−7	1.91(.16/.14)E−6	4.92(.33/.30)E−8
	20	1550	3600	4.34(.60/.50)E−7	7.97(.61/.55)E−7	2.80(.19/.18)E−8
	20	1650	2400	1.23(.07/.07)E−6	1.79(.08/.07)E−7	7.98(.68/.59)E−8
Di0	6	1400	7200	—	—	1.74(.29/.19)E−10
	9.5	1400	7200	—	—	2.74(.40/.27)E−10
	10	1550	5700	—	—	2.18(.12/.10)E−9
	16	1400	7200	—	—	5.23(.56/.40)E−10
	16	1550	6000	—	—	3.23(.35/.32)E−9
	16	1650	3600	—	—	8.52(.89/.77)E−9
	20	1400	7200	—	—	8.97(.74/.60)E−10
	20	1550				5.26E−9*

*Calculated from the value at 20kb, 1400°C, and an activation energy of 72 kcal/mol.

◁

Fig. 1. Isotope ratios of Mg(25/24), Si(30/28), and Ca(44/40) measured as a function of position in typical diffusion profiles obtained for the diopside liquid run at 14 kbar and 1650°C. Arrows indicate locations of initial interface determined as x corresponding to $C = (C_1 + C_2)/2$. Note that the locations determined independently for three elements agree within the spacing of analyzed spots. Note also that isotope ratios at both ends of the charge demonstrate that a diffusion model of a composite of two semi-infinite media is valid for all elements. A precision of 0.5% results in error bars smaller than the size of the symbol.

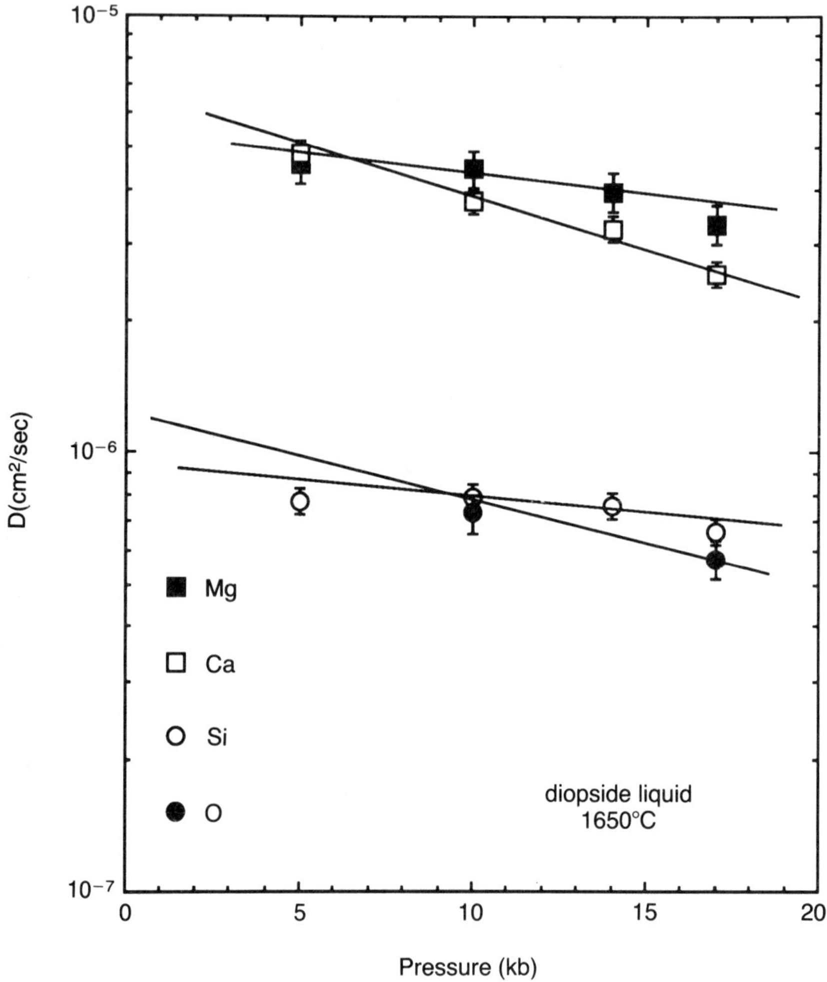

Fig. 2. Effect of pressure on diffusivities of Mg, Ca, Si, and O in the diopside liquid at 1650°C. Data for oxygen are from Shimizu and Kushiro (1984).

(1) The Diopside Liquid (Di100)

Combined with those of Shimizu and Kushiro (1984), the present results show that mobilities of Ca and Mg are generally 5 to 6 times greater than those of Si and O in the diopside liquid. Fig. 2 illustrates that diffusivity of these elements decreases with increasing pressure and that the pressure effect appears to be smallest on Si, the smallest ion, and largest on Ca, the largest ion. Although the physical meaning of the pressure derivative of isothermal diffusivity (i.e., activation volume) is unclear (for instance, see Watson, 1979), the observed parallelism between the magnitude of the pressure effect and ionic size may indicate that the pressure derivative could in some way be related to the size of the diffusing species, particularly if individual ions are diffusing species in the diopside liquid. It has been argued that the single oxygen ion (O^{2-}) is the dominant diffusing species in diverse silicate liquids, including the diopside liquid (e.g., Dunn, 1982; Shimizu and Kushiro, 1984; Dingwell, 1989). The decrease in diffusivity with increasing pressure coincides with an increase in viscosity (Scarfe et al., 1979). Extrapolating trends in Fig. 2, it is concluded that at atmospheric pressure, $D_{Ca} > D_{Mg} > D_O > D_{Si}$. Oishi et al. (1975) reported $D_O > D_{Si}$ in liquids in the system $CaO-Al_2O_3-SiO_2$. This sequence is the same as that obtained by the ion dynamics simulations on the diopside liquid at 2.6 kbar at 6000 K (Angell et al.,

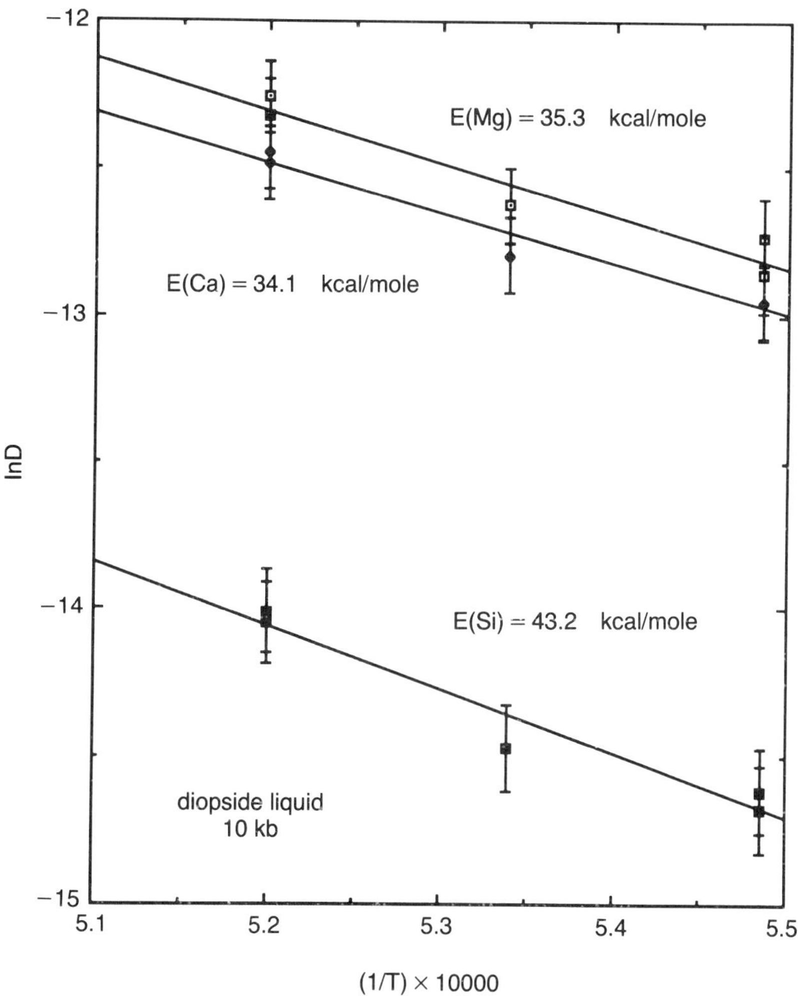

Fig. 3. Arrhenius diagram showing the effect of temperature on diffusivities of Mg, Ca, and Si in the diopside liquid at 10 kbar. Activation energies calculated from the slope are also shown.

1987). However, the effect of pressure on D_{Mg} is not consistent; D_{Mg} increases to about 200 kbar in the simulations, whereas it decreases in the present experiments.

The effect of temperature on diffusivity at 10 kbar is shown in Fig. 3. The Arrhenius lines for Mg, Ca, and Si are subparallel with activation energies ranging from 34 kcal/mol for Ca to 43 kcal/mol for Si. Watson (1979) reported an activation energy value for Ca (34 kcal/mol) at 10 kbar, identical to the present result in a Na–Ca aluminosilicate liquid. In addition, the activation energy for viscous flow of the diopside liquid at 1 atm is 37 kcal/mol (Scarfe et al., 1979), similar to the values obtained above.

(2) The Jadeite Liquid (Di0)

Fig. 4 shows the increase in diffusivity of both Si and O (data from Shimizu and Kushiro, 1984), with increasing pressure. This increase coincides with a decrease in viscosity (Kushiro, 1976). The present results are qualitatively consistent with those of Angell et al. (1982), who reported, on the basis of the ion dynamics simulation, that diffusivities of Si and O (together with Al) in the jadeite liquid increase with pressures up to 200 kbar at 6000 K. Fig. 4 also shows that the pressure effect is much greater on Si than on O, and that the magnitude of the pressure dependence of Si diffusion is similar between 1400 and 1550°C.

The pressure-induced increase in the mobility of network forming elements is coupled with a decrease in viscosity and represents a major characteristic of silicate liquids with an open network structure (e.g., Angell et al., 1982). Although it appears "anomalous" in terms of a notion that pressure reduces interatomic distances, thereby reducing the mobility of atoms, the pressure-induced decrease in viscosity occurs commonly in silicate liquids of geologic interest, including natural basalts and andesites (Scarfe et al., 1979).

The magnitude of pressure effects on diffusion is generally related to the molar volume of diffusing ions. For instance, Watson (1979) pointed out that the activation volume of Ca self-diffusion in a Na-Ca aluminosilicate liquid is close to the molar volume of Ca^{2+} ion. Angell et al. (1987) showed that the ion dynamics computer simulation produced a pressure dependence of Ca diffusion in the diopside liquid consistent with the notion that Ca^{2+} is the diffusing ion species. As noted earlier, the present result on the diopside liquid extends this notion to other major element constituents.

However, the physical significance of pressure effect is particularly unclear when the activation volume is negative as in Si and O diffusion in the jadeite liquid. Shimizu and Kushiro (1984) indicated that the absolute value of the activation volume for O diffusion in the jadeite liquid at 1400°C is close to the molar volume of O^{2-}. The present results yield an activation volume of Si diffusion as 16cc/mol at 1400°C and 13 cc/mol at 1550°C, again comparable to 11 cc/mol for the molar volume of $SiO_4{}^{4-}$, based on the ionic radius of oxygen of 1.4 Å and the O–O distance of 2.56 Å (Soules, 1979).

Fig. 5 compares the effect of temperature on oxygen and silicon diffusion in

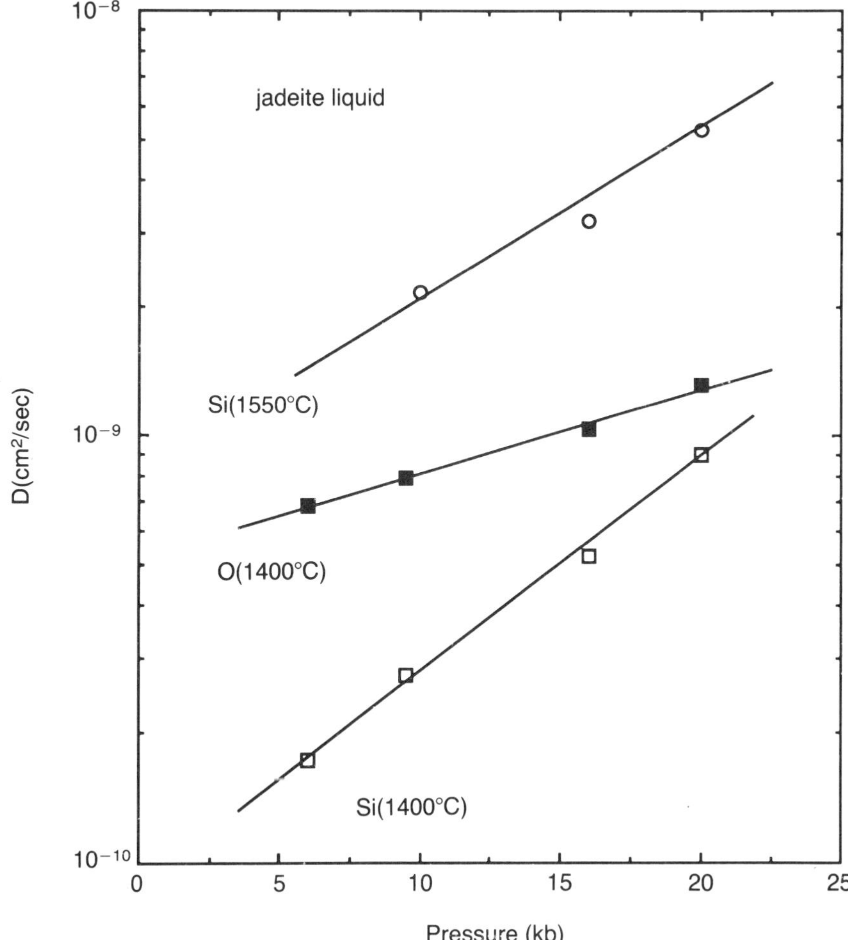

Fig. 4. Effect of pressure on diffusivities of O and Si in the jadeite liquid. Note subparallel pressure effects on Si diffusion at 1400°C (open squares) and 1550°C (open circles). Also note that the pressure effect is greater for Si than for O. Data for oxygen are from Shimizu and Kushiro (1984).

the jadeite liquid. As shown previously (Shimizu and Kushiro, 1984), the activation energies of viscous flow (67.3 ± 2.4 kcal/mol) and of oxygen diffusion (61 ± 2 kcal/mol) at 15 kbar, similar to that of silicon diffusion (72 kcal/mol) at 16 kbar.

(3) Intermediate Compositions (Di60, Di40 and Di20)

It is clear from Table 1 that mobilities of elements decrease as the composition varies from diopside to jadeite. Since the degree of polymerization increases

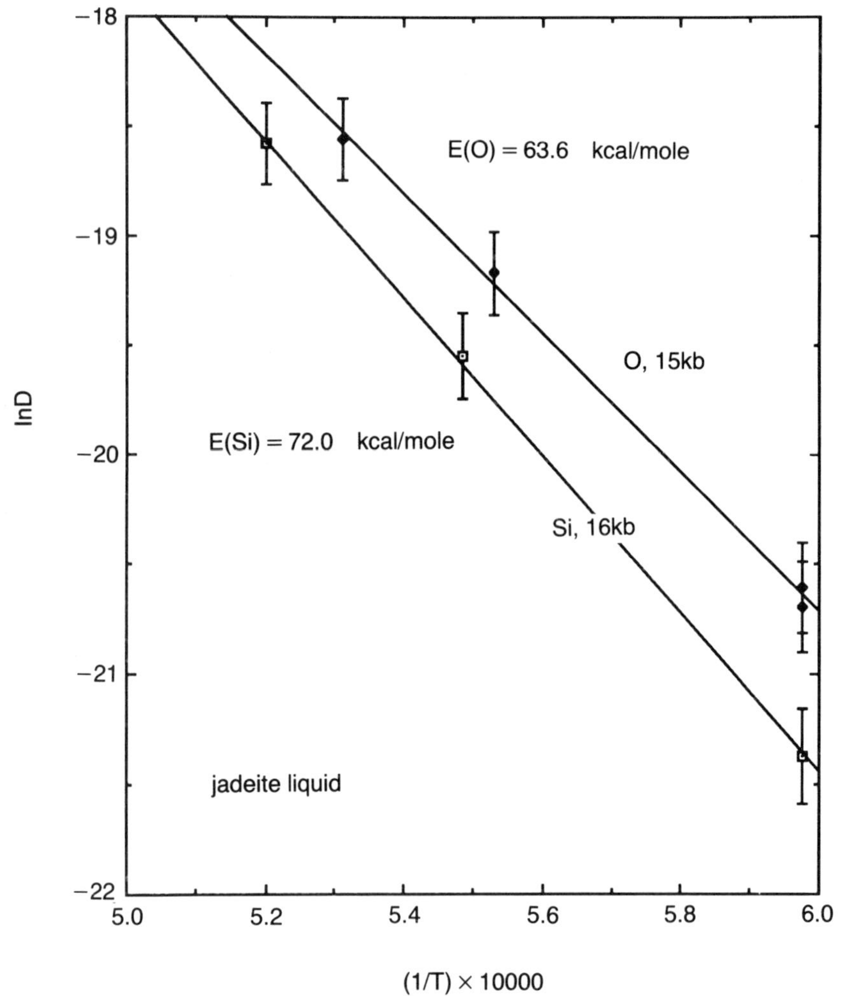

Fig. 5. Arrhenius diagram showing the effect of temperature on diffusivities of O (at 15 kbar) and Si (at 16 kbar) in the jadeite liquid. Data for oxygen are from Shimizu and Kushiro (1984).

toward jadeite, the present results are understood as a decrease in element mobility with increasing network polymerization. The result is consistent with the conclusion based on comparison of diffusivities of various elements between basic and silicic liquids (e.g., Henderson et al., 1985).

The composition-dependent variation in mobilities can be examined in more detail based on the isothermal (1550°C) diffusion data at 10 and 20 kbar. Fig. 6 shows that at 10 kbar, Si diffusion is most (by a factor of 200 between end-members) and Ca diffusion is least dependent (by a factor of 2 from Di100 to

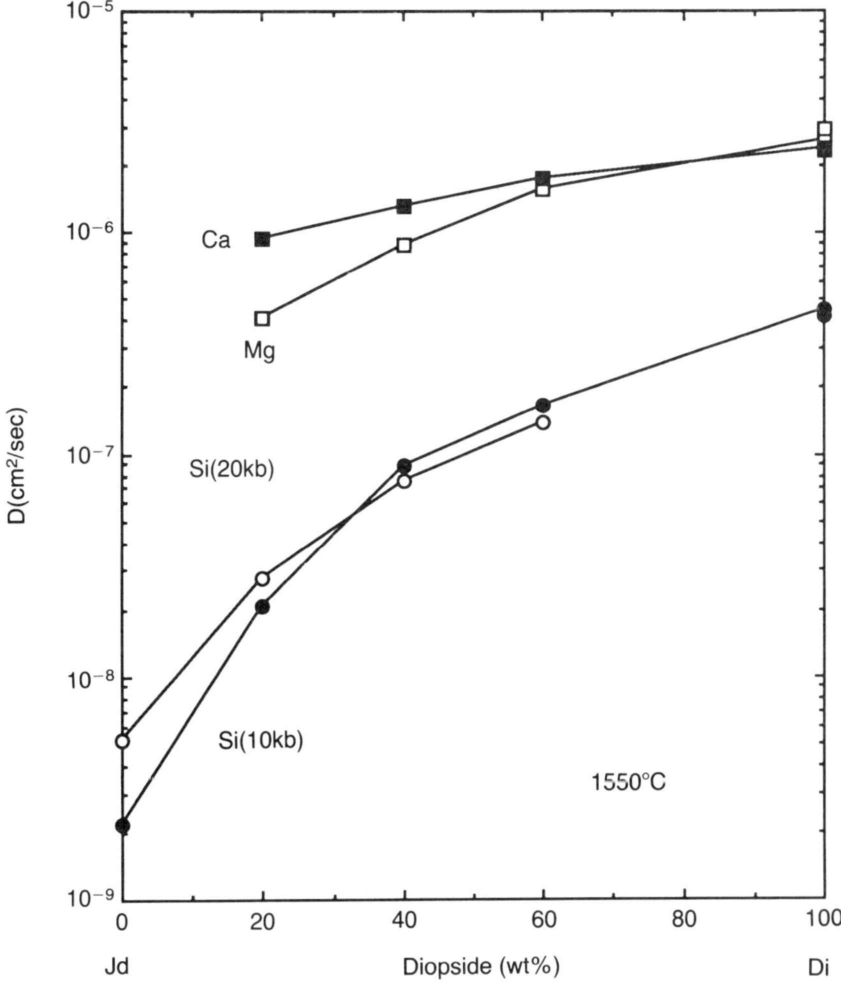

Fig. 6. Diffusivities of Mg, Ca and Si as a function of composition in the diopside–jadeite system at 1550°C. Data for Si at 20 kbar are also shown to illustrate an inversion of the pressure dependence around Di30Jd70 composition. Note that Si diffusivities for the intermediate compositions are greater than those expected by a linear variation between the Di and Jd liquids. Also note that significant difference exists in behavior of Mg and Ca diffusivities as a function of composition.

Di20) on composition. As a result, the total range of diffusivities among elements increases toward jadeite composition with increasing network polymerization from a factor of 5 for Di100 to a factor of 50 for Di20. Henderson et al. (1985) noted a much larger increase in diffusivity ranges from basalt to pitchstone and pantellerite liquids.

The variation of Mg diffusivity as a function of composition is noticeably different from that the Ca. For range from Di100 to Di60, Mg diffusivity decreases slightly more than that of Ca, but further toward jadeite composition it decreases much more strongly than Ca, being almost parallel to the decrease of Si diffusivity. This finding is in contrast to a common notion that Ca and Mg are network modifiers occupying network channel (or cavity) position.

As diffusivities decrease toward jadeite, activation energies increase. At 10 kbar, the activation energy for Ca diffusion increases from 34 kcal/mol at Di100 and 41 kcal/mol at Di60 to 56 kcal/mol at Di20. For Si diffusion the energy increases from 43 kcal/mol at Di100 and 50 kcal/mol at Di60 to 67 kcal/mol at Di20. In contrast, virtually no change was observed for the activation energy of Mg diffusion at around 50 kcal/mol. Fig. 7 shows the Arrhenius relationships for Ca, Mg, and Si diffusion for the Di20 composition at 10 kbar. At 20 kbar, it appears that the effect of composition on the activation energy is enhanced; "2-point Arrhenius lines" indicate increases in activation energies from 45 to 73 kcal/mol for Mg, 34 to 57 kcal/mol for Ca and 62 to 74 kcal/mol for Si as composition varies from Di60 to Di20.

It was noted earlier that the effect of pressure on diffusivities of Si and O has opposite signs for Di and Jd liquids. The Si diffusion data at 1550°C in Fig. 6 show that an inversion of the pressure effect occurs approximately at Di30Jd70 composition. In an analogous system, diopside-albite, Brearley et al. (1986) reported occurrence of a similar inversion in the pressure dependence of viscosity between Ab25Di75 and Ab50Di50 compositions. Furthermore, Kushiro (1981) also showed the existence of an inversion for the pressure dependence of viscosity at anorthite composition in the system $CaAl_2O_4$–SiO_2. Although it is likely that these inversions (viscosity and diffusion of network-forming elements) are related to a critical change in the liquid structure, the exact nature of the change is difficult to define (see discussion below).

It is important to note that an inversion also occurs for Mg diffusion; at 1650°C, Mg diffusivity for Di20 *increases* with pressure from 8.52×10^{-7} cm^2/sec at 10 kbar to 1.23×10^{-6} cm^2/sec at 20 kbar, whereas for Di60 it *decreases* from 3.30×10^{-6} cm^2/sec at 10 kbar to 2.63×10^{-6} cm^2/sec at 20 kbar.

The composition dependence of Si diffusion (Fig. 6) shows a positive deviation from simple additivity, i.e., Si diffusivity for intermediate compositions are greater than those predicted by a linear additivity. This dependence is analogous to the mixed alkali effect (Isard, 1969; Day 1976), composition dependence of viscosity in alkali silicate liquids that can be interpreted based on the configurational entropy theory of Adam and Gibbs (1965). The presence of two different alkali elements increases the number of configurations of "structural units", thereby increasing the configurational entropy of the system. This change results in an increase of probability of cooperative rearrangements of subsystems, and hence viscosity decreases. Furthermore, the magnitude of this effect is expected to increase from one endmember toward the middle of the binary composition and then decrease toward the other endmember. In fact, Richet (1984) reported that a negative deviation of viscosity in liquids in the system $K_2Si_3O_7$–$Na_2Si_3O_7$ at a given temperature is caused by contributions of the configurational entropy

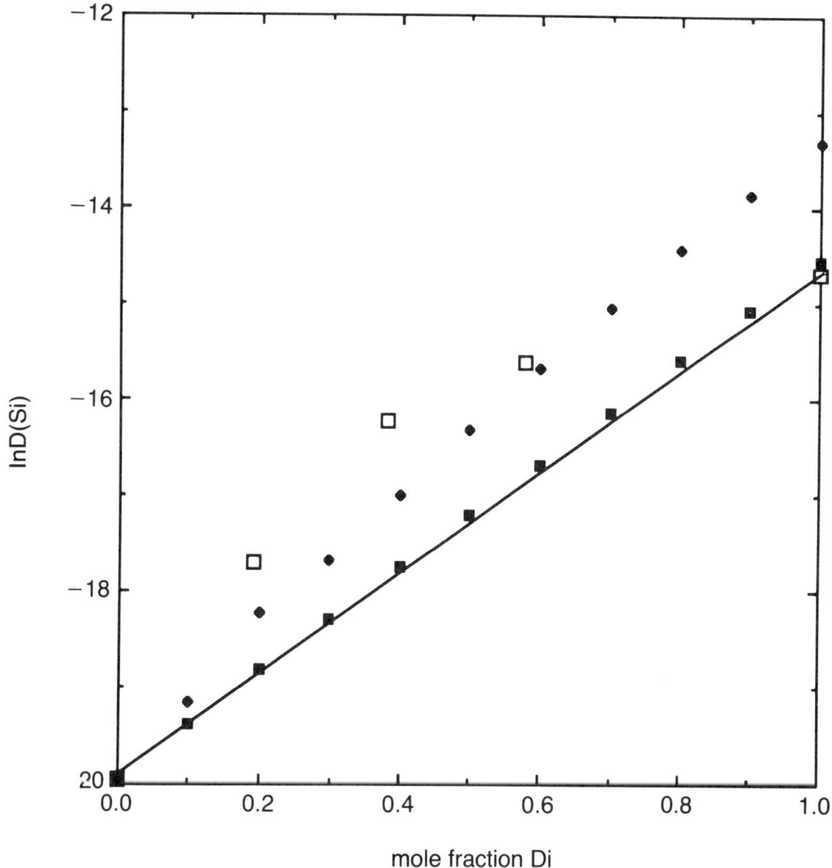

Fig. 7. Effect of composition on Si diffusivity as predicted by equation (3) using multicomponent configuration of anion species given in Table 2 as a correction term. Small squares denote values with $A = 0.1$, approximating a simple linear relationship. Solid diamonds denote values with $A = 1.0$, showing increasingly large deviations from the linearity toward Di, and a poor fit to the observed values (large open squares).

to the entropy of mixing. Indeed calculated viscosities for intermediate composition involving this term mimic the measured viscosities well.

Brearley et al. (1986) reported a similar observation on the composition dependence of viscosity in the system diopside-albite.

Interpretation of the composition dependence of Si diffusivity in the present system is complicated. Virgo et al. (1979) suggest that liquids in the diopside-forsterite-nepheline-silica system have a domain with three-dimensional network similar to the structure of liquids in the nepheline-silica system, and a domain with more depolymerized network. A similar domain structure could be expected of liquids in the present system, and proportions of domains change as composition changes from diopside toward jadeite. A tentative interpretation of the observed composition dependence of Si diffusion will be presented below.

In summary, the present results provide partial and provisional answers to some of the questions raised earlier. As to the effect of network polymerization on diffusivity of network elements, it is concluded that polymerization not only reduces mobility of network elements significantly, but also increases the activation energy. Also noticeable is that signs of the pressure dependence of Si diffusion are opposite between polymerized and depolymerized liquids, with the former showing an "anomalous" pressure induced enhancement of Si mobility.

The effect of network polymerization on diffusion of a network modifier (Ca) is small as shown in Fig. 6. Henderson et al. (1985) show that except for Li and Na, Sr diffusion is least affected by the liquid chemistry/structure, indicating consistency with the present results if Ca and Sr have similar structural roles. The results on Mg diffusion are, however, unexpected and a speculative interpretation will be presented below.

Interpretations of Results in Relation to the Structure of Liquids

Some of the questions raised at the beginning of this chapter concern relating mobility of elements to the structure of liquids. The purpose of this section is to examine whether structural parameters of liquids in the present system explain salient features of this study. Structural parameters are abundances of network-forming anion species and network-modifying cations assigned to individual species to satisfy charge neutrality as calculated from the chemical composition. The concept behind calculations is aptly summarized in Mysen's (1988) book and is based on the massive recent effort on diverse silicate systems using spectroscopic techniques including Raman and NMR. The reader is referred to individual contributions given in Mysen (1988).

A particular interest is in the composition dependence of Si mobility at constant P (10 kbar) and T (1550°). It is clear from Fig. 6 that Si mobility decreases from Di100 to Di0 with "increasing polymerization." A salient feature, however, is, as noted earlier, that $\ln D_{Si}$ displays deviation from linearity, which is greater toward the middle of the system. Assuming that viscous flow is intimately associated with network motion, it is noticeable that there is a strong resemblance between the present results and those of Brearley et al. (1986) on viscosity variation in the system diopside-albite. These authors showed that viscosity for intermediate compositions was lower than that expected from linearity.

Table 2 contains abundances of network anion species computed by the Mysen (1988) method, which are used as structural parameters, and shows variations of anion species as a function of composition in the present system. As the composition varies from diopside to jadeite, three-dimensional network

Table 2. Mole fractions of network-forming anion species as a function of composition expressed as mol% diopside

comp.	3-D network (TO_2)	Monomers (TO_4)	Dimers (T_2O_7)	Chains (TO_3)	Sheets (T_2O_5)	$\sum x_i \ln x_i$
Di100	0.266	0.192	0.156	0.379	0.007	−1.361
90	0.349	0.150	0.128	0.358	0.015	−1.346
80	0.433	0.118	0.104	0.324	0.020	−1.293
70	0.516	0.095	0.083	0.283	0.023	−1.216
60	0.597	0.077	0.064	0.239	0.023	−1.110
50	0.674	0.062	0.046	0.195	0.023	−0.986
40	0.747	0.050	0.030	0.153	0.021	−0.841
30	0.815	0.039	0.015	0.114	0.017	−0.673
20	0.880	0.028	0.003	0.077	0.013	−0.669
10	0.944	0.016	0.000	0.034	0.006	−0.266
0	1.000	0.000	0.000	0.000	0.000	0.000

Abundances of anion species were provided by Dr B.O. Mysen, Carnegie Inst. Washington.

species (TO_2) increase whereas abundances of chains (TO_3), monomers (TO_4) and dimers (T_2O_7) decrease. Abundances of sheets (T_2O_5) remain small (up to 2 mole %) increasing from diopside to the middle of the system, then declining toward jadeite.

The fundamental difficulty of relating Si mobility to the structure however, is that the degree of polymerization, as represented by the abundance of the three-dimensional network species, increases only linearly with increasing jadeite component and the marked non-linear behavior cannot be explained by the abundance of this species alone. In fact, only a mediocre relationship with a correlation coefficient of $r^2 = 0.82$ was obtained between $\ln D_{Si}$ and TO_2. Relationships with other anion species were also found to be mediocre.

An attempt was then made to compute contributions of the configurational entropy to the non-linear behavior by analogy to viscosity variations in binary silicate systems (e.g., Richet, 1984; Brearley et al., 1986). Based on the abundances of anion species for a given composition, a quantity, $x \ln x$ was computed for each species and the sum for all species was obtained. It was found that the function $\sum x_i \ln x_i$ showed a monotonous increase from jadeite toward diopside (Table 2). Therefore, if this function is used as a "correction" term added to the linearity function, such that:

$$\ln D_{Si} = x \ln D_{Si}(\mathrm{Di100}) + (1 - x) \ln D_{Si}(\mathrm{Di0}) - A \sum x_i \ln x_i \qquad (3)$$

where A is a constant and x is mole fraction of diopside component, Si mobility in the present system varies as a function of composition as shown in Fig. 7. In general, departure from the binary linearity increases as $\sum x_i \ln x_i$ increases toward diopside (Table 2), and the $\ln D$ curve approaches a simple binary linearity when A is sufficiently small. For finite A, it is clear that the above equation produces only a very poor fit to the observed values (Fig. 7). It is concluded that configuration of the anion species as calculated in Table 2 does not explain the observed non-linearity.

In contrast to the failure of the attempt above, the observed Si mobility (D_{Si}) for the intermediate compositions for 10 kbar and 1550° is adequately calculated by a formula:

$$\ln D_{Si} = x \ln D_{Si}(Di100) + (1 - x) \ln D_{Si}(Di0) - B[x \ln x + (1 - x) \ln (1 - x)] \quad (4)$$

where x is mole fraction of diopside component, and $B = 2.0$. Note that this formula differs from equation (3) only in that the "correction" term (the third term) has a form used in explaining viscosity variation in a binary alkali silicate system based on the configurational entropy theory (Richet, 1984). Fig. 8 com-

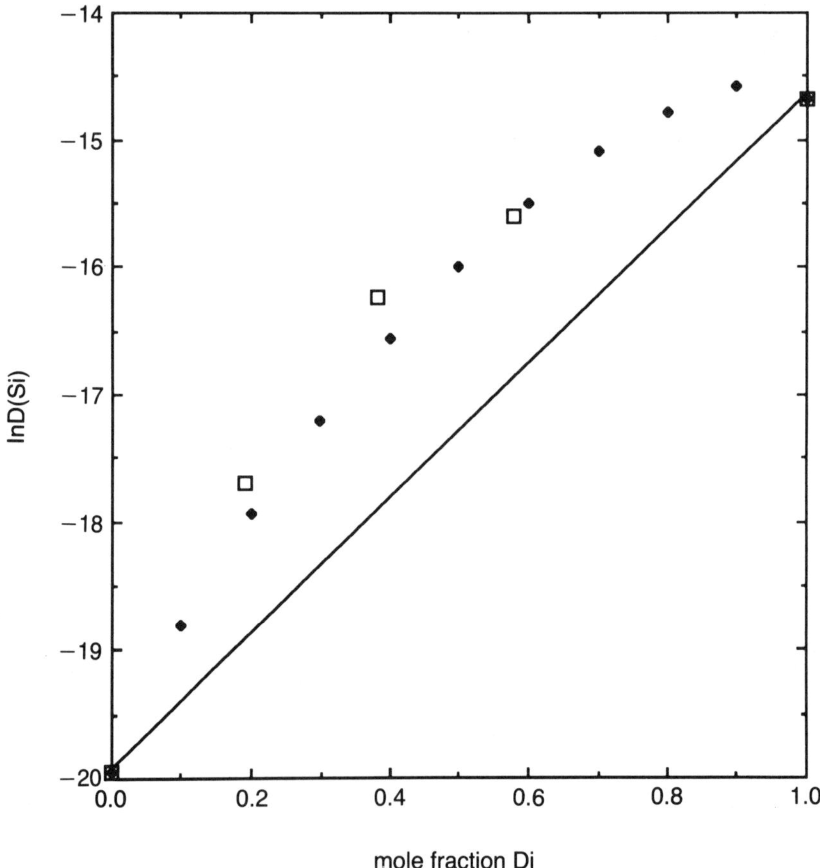

Fig. 8. Effect of composition on Si diffusivity as predicted by equation (4) using the binary configuration as a correction term. Diamonds denote calculated values with $B = 2.0$ (approximately equal to the gas constant of 1.98). Open squares are the observed values.

pares Si mobility computed by equation (4) with those actually measured. The results suggest that the observed behavior of Si mobility is better described as a binary system with a "mixed alkali effect"-like correction term, rather than as a system with multiple anion species to form a similar mixing term. It is noteworthy that viscosity variations in the system diopside-albite is also well described by a similar binary treatment (Brearley et al., 1986), although anion speciation of this system undoubtedly varies in a complex multi-component fashion as in the diopside-jadeite system.

Thus, there is an apparent conflict between calculated structural parameters (i.e., abundances of anion species and assigned cations) and the observed behavior of Si mobility in the diopside-jadeite system. A similar discrepancy is indicated for the system diopside-albite in terms of viscosity.

This discrepancy seems to suggest that transport phenomena such as viscous flow and diffusion of network elements could be governed by ionic motions that occur in molecular units different from the anionic speciation. In fact, a way to reconcile this apparent discrepancy is through a notion that transport phenomena could be governed more importantly by a microscopic (or local) environment than by macroscopic, individual anion species. In other words, the apparently binary behavior of Si mobility does not indicate that the system is structurally binary; rather that Si diffusion occurs in subsystems, in the presnet example in two different subsystems with each representing local environment for end-member structure. Co-existence of these subsystems in the intermediate compositions increases the number of configurations of Si-bearing subsystems and thus enhances mobility of Si over the simple additivity. The subsystems are smaller in molecular unit than anion species, and their identity remains unchanged with varying abundances of anion species.

Another salient feature of the present study is an inversion of the pressure dependence of Si mobility occurring around Di30 composition. In liquids with higher Di content, Si mobility decreases with increasing pressure, whereas in those with lower Di content Si mobility increases with pressure. Based on the results of Shimizu and Kushiro (1984), an inversion is also expected for oxygen diffusion. Brearley et al. (1986) report a similar inversion for the pressure dependence of viscosity in the system diopside-albite and Kushiro's (1981) results show that an inversion also occurs for the pressure dependence of viscosity in the system $CaAl_2O_4$–SiO_2 around anorthite composition. Assuming again that viscous flow is intimately associated with network motion, the observation made here, and those on viscosity by other investigators, reflect critical differences in liquids on both sides of the inversion particularly in terms of their network structure. It is noticeable, however, that the inversion appears to be independent of NBO/T, a parameter generally considered as a measure of network polymerization. The inversion occurs at $NBO/T = 0.445$ in the diopside-jadeite system, ~ 1.2 in the diopside-albite, and 0 for the system $CaAl_2O_4$–SiO_2 (all liquids in this system have $NBO/T = 0$). It is suggested that the pressure dependence of viscous flow, and Si and O diffusion is determined by the nature of subsystems in which network motion takes place, rather than macroscopic anion speciation as indicated by NBO/T.

Raman spectroscopic studies of Sharma et al. (1979) and Mysen et al. (1980) show that the spectra of the jadeite liquid were essentially identical at various pressures up to 38 kbar, indicating that no constructive changes of the structure occurred. It follows that abundances of anion species for a given liquid would remain unchanged as a function of pressure, because the anion speciation is primarily based on deconvolution of Raman spectra. It is also noted that for Di20 composition, in which pressure-induced increase in Si diffusion occurs, activation energy of Si diffusion increases slightly from 67 kcal/mol at 10 kbar to 74 kcal/mol at 20 kbar. This increase would not be expected if weakening of T-O-T bond occurs with pressure and if it is reflected directly on activation energy of Si diffusion.

Causes and mechanisms for the pressure-induced increase in network element diffusion (or decrease in viscosity) are still unclear; the present results and arguments presented above indicate, however, that the transport phenomena are controlled by motion of ionic units much smaller than anion species recognizable by the Raman spectroscopy. Identification of the subsystems is difficult, but their small size may be implied by monatomic diffusion species inferred in the diopside liquid from the pressure dependence of diffusivities of O, Si, Ca, and Mg, diffusivities of O in the jadeite liquid and from a prevalent view of viscous flow being rate-controlled by oxygen self-diffusion (e.g., Oishi et al., 1975; Dunn, 1982; Shimizu and Kushiro, 1984; Dingwell and Webb, 1989).

It is much more difficult to relate Ca and Mg mobilities to the structure of liquids. Although a general conclusion is that cation mobilities decrease with increasing network polymerization, Mg mobility is more strongly affected by it than Ca mobility (see Fig. 6). In fact, the composition dependence of Mg mobility shows a significant resemblance to that of Si especially for compositions close to jadeite. Another feature worth reiterating here is that there is an inversion of the pressure dependence of Mg mobility between Di60 and Di20, very similar to that of Si. It was argued above that Si mobility of the present system appears to be controlled by the nature of molecular units smaller than anion species. It may then be argued that the observed similarities between Mg and Si mobilities indicate that Mg mobility is also independent of the anion speciation and of the assignment of Mg^{2+} to certain species for charge balancing.

Inversion of the pressure dependence of Mg mobility is unexpected and in strong contrast to the fact that Ca mobility decreases with increasing pressure irrespective of the composition. This inversion suggests firstly that Mg and Ca occupy significantly different "structural environment", and secondly that the environment for Mg resembles that of Si. With the present diffusion data alone, it is not possible to conclude that Mg has a tetrahedral coordination, but in this context, it is noteworthy that Angell et al. (1987) concluded that Mg is tetrahedrally coordinated in the diopside liquid up to pressures above 50 kbar at 4000K.

However, the similarity between Mg and Si does not appear to be universal. Our preliminary data on self-diffusion of Mg, Ca, Si, Ti, Cr and Zr in a synthetic basalt (SiO_2 = 52.5 wt%; Al_2O_3 = 16.5; TiO_2 = 1.5; MgO = 12.5; CaO = 11.5; Na_2O = 2.5; K_2O = 1.0; Cr_2O_3 = 1.0; ZrO_2 = 1.0) show that at 1550°C Mg mobility decreases from 8.5 kbar to 15 kbar pressure together with Ca and Cr, whereas Si, Ti, and Zr mobilities increase significantly.

In summary, we note that relating transport phenomena (diffusion, viscous flow) to the structure of liquids is not simple beyond a first-order conclusion that network polymerization reduces transport. The difficulty arises probably because transport phenomena are controlled by motion in subsystems that have smaller molecular units than the anion speciation recognized by Raman spectroscopy. In this sense, it may be concluded that mobilities of elements in a given liquid reflect ionic interaction on local scales, especially in terms of the hindrance to ionic motion exerted by the Coulomb interaction. Therefore, mobility data could provide an "intimate" look at the structures of silicate liquids, complementing ion-dynamics simulation, viscosity measurements, and other types of approaches.

Acknowledgments

Thanks are due to K.D. Burrhus for keeping the ion microprobe at the highest level of performance, and to colleagues at MIT, WHOI and Tokyo for discussion and advice. Special thanks are due to Bjorn Mysen of Carnegie Institution of Washington for providing a thorough review of an early version of this paper with extensive friendly criticism. He also provided anion speciation data from which Table 2 was made. This research was partly supported by a grant from NSF, EAR-8805221.

References

Adam, G., and Gibbs, J.H. (1965) On the temperature dependence of cooperative relaxation properties in glass-forming liquids. *J. Chem. Phys.*, **43**, 139–146.

Angell, C.A., Cheeseman, P.A., and Tamaddon, S. (1982) Pressure enhancement of ion mobilities in liquid silicate from computer simulation studies to 800 kbars *Science*, **218**, 885–887.

Angell, C.A., Cheeseman, P.A., and Kadiyala, R.R. (1987) Diffusivity and thermodynamic properties of diopside and jadeite melts by computer simulation studies. *Chem. Geol.*, **62**, 83–92.

Brearley, M., Dickinson, J.E., Jr., and Scarfe, C.M. (1986) Pressure dependence of melt viscosities on the join diopside-albite. *Geochim. Cosmochim. Acta*, **50**, 2563–2570.

Cooper, A.R. (1974) Vector space treatment of multicomponent diffusion, in *Geochemical Kinetics and Transport*, edited by A.W. Hofmann, B.J. Giletti, H.S. Yoder, Jr., and R.A. Yund, pp. 15–30. Carnegie Institute, Washington.

Crank, J. (1975) *Mathematics of Diffusion*, 2nd ed. Oxford University Press, 414 pp.

Day, D.E. (1976) Mixed alkali glasses-their properties and use. *J. Noncrystall. Solids*, **21**, 343–372.

Dingwell, D.B. (1989) Effects of structural relaxation on cationic tracer diffusion in silicate melts. *Chem. Geol.* (in press).

Dinwell, D.B., and Webb, S.L. (1989) Structural relaxation in silicate melts and non-Newtonian melt rheology in geologic processes. *Phys. Chem. Mineral.*, **16**, 508–516.

Dunn, T. (1982) Oxygen diffusion in three silicate melts along the join diopside-anorthite. *Geochim. Cosmochim. Acta*, **46**, 2293–2299.

Henderson, P., Nolan, J., Cunningham, G.C., and Lowry, R.K. (1985) Structural controls and mechanisms of diffusion in natural silicate melts. *Contrib. Mineral. Petrol.*, **89**, 263–272.

Isard, J.O. (1969) The mixed alkali effects in glass *J. Noncrystall. Solids*, **1**, 235–261.

Kushiro, I. (1976) Changes in viscosity and structure of melt of $NaAlSi_2O_6$ composition at high pressures. *J. Geophys. Res.*, **81**, 6347–6350.

Kushiro, I. (1981) Change in viscosity with pressure of melts in the system $CaO–Al_2O_3–SiO_2$, *Carnegie Inst. Wash. Yearbook*, **80**, 339–341.

Kushiro, I. (1983) Effect of pressure on the diffusivity of network-forming cations in melts of jadeitic compositions. *Geochim. Cosmochim. Acta*, **47**, 1415–1422.

Mysen, B.O. (1988) *Structure and Properties of Silicate Melts.* Elsevier, 354 pp.

Mysen, B.O., Virgo, D., and Scarfe, C.M. (1980) Relations between the anionic structure and viscosity of silicate melts-a Raman spectroscopic study. *Amer. Mineral.*, **65**, 690–710.

Mysen, B.O., Virgo, D., and Seifert, F.A. (1982) The sturture of silicate melts: Implications for chemical and physical properties of natural magma, *Rev. Geophys. Space Phys.*, **20**, 353–383.

Oishi, Y. (1965) Analysis of ternary diffusion: solutions of diffusion equations and calculated concentration distribution. *J. Chem. Phys.*, **43**, 1611–1620.

Oishi, Y., Terai, R., and Ueda, H. (1975) Oxygen diffusion in liquid silicates and relation to their viscosity. in: *Mass Transport Phenomena in Ceramics*, edited by A.R. Cooper and A.H. Heuer. Plenum Press, New York, pp. 297–310.

Richet, P. (1984) Viscosity and configurational entropy of silicate melts. *Geochim. Cosmochim. Acta*, **48**, 471–483.

Scarfe, C.M., Mysen, B.O., and Virgo, D. (1979) Changes in viscosity and density of melts of sodium disilicate, sodium metasilicate, and diopside composition with pressure. *Carnegie Inst. Wash. Yearbook*, **78**, 547–551.

Sharma, S.K., Virgo, D., and Mysen, B.O. (1979) Raman study of the coordination of aluminum in jadeite melts as a function of pressure. *Amer. Mineral.*, **64**, 779–788.

Shimizu, N., and Hart, S.R. (1982a) Applications of the ion microprobe to geochemistry and cosmochemistry. *Ann. Rev. Earth Planet. Sci.*, **10**, 483–526.

Shimizu, N., and Hart, S.R. (1982b) Isotope fractionation in secondary ion mass spectrometry. *J. Appl. Phys.*, **53**, 1303–1311.

Shimizu, N., and Kushiro, I. (1984) Diffusivity of oxygen in jadeite and diopside melts at high pressures. *Geochim. Cosmochim. Acta*, **48**, 1295–1303.

Soules, T.F. (1979) A molecular dynamic calculation of the structure of sodium silicate glasses. *J. Chem. Phys.*, **71**, 4570–4578.

Taylor, M., and Brown, G.E., Jr. (1979) Structure of mineral glasses-II. The $SiO_2–NaAlSiO_4$ join. *Geochim. Cosmochim. Acta*, **43**, 1467–1473.

Virgo, D., Mysen, B.O., and Seifert, F. (1979) Structures of quenched melts in the system $NaAlSiO_4–CaMgSi_2O_6–Mg_2SiO_4–SiO_2$ at 1 atm. *Carnegie Inst. Wash. Yearbook*, **78**, 502–506.

Watson, E.B. (1979) Calcium diffusion in a simple silicate melt to 30 kbar, *Geochim. Cosmochim. Acta*, **43**, 313–322.

York, D. (1969) Least-squares fitting of a straight line with correlated erros. *Earth Planet. Sci. Lett.*, **5**, 320–324.

Chapter 7
Thermodynamic Properties of Silicate Liquids at High Pressure and Their Bearing on Igneous Petrology

Yan Bottinga

Introduction

Knowledge of the melting relations in simple binary or ternary systems containing silica is required for the understanding of igneous petrology. Due to the pioneering efforts of petrologists like Bowen, Greig, or Kracek, we possess a substantial amount of information for such systems at 1 atm pressure. But at high pressure much less experimental data have been acquired. This paper discusses the application of classical thermodynamics to the few observations available on high pressure melting in relatively simple systems containing silica.

In general the thermodynamic properties of silicate minerals are far better known than are those of liquids with the same composition. Data that is available for several minerals for $P = 1$ atm. include:

The entropy, enthalpy and volume;

An isothermal equation of state, like the Birch–Murnaghan equation, that gives implicitly the specific volume and the isothermal compressibility as a function of pressure;

The coefficient of thermal expansion and the isobaric specific heat as functions of temperature.

The Birch–Murnaghan equation may be written as

$$P = 1.5K_0(r^7 - r^5)\{1 - 0.75(4 - K'_0)(r^2 - 1)\} \tag{1}$$

where the constants K_0 and K'_0 are the isothermal bulk modulus at 1 atm and its pressure derivative, $r = (V(0, T)/V(P, T))^{1/3}$, with $V(0, T)$ being the specific or molar volume at 0.001 kbar and T K. If the temperature dependence of K_0 is known then it is possible to evaluate:

The temperature dependence of the compressibility ($\kappa = -1/V(\partial V/\partial P)_T$);

The pressure dependence of the thermal expansion ($\alpha = 1/V(\partial V/\partial T)_P$), because

$$(\partial a/\partial P)_T = -(\partial \kappa/\partial T)_P \tag{2}$$

The pressure dependence of the heat capacity ($C_P = T(\partial S/\partial T)_P$), because

$$(\partial C_P/\partial P)_T = -TV(\alpha^2 + \partial \alpha/\partial T) \tag{3}$$

Calculation with reasonable accuracy of the thermodynamic properties of crystalline silicate compounds in the temperature interval 300 to 2500 K and for pressures ranging from 0 to 200 kbar can now be performed. But for silicate liquids, the situation is totally different, because the only information available is for $P = 1$ atm., and is limited to the molar volume V_L, the entropy S_L (only in favorable instances), the coefficient of thermal expansion (α_L) and the heat capacity (C_{PL}). Exceptionally α_L and C_{PL} are known as functions of temperature.

Direct observation of the high pressure thermodynamic properties of liquid silicates is difficult because of the required high temperatures. But this type of information can be obtained indirectly from the melting temperature variation with pressure of crystalline silicates for which the above-mentioned data already exists. Direct observation, which has been used frequently before, by among others: Carmichael et al. (1977), Ohtani (1983), Navrotsky et al. (1982) and Bottinga (1985) was used for this work. This paper discusses the properties of liquid Mg_2SiO_4, Fe_2SiO_4, $MgSiO_3$, $Mg_3Al_2Si_3O_{12}$ and SiO_2, along their respective melting curves in $P-T$space. The results for these stoichiometric compounds will be used to calculate the variation with pressure of the eutectic temperature in the system $MgSiO_3$–SiO_2, and of the partitioning of Fe and Mg between olivine and liquid. The final subject to be discussed is the hypothesis that silicate liquids produced by partial melting in the upper mantle may have a larger density than the residual solid material.

Evaluation of the Thermodynamic Properties of Silicate Liquids Along the Melting Curve

The Clausius–Clapeyron equation

$$(dT/dP)_f = \Delta_f V/\Delta_f S \tag{4}$$

may be used to calculate the thermodynamic properties of liquids produced by congruently melting crystalline substances. In equation (4) $(dT/dP)_f$ is the slope of the melting point curve, and $\Delta_f V$ and $\Delta_f S$ are the volume and entropy of melting, respectively. An analytical expression of the melting point curve in $P-T$ space facilitates the usage of this equation considerably. The Simon equation

$$P = a\{(T/T_0)^b - 1\} + P_0 \tag{5}$$

also is often used for the same purpose. In equation (5) a and b are constants, T and P are melting point coordinates, and T_0 and P_0 are the initial values of P and T. The Simon equation is an empirical relation that should not be used outside the P–T interval for which the melting curve is known; see also Wolf and Jeanloz (1984). The constants a and b are obtained by fitting equation (5) to melting temperature measurements at different pressures. The melting pressure variation with temperature is given by

$$(dP/dT)_f = (ab/T_0)(T/T_0)^{b-1} \tag{6}$$

The entropy of fusion along the melting curve can be obtained by integrating

$$\Delta_f S(P, T) = \Delta_f S(P', T') + \int_{T'}^{T} \{\Delta C_P/T - \Delta(\alpha V)(dP/dT)_f\}\, dT \tag{7}$$

where $\Delta C_P = C_{PL} - C_{PC}$ and

$$\Delta(\alpha V) = \Delta(\alpha V') + \Delta(\alpha^2 V')\Delta T - \Delta(\alpha\kappa V')\Delta P \tag{8}$$

In equation (8) $\Delta(\alpha V) = \alpha_L V_L - \alpha_C V_C$, $\Delta(\alpha^2 V') = \alpha_L^2 V_L' - {}^{a2}_{\ C} V_C'$, $\Delta(\alpha\kappa V') = \alpha_L \kappa_L V_L' - \alpha_C \kappa_C V_C'$, $\Delta T = T - T'$ and $\Delta P = P - P'$; the subscripts L and C stand for liquid and crystal and the prime indicates an initial, known value. Using equations (4, 6–8), $\Delta_f V(P, T)$ can be calculated and with equation (9) $V_L(P, T)$ can be computed

$$V_L(P, T) = \Delta_f V(P, T) + V_C(\mathrm{O}, T)\left(1 + \int_{T_0}^{T} \alpha_C\, dT\right) r_C^{-3} \tag{9}$$

where r_C has been obtained with equation (1).

Initially at $P' = P_0$ and $T' = T_0$, α_L, V_L, and S_L, are known and a ΔT is selected for which it may be supposed that the variation of κ_L is negligible. Equation (6) allows calculation of ΔP and V_L can be calculated with equation (9). But to do this, a starting value for κ_L, which is needed in equation (8) must be assumed. Using the calculated $V_L(P, T)$ a new value for κ_L is obtained from equation (10).

$$\kappa_L = \{\ln(V_L(P', T')/V_L(P, T)) + \alpha_L \Delta T\}/\Delta P \tag{10}$$

Iteration will give a κ_L value (eq. 10) that differs only within permissible bounds from the value used in equation (8).

The calculations began with fitting of melting data for the minerals forsterite, fayalite, enstatite, pyrope and β-quartz to the Simon equation (5). The computed constants a and b, the values for P_0 and T_0, and the quality of the fits are given in Table 1. Noteworthy is that of these five minerals only forsterite shows the melting behavior of a simple well-behaved crystalline substance; fayalite melts slightly incongruently at $P = 1$ atm., enstatite and pyrope melt also incongruently at $P = 1$ atm., and β-quartz becomes liquidus mineral at $P > 6$ kbar. The sources for the melting data also are given in Table 1. The quality of the fits to the data is expressed as the root mean square difference (RMS) between the fitted values and the observations. The RMS values are smaller or comparable to the uncertainty in the experimental data.

Experimental temperatures are believed to be correct to ± 10 K (Jackson, 1976; Davis and England, 1964) and the uncertainty in the measured pressure

Table 1. Simon-equation parameters for silicate mineral melting curves.

Mineral	Reference	T_0 (K)	T(max) (K)	P_0 (kbar)	P(max) (kbar)	a (kbar)	b	RMS[a] (kbar)
Forsterite	Ohtani and Kumazawa (1981)	2163	2736	0[d]	150	108.33	3.7	0.6
	Davis and England (1964)							
Faylite	Bowen and Schairer (1932)	1490	1870	0	70	157.80	1.59	1.3
	Akimoto et al. (1967)[c]							
	Ohtani (1979)							
Pyrope	Ohtani et al. (1981)	2073	2410	40	100	19.79	9.25	1.7
Enstatite	Boyd et al. (1964)	1830[b]	2220	0	47	28.65	5.01	1.0
β-quartz	Jackson (1976)	2003	2511	7	25	15.99	3.34	0.1

[a] $\mathrm{RMS} = \sum_{i}^{N} \{(P_{calc} - P_{obs})_i^2/N\}^{.5}$, i indexes the observations.

[b] Estimated

[c] Omitted the point at 1773K

[d] i.e, 0.001 kbar

amounts to $\pm 5\%$ for $10 < P < 50$ kbar. However, at higher pressures, estimation of the uncertainty in pressure becomes difficult because the equilibria used, for calibrations are not always well established. For instance Ohtani (1979) and Ohtani and Kumazawa (1981) have calibrated their pressure measurements by means of the Fayalite-Fe spinel equilibrium, determined by Akimoto et al. (1977), for $P < 100$ kbar. But the observations by the latter authors are now believed to be incorrect (Navrotsky and Akaogi, 1984).

Ohtani et al. (1982) estimated that the temperature uncertainty in their data below 100 kbar amounted to ± 50 K and above 100 kbar to ± 100 K. For uncertainties in their pressure measurements, they reported ± 3 kbar at $P = 70$ kbar and ± 6 kbar at $P = 140$ kbar.

The thermodynamic input data are listed in Table 2. Frequently it has been found that the independently determined quantities $\Delta_f V$, $\Delta_f S$ and $(dP/dT)_f$ are not consistent with the Clausius–Clapeyron equation. There are probably several reasons for this disagreement.

$\Delta_f V$ is a small difference between two independently and often indirectly, obtained quantities namely, V_C and V_L. V_C may not be known too accurately close to the melting point because of high temperature polymorphism, which is quite common. In addition, the coefficient of thermal expansion is usually unknown for $T > 1100$ K and its variation with temperature may be nonlinear. Usually V_L has not been measured but calculated from partial molar volume data (Bottinga et al., 1982; Mo et al., 1982). The thermal expansion coefficient for liquid silicates may be in error by a factor of two (Bottinga et al., 1983). As a net result the accuracy of V_L is not better than 1%. But even such a small error may cause a significant error in $\Delta_f V$, because $\Delta_f V$ ranges from 8 to 16% of V_C.

The initial value of $(dP/dT)_f$ is often poorly constrained by the available experimental observations, as is well illustrated by the controversy about the melting point of albite and the variation of the albite melting point with pressure for $0 < P < 10$ kbar (Navrotsky et al., 1982; Boettcher et al., 1982).

$\Delta_f S$ may be not too well known because of premelting phenomena, as has been observed by Orr (1953) for fayalite. The uncertainty in $\Delta_f S$ may amount to 10%, depending on the size of $\Delta_f S$.

This disagreement among the quantities occurring in the Clausius–Clapeyron equation is not limited to silicates but has also been observed for molten salts, see for example Clark (1959) and Kim et al. (1972).

The coefficients in the heat-capacity equation have been recalculated (see footnote to Table 2) for fayalite, because the coefficients listed in Robie et al. (1978) produced a larger heat capacity at high temperature than has been observed for liquid fayalite.

The computations were done in steps of 10 K along the fusion curve. With the Simon equation, the ΔP corresponding to this temperature interval was calculated, and the Birch–Murnaghan equation was used to calculate $V_C(P, T)$. The next step was the calculation of $\Delta_f S(P, T)$ by integrating equation (7), in which were inserted equations (6) and (8). The specific heat and the coefficient of thermal expansion of the liquid were assumed to be temperature and pressure

Table 2. Thermodynamic data

	Mg_2SiO_4 Fo	Fe_2SiO_4 Fa	$Mg_3Al_2Si_3O_{12}$ Py	$MgSiO^{33}$ En	SiO^{24} Si	Dimension
P_0	0.001	0.001	40	0.001	7	kbar
T_0	2163[a]	1490[a]	2073[q]	1830[aa]	2003[gg]	K
Crystalline phases						
$V(P_0, T_0)$	47.13[b]	48.42[i]	115.72[r]	33.41[bb]	23.45[hh]	cm^3/gfw
α_1^1	4.39[b]	2.38[j]	2.338[s]	1.391[s]	0.1[ii]	$\times 10^{-5}$ 1/K
α_2		1.1	0.571	2.544		$\times 10^{-8}$
α_3			−0.492	0.1282		
$S(0, T_0)$	422[a]	425.7[a]	1149.7[t]	269.1[cc]	166.3[jj]	J/(gfw.K)
C_1^2	227.98[a]	155.11[k]	544.95[u]	205.56[cc]	65.28[jj]	J/(gfw.K)
C_2	.34139	3.6446	2.068	−1.28	5.529	$\times 10^{-2}$
C_3	−0.894	−2.9456	−8.3312	1.1926	−1.8463	$\times 10^{6}$
C_4	−1.7446		−2.2830	−2.2977		$\times 10^{3}$
$K_0(T_0)$	0.8372[c]	0.9642[l]	1.43[v]	0.7184[dd]	0.5631[kk]	Mbar
K_0'	5.33[d]	5.40[l]	4.5[w]	9.6[ee]		
$(\partial K_0/\partial T)_P$	−0.16[c]	−0.23[m]	−0.18[s]	−0.17[s]		kbar/k
Liquid phases						
$V(P_0, T_0)$	50.91[e]	52.10[n]	134.05[x]	38.77[x]	25.92[ll]	cm^3/gfw
α	8.2[f]	10.7[o]	5.65[f]	6.20[f]	1.0[ll]	$\times 10^{-5}$ 1/K
$S(0, T_0)$	492[g]	487.5[p]	1311.7[y]	311.1[ff]	172.9[jj]	J/(gfw.K)
C_P	252.93[h]	239.25[h]	682.21[z]	167.15[h]	81.37[jj]	J/(gfw.K)

Notes

[1] $\alpha = \alpha_1 + \alpha_2 T + \alpha_3/T^2$. [2] $C_P = C_1 + C_2 T + C_3/T^2 + C_4/T^{.5}$. [3] All solid data are for clinoenstatite. [4] All solid data are for β-quartz.

[a] Robie et al. (1978).

[b] Takeuchi et al. (1984), used V (1837 K) and extrapolated to 2163 K with the expansion coefficient calculated from the reported V (1873 K) and V (1673 K).

[c] Sumino et al. (1979).

[d] Kumazawa and Anderson (1969).

[e] Calculated with the Clausius–Clapeyron equation.

[f] Calculated with the model of Bottinga et al. (1982).

[g] Calculated with the $\Delta_f S(P_0, T_0) = 70$J/(gfw.K), estimated by Stebbins et al. (1984).

[h] Calculated with the model of Richet and Bottinga (1985).

[i] Used the value $V(0, 300) = 46.28$ cm^3/gfw listed by Watanabe (1982) and the expansion data of Suzuki et al. (1981) to calculate $V(0, 1490) = 48.15$ cm^3/gfw. To obtain agreement with the Clausius–Clapeyron equation (see text), the latter value was increased by 0.55%.

[j] Derived from the high-temperature expansion data of Suzuki et al. (1981).

[k] Coefficients derived from the enthalpy measurements of Orr (1953), excluding the data at $T > 1371$ K, because of premelting phenomena.

[l] Refitted the experimental data of Yagi et al. (1975) to the Birch–Murnaghan equation, and used the listed temperature derivative to calculate $K_0(T_0)$.

[m] Sumino (1979).

[n] Calculated with the model of Bottinga et al. (1982): $V(0, 1490) = 52.36$ cm^3/gfw, reduced this value by 0.51% to obtain agreement with the Clausius–Clapeyron equation.

[o] From Bottinga and Weill (1970).

[p] From $\Delta_f S(P_0, T_0) = 61.77$ J/(gfw.K) (Orr, 1953), which is essentially the same as the more recent value of Stebbins and Carmichael (1984).

Table 2 (*continued*)
[q] Ohtani et al. (1981).
[r] Calculated from the data compiled by Watanabe (1982).
[s] Watanabe (1982).
[t] Used $S(0, 298) = 266.3$ J/(gfw.K) (Haselton and Westrum, 1980) and the listed C_p coefficients to calculate $S(0, 2073)$.
[u] Holland (1981).
[v] Calculated from K_0 measured by Levien et al. (1979) and the listed temperature derivative.
[w] Levien et al. (1979).
[x] Calculated with Clausius–Clapeyron equation.
[y] From data for crystalline pyrope and $\Delta_f S$ (Richet and Bottinga, 1984).
[z] Richet and Bottinga (1984).
[aa] Incongruent melting point; the melting of enstatite becomes congruent at $P < 2$ kbar (Chen and Presnall, 1975).
[bb] The data of Skinner (1966) give V (0, 1830) = 32.92; to obtain agreement with the Clausius–Clapeyron equation this value was increased by 1.6%.
[cc] Calculated with data from Robie et al. (1978).
[dd] Calculated from $K_S = 1.08$ Mbar (Weidner et al., 1978).
[ee] Frisillo and Barsch (1972).
[ff] Calculated with $\Delta_f S$ of Stebbins et al. (1984).
[gg] Jackson (1976).
[hh] Caclulated from data listed in Birch (1966), Robie et al. (1966) and Skinner (1966).
[ii] Estimated.
[jj] Richet et al. (1982).
[kk] Birch (1966)
[ll] Calculated from data by Brückner (1970) and Bucaro and Dardy (1976).

independent. The temperature independence is compatible with currently available data. The pressure independence is an unavoidable assumption because the thermodynamic data for liquid and crystalline silicates are not precise enough to permit a self-consistent computation of both α_L and κ_L along the fusion curve. Attempts were made to do this, but the calculations do not converge to physically acceptable variations with pressure and temperature of both α_L and κ_L. In other words, available information allows evaluation of V_L and κ_L, but second derivatives of V_L could not be computed. For the solid phase both α_C and C_{PC} were corrected for pressure, using equations (2, 3) and $\partial\kappa/\partial T$ was derived from the equation of state (eq. 1). Hence knowing $\Delta_f S(P, T)$, $\Delta_f V$ was calculated and subsequently $V_L(P, T)$ with equation (9). Finally, κ_L was obtained from equation (10).

A Birch–Murnaghan equation was fitted to the calculated $V_L(P, T)$ values; this fit gave the constants K_0 and K'_0 and the derivative $\partial K_0/\partial T$ for a liquid composition. Differentiation of this Birch–Murnaghan equation provides a second way to compute κ_L (see Bottinga, 1985). The isothermal compressibilities obtained in this way were in perfect harmony with those computed with equation (10).

The computations are most seriously affected by errors in $(dP/dT)_f$ and $\Delta_f V$. As a result the calculated compressibilities may be off by a factor of two and the $\Delta_f V$ values (Table 3) by about 30%. However the $\Delta_f S$, S_L and V_L values should

Table 3. Thermodynamic properties of liquids along their freezing curves.

T (K)	P (kbar)	V_L (cm^3/gfw)	$\Delta_f V$	S_L J/(gfw.K)	$\Delta_f S$	K_L[b] Mbar^{-1}
Mg_2SiO_4						
2163	0.	50.9	3.8	492	70	1.9
2173	1.9	50.8	3.7	492	70	1.9
2273	21.8	49.4	3.2	496	68	1.7
2373	44.3	48.1	2.7	498	64	1.5
2473	69.5	46.8	2.2	498	59	1.3
2573	97.6	45.6	1.8	498	53	1.2
2673	128.8	44.4	1.4	496	44	1.1
2723	145.6	43.8	1.2	494	40	1.0
Fe_2SiO_4						
1490	0.	52.1	3.7	488	62	2.0
1570	13.7	51.3	3.4	493	58	1.7
1670	31.3	50.3	3.0	498	53	1.7
1770	49.6	49.4	2.5	502	47	1.6
1870	68.5	48.4	2.1	505	41	1.5
$MgSiO_3$						
1830	0.	38.8	5.4	311	42	4.5
1870	3.3	38.3	5.0	314	43	4.0
1970	12.8	37.4	4.2	320	44	3.0
2070	24.5	36.5	3.5	326	46	2.3
2170	38.6	35.7	3.0	331	46	1.8
2210	45.1	35.4	2.8	332	47	1.7
$Mg_3Al_2Si_3O_{12}$						
2073	40.	134.1	18.3	1296	162	8.8
2173	50.8	127.8	12.6	1318	164	3.7
2273	66.6	123.2	8.7	1337	164	2.1
2373	89.3	119.3	6.0	1349	161	1.4
2393	94.9	118.5	5.6	1351	161	1.3
SiO_2						
2003	7.	25.9	2.5	n.c.[a]	6.6	3.6
2109	10.	25.7	2.3		6.8	3.1
2262	15.	25.4	2.0		7.0	2.2
2394	20.	25.2	1.7		7.1	1.8
2511	25.	25.0	1.6		7.1	1.6

[a] Not calculated.

[b] Isothermal compressibility of the liquid.

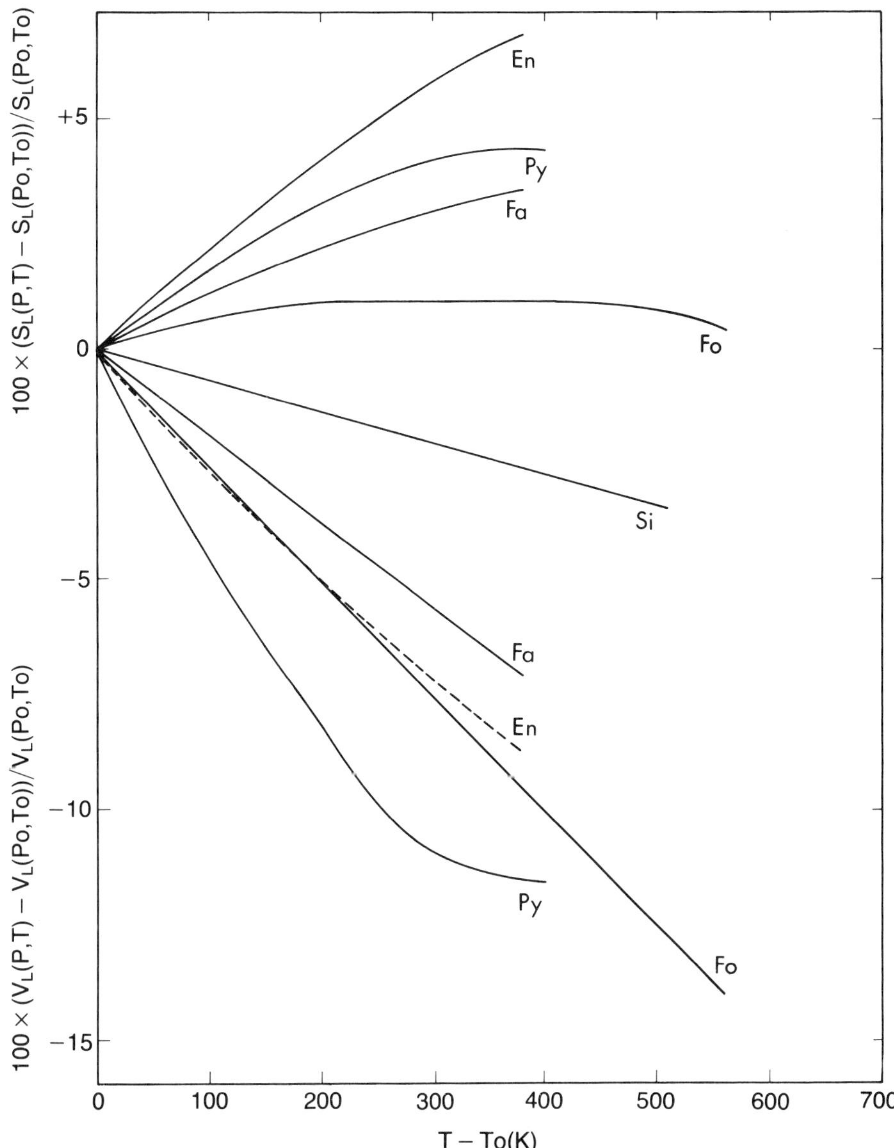

Fig. 1. Relative variation (%) of the molar entropy (top part of the diagram) and molar volume of several silicate liquids with temperature along the melting curve. P_0 and T_0 are the initial melting point coordinates for which the calculations were done, see Table 2. The liquid compositions are: forsterite (Fo, Mg_2SiO_4), fayalite (Fa, Fe_2, SiO_4), enstatite (En, $MgSiO_3$), pyrope (Py, $Mg_3Al_2Si_3O_{12}$), and quartz (Si, SiO_2).

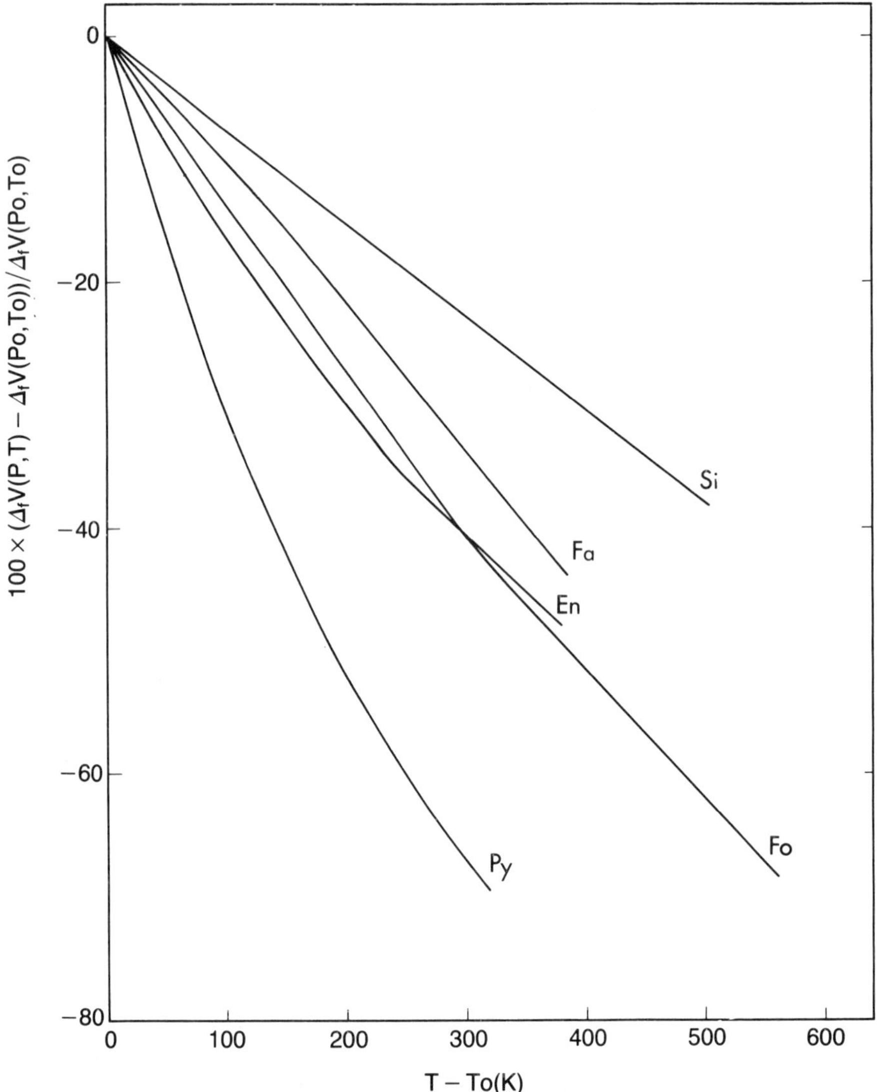

Fig. 2. Relative variation (%) of the molar volume of melting along the melting curve as a function of temperature, see Fig. 1.

be accurate to 4%. The computed κ_L, $\Delta_f V$, $\Delta_f S$, S_L and V_L values along the melting curves are listed in Table 3 and plotted in Fig. 1 through 3. The compressibilities have already been discussed (Bottinga, 1985), except for those of Fe_2SiO_4 and SiO_2. For the latter substance, the logarithmic compressibility has been plotted against pressure in Fig. 4 together with the determination at 1 atm by Bucaro and Dardy (1976). The agreement between the observed value and those calculated is good.

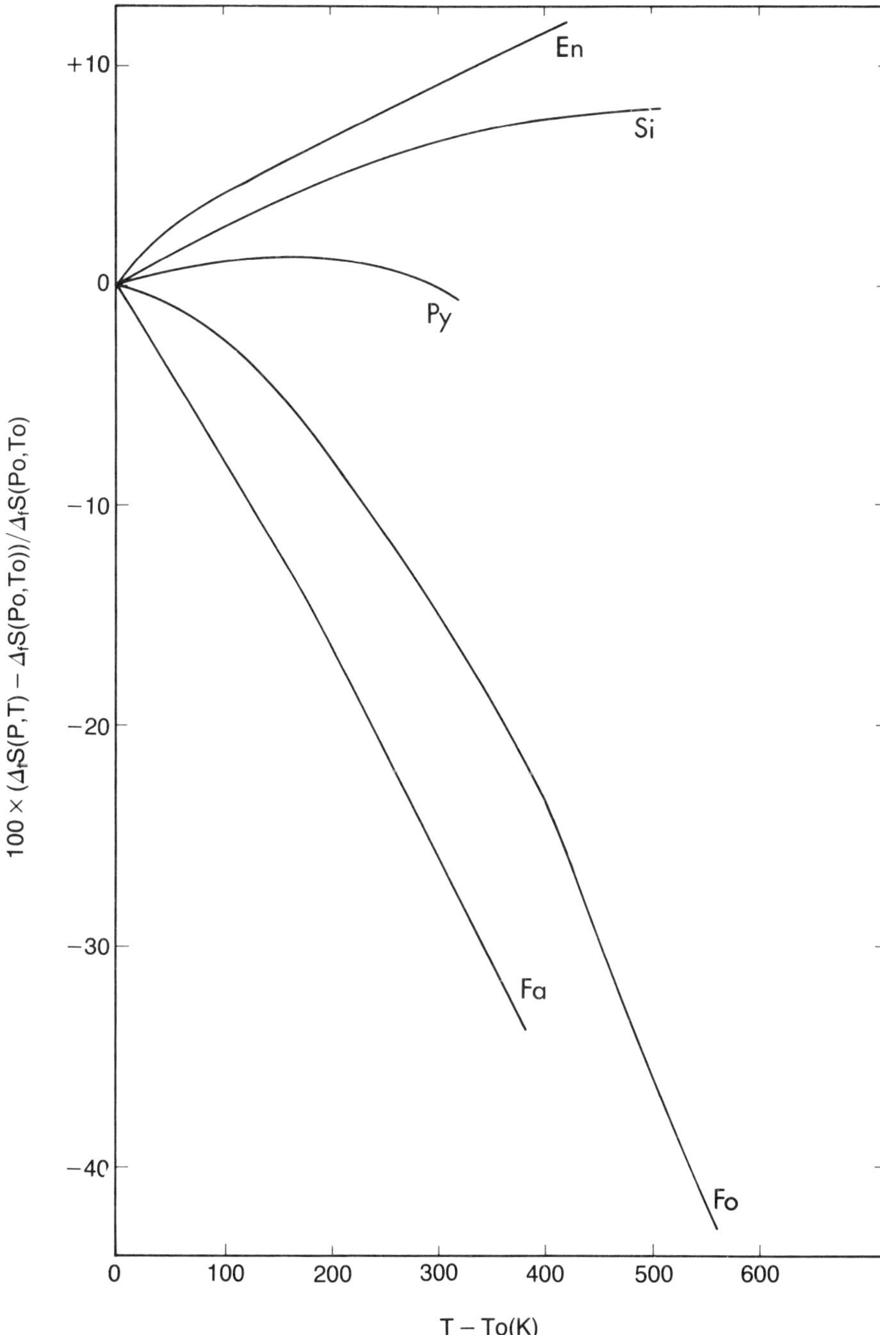

Fig. 3. Relative variation (%) of the molar entropy of melting along the melting curve as a function of temperature, see Fig. 1.

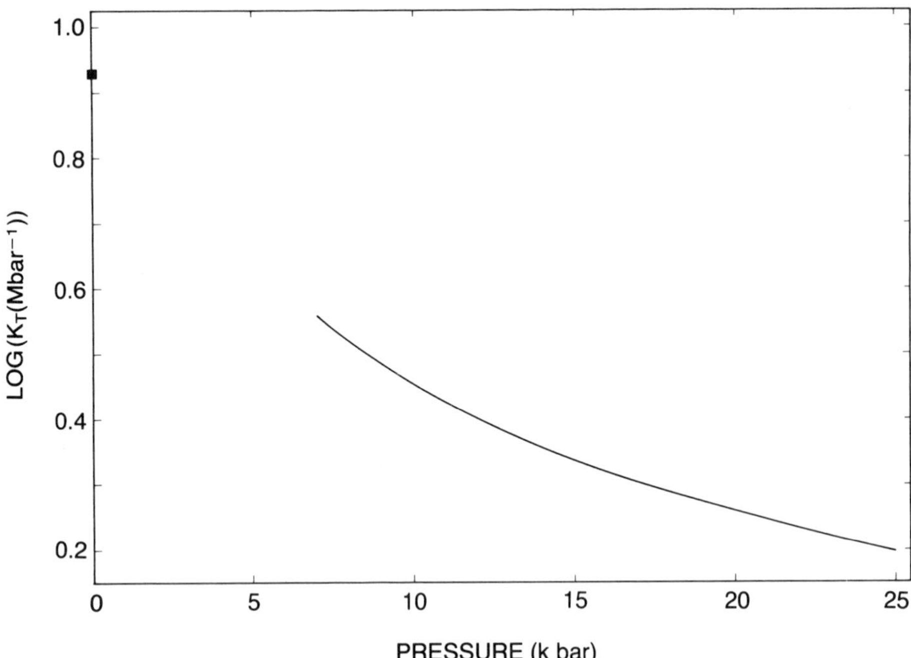

Fig. 4. Logarithmic isothermal compressibility ($Mbar^{-1}$) plotted versus pressure (kbar) for liquid SiO_2. The point at P = 0.001 kbar was determined by Bucaro and Dardy (1976) by means of a light scattering technique for the temperature interval glass transition (1470 K) to melting point of cristobalite (2003); no dependence on temperature was noted. The calculated results are for 2290 K.

It is noteworthy that S_L is relatively constant over a large pressure interval (see Fig. 1) and V_L always decreases with pressure along the melting curve. $\Delta_f S$ decreases only for completely depolymerized melts, i.e. Mg_2SiO_4 and Fe_2SiO_4, see Fig. 2 and 3. The physical interpretation of the observations will be the subject of a fourth publication. Geological applications of these results are discussed in sections III and IV of this chapter.

Table 4 contains a compilation of the Birch-Murnaghan parameters obtained for the five liquid compositions. These values should not be used outside the $P-T$ region for which they were derived (see Table 1).

Eutectic Melting

From a geological point of view, eutectic melting is far more important than congruent melting. Partial melting in the upper mantle is in general approximated by eutectic melting. The two important aspects of eutectic melting are:

Table 4. Birch–Murnaghan parameters for various silicate melts.

Liquid composition	$K_0(T_0, P_0)$ (kbar)	K'_0	$\partial K_0/\partial T$ (kbar/K)	RMS[a] (kbar)
Mg_2SiO_4	524.4	4.40	-8.2×10^{-2}	4.6×10^{-3}
Fe_2SiO_4	549.3	2.82	-8.6×10^{-2}	1.1×10^{-2}
$Mg_3Al_2Si_3O_{12}$	112.9	25.17	-4.3×10^{-6}	3.2×10^{-1}
$MgSiO_3$	217.8	10.98	-3.9×10^{-7}	5.0×10^{-2}
SiO_2[c]	274.9	28.7	n.c.[b]	5.5×10^{-2}

[a] See footnote Table 1
[b] Not calculated
[c] Results are for $T = 2290$ K

The variation of the eutectic temperature with pressure;
The variation of the eutectic melt composition with pressure, while remaining on the eutectic melting curve.

For the discussion of the first point, a Gibbs–Duhem equation is to be written for each of the phases in equilibrium at the eutectic point, namely k crystalline solids and one melt phase. This work will give $k + 1$ equations

$$\sum_{i=1}^{k} x_{ij} d\mu_{ij} + S_j dT - V_j dP = 0 \tag{11}$$

X_{ij} is the mole fraction of crystalline phase i dissolved in phase j. The crystalline phases are supposed to be stoichiometric compounds, hence all $X_{ij} = 0$ except for $X_{ii} = 1$. For the melt phase: $0 < X_{ij} < 1$, and μ_{ij} is the chemical potential of crystalline phase i dissolved in phase j. S_j and V_j are the mean molar entropy and volume of phase j. The $k + 1$ equations (11) contain $k + 1$ unknowns, namely the variation of the eutectic pressure with temperature $(dP/dT)_e$ and the variation with temperature of the chemical potentials $(\partial\mu_i/dT)_e$ along the eutectic line in P–T space. Solving the equations for $(dP/dT)_e$ we obtain

$$\begin{aligned}(dP/dT)_e &= \left(-S_L + \sum_i X_{iL} S_{iC}\right) \Big/ \left(-V_L + \sum_i X_{iL} V_{iC}\right) \\ &= \sum_i X_{iL}(\bar{S}_{iL} - S_{iC}) / \sum_i X_{iL}(\bar{V}_{iL} - V_{iC})\end{aligned} \tag{12}$$

$\bar{V}_{iL}$ is the partial molar volume of phase i dissolved in the melt. At 1 atm the partial molar volume of a stoichiometrically dissolved crystalline silicate phase in a melt is well approximated by the molar volume of the molten phase i at the same temperature (Bottinga and Weill, 1970). The same approximation will be used at high pressure and the P–T slope of the eutectic line will be calculated for the enstatite-quartz system. The partial molar volumes are approximated by the calculated molar volumes, obtained in the previous section, at the selected pressures and corrected for temperature with the known thermal expansion

Table 5. Eutectic temperatures and pressures for the system $MgSiO_3$–SiO_2

		SiO_2		$MgSiO_3$		$(dP/dT)_e$	
P^a (kbar)	T^a (K)	ΔV^b (cm^3/gfw)	ΔS^c J/(gfw.K)	ΔV^b (cm^3/gfw)	ΔS^c [J/(gfw.K)]	calc. bar/K	obs. bar/K
0	1820	−0.54	6.50	5.36	41.9	83	76
12	1979	1.92	8.98	4.27	44.7	99	143
25	2070	1.58	9.51	3.52	45.6	123	

[a] Observed by Chen and Presnall (1975)
[b] $\Delta V = V_{iL} - V_{iC}$
[c] $\Delta S = S_{iL} - S_{iC}$

coefficients in Table 2. The partial molar entropies were calculated with equation (12)

$$\bar{S}_{iL} = S_{iL} + \Delta_m \bar{S}_{iL} \tag{13}$$

where S_{iL} is the molar entropy of pure liquid compound i and $\Delta_m \bar{S}_{iL}$ is the partial molar entropy of mixing for component i in the melt. This contribution to the partial molar entropy was evaluated by assuming ideal mixing of compositional units proposed by Ghiorso et al. (1983).

The observations by Chen and Presnall (1975) and the author's results are listed in Table 5. The eutectic composition (19.1 mol% SiO_2) is independent of pressure for $0 < P < 25$ kbar according to Chen and Presnall (1975). A minor complicating feature is that at 1 atm. enstatite does not melt congruently, but this aspect is ignored in this work. The agreement between observations and calculated results, shown by the entries in Table 5, is reasonable.

The only other system with a well defined eutectic and for which there are experimental observations, is forsterite–pyrope. Combining the observations by Davis (1964) at 40 kbar and those by Ohtani et al. (1982) at 50 kbar provides a variation of the eutectic pressure with temperature of about −1000 bar/K. The author's results give +110 bar/K at 40 kbar and +150 bar/K at 50 kbar, although the calculated gradients are quite different from the experimental one, they are tolerated by the experimental observations because of the uncertainties in measured pressures and temperatures (see section II).

An interesting aspect is that a good knowledge of the P–T coordinates of a eutectic provides a direct means to verify proposed models for the calculation of the entropy of mixing.

The equation for computing the shift of the eutectic composition with pressure is more complicated. But for a two-component eutectic system it is easy to evaluate how the eutectic melt composition changes qualitatively with pressure, with equation (14)

$$\left(\frac{dX_2}{dP}\right)_e = \frac{X_1 \Delta S_1 \Delta S_2}{RT(\partial \ln X_2\gamma_2/\partial X_2)(X_1\Delta S_1 + X_2\Delta S_2)}\left(\frac{\Delta V_1}{\Delta S_1} - \frac{\Delta V_2}{\Delta S_2}\right) \tag{14}$$

Where $\Delta S_1 = \bar{S}_{1L} - S_{1C}$, likewise for ΔV_1; the subscript L has been omitted, and the remaining symbols have their conventional meaning. The usefulness of equation (14) resides in the fact that the sign of $(dX_2/dP)_e$ is determined by the term $(\Delta V_1/\Delta S_1) - (\Delta V_2/\Delta S_2)$, as was pointed out by Prigogine and Defay (1962), because all other terms are positive. A good application of equation (14) has not been found among the silicate-liquid eutectic systems investigated at high pressure.

Pressure Dependence of Partition Coefficients

The olivine-melt Fe/Mg partition coefficient (K_D) plays an important role in igneous petrology.

$$K_D = (\mathrm{Fe/Mg})_{\mathrm{Ol}}/(\mathrm{Fe/Mg})_{\mathrm{L}} \tag{15}$$

where $(\mathrm{Fe/Mg})_{\mathrm{Ol}}$ is the atomic abundance ratio in olivine. The value of K_D at $P = 1$ atm varies little with temperature and liquid composition (Roeder and Emslie, 1970; Takahashi, 1978). The effect of pressure on K_D has been determined by Takahashi and Kushiro (1983) and Mysen and Kushiro (1977).

With the results obtained in section II, it is possible to compute the variation of K_D with pressure for the system forsterite–fayalite. The exchange of Fe and Mg between an olivine crystal and an olivine liquid can be represented by

$$(\mathrm{Fe_2SiO_4})_L + (\mathrm{Mg_2SiO_4})_{\mathrm{Ol}} = (\mathrm{Mg_2SiO_4})_L + (\mathrm{Fe_2SiO_4})_{\mathrm{Ol}} \tag{16}$$

Under equilibrium conditions

$$\Delta\mu_{Fo} - \Delta\mu_{Fa} + RT\ln(a_{\mathrm{Fa}}/a_{\mathrm{Fo}})_{\mathrm{Ol}}/(a_{\mathrm{Fa}}/a_{\mathrm{Fo}})_L = 0 \tag{17}$$

where $\Delta\mu_{\mathrm{Fa}}$ is the difference between the chemical potentials of pure liquid fayalite and pure crystalline fayalite, and a_{Fa} is the activity of fayalite component. It is well known that to a good approximation

$$(a_{\mathrm{Fa}})_{\mathrm{Ol}} = (X_{\mathrm{Fa}})^2_{\mathrm{Ol}} \tag{18}$$

and it is assumed that for the liquid phase

$$(a_{\mathrm{Fa}})_L = (X_{\mathrm{Fa}})^2_L \tag{19}$$

From equations (15 and 17 through 19)

$$\ln K_D = (\Delta\mu_{\mathrm{Fa}} - \Delta\mu_{\mathrm{Fo}})/2RT \tag{20}$$

and the pressure dependence of K_D is given by

$$\partial \ln K_D/\partial P = \{(V_L - V_C)_{\mathrm{Fa}} - (V_L - V_C)_{\mathrm{Fo}}\}/2RT \tag{21}$$

where $(V_L - V_C)_{\mathrm{Fo}}$ is the difference between the molar volumes of pure liquid forsterite and pure crystalline forsterite.

Applying these equations and using the data of Table 2 and the results of section II, values are obtained for $(\partial \ln K_D/\partial P)_{1700\ K}$ of 5.7 $Mbar^{-1}$ at $P = 1$ atm and 4.3 $Mbar^{-1}$ for $P = 35$ kbar. From the fit to the observations given by Takahashi and Kushiro (1983) it is possible to derive that $\partial \ln K_D/\partial P$ equals 6 $Mbar^{-1}$ at 1 atm and about 5.1 $Mbar^{-1}$ at 35 kbar. Taking into account the approximate character of these calculations, the agreement between observations and calculations is judged to be acceptable. For $P = 67$ kbar and $T = 1700$ K the author's calculations give $\partial \ln K_D/\partial P = 2.7$ $Mbar^{-1}$. Hence it seems that to a depth of 200 km in the upper mantle, the sympathetic variation of K_D with pressure decreases slowly from about 6% per 10 kbar at 1 atm to about 3% per 10 kbar at 67 kbar.

Density of Silicate Liquids at High Pressure

Several authors, including Ohtani (1983), Walker et al. (1978), Stolper et al. (1981) and Herzberg (1984) have argued in favor of the hypothesis that liquids produced by partial melting in the upper mantle below a certain depth are more dense than the residual solid material. Rigden et al. (1984) have deduced from their shock-wave experimental data for a liquid consisting of 36 mol% $CaAl_2Si_2O_8$ and 64 mol% $CaMgSi_2O_6$ at $T > 1673$ K that this liquid will be more dense than an equilibrium assemblage of crystalline phases with the same bulk composition at pressures greater than 150 kbar. The far-reaching consequences of this hypothesis are impressive, see Rigden et al. (1984).

Besides the experiments by Rigden et al. the only independent evidence in favor of this hypothesis is that the Birch–Murnaghan equation for liquid basalt, proposed by Stolper et al. gives liquid densities at depths greater than about 150 km in the upper mantle that are in excess of seismologically and mineralogically derived values for that region. From a thermodynamic point of view it is noted that the melting curves for all investigated mantle minerals show a positive $(dT/dP)_f$ slope; for these volatile-free minerals, $\Delta_f S > 0$ and hence $\Delta_f V > 0$. These minerals include forsterite (0–150 kbar) (Ohtani and Kumazawa, 1981; Davis and England, 1964), fayalite (0–70 kbar) (Ohtani, 1979), Fe-spinel (70–140 kbar) (Ohtani, 1979), enstatite (0–50 kbar) (Boyd et al., 1964), pyrope (40–100 kbar) (Ohtani et al., 1981), diopside (0–50 kbar) (Boettcher et al., 1982; Williams and Kennedy, 1969) and jadeite (23–49 kbar) (Bell and Roseboom, 1969).

However, partial melting in the mantle is not congruent, and incongruent melting offers at least two ways for producing melts denser than the solid phases with which they are in equilibrium. In eutectic systems with solid components having positive sloping melting curves, the eutectic may have a negative slope in P–T space if the eutectic temperature is much smaller than the melting point(s) of one or more of the solid component(s). In such a system, in spite of the positive volume of melting of the solid components, the difference between the partial molar volume of one of the components in the liquid phase and its molar volume as a crystalline phase in equilibrium with the melt is negative. This possibility exists because the thermal expansion of the liquid is nearly always significantly greater than that of the solid.

Partial fusion of mineral solid solutions may also give rise to liquids which are denser than the residual solid material in equilibrium with the liquid. This condition may be the result of Fe/Mg disproportionation between the liquid and the solid phases causing the liquid to be enriched in Fe. This phenomenon takes place in the earth. Moreover Fe-containing liquids are more compressible than liquids in which all Fe is replaced by Mg. The compressibility of liquid fayalite is greater than that of liquid forsterite at the same pressure and temperature, as can be seen from the results given here for fayalite and forsterite (Table 3).

Conclusions

One aim of this chapter was to show that experimental observations on simple well defined systems are really indispensable for the acquisition of general knowledge of petrological interest. Besides the essential thermodynamic information discussed here it is also possible to obtain structural information indirectly, as was shown in Bottinga (1985). In this connection, a precise determination of the anorthite melting curve for $0 < P < 15$ kbar would be very helpful for the study of possible changes in the coordination of aluminum in silicate liquids.

The author's survey of the petrological literature has shown that very few measured data are available on the P, T dependence of eutectic points in simple silicate systems. The importance of studying eutectic systems has already been pointed out by the late George Kennedy, see for example Kim et al. (1972). Here, attention has been drawn to the fact that such a study can be very useful for verifying entropy of mixing models for liquid silicates (see section III).

It is now common for authors of papers on experimental metamorphic petrology to deal also with the thermodynamic consistency of their observations. Prior to 1960 such discussions were rare, but in igneous petrology this aspect is still frequently neglected. Typical of such discussions are the observations on the fayalite-spinel transformation in section II.

Acknowledgments

The constructive criticism by P. Richet was greatly appreciated and the authors thank the C.N.R.S. interdisciplinary research program P.I.R.P. S.E.V. for financial support.

References

Akimoto, S., Yagi, T., Inoue, K. (1977) High temperature-pressure phase boundaries in silicate systems using in situ X-ray diffraction. In: *High Pressure Research*, pp. 535–602. Academic Press, New York.

Akimoto, S., Komada, E., and Kushiro, I. (1967) Effect of pressure on the melting of spinel polymorph of Fe_2SiO_4. *J. Geophys. Res.*, **72**, 679–686.

Bell, P.M. Roseboom E.H., Jr. (1969) Melting relationships of jadeite and albite to 45 kbars with comment on melting diagrams of binary systems at high pressures. *Mineral. Soc. Amer. Spectr.* Paper 2, 151–162.

Birch, F. (1966) Compressibility: elastic constants. In: *Handbook of Physical Constants. Geol. Soc. Amer.*, **97**, 107–173.

Boettcher, A.L., Wayne Burnham, C., Windom, K.E., Bohlen, S.R. (1982) Liquids, glasses and melting of silicates to high pressures. *J. Geol.*, **90**, 127–138.

Bottinga, Y. (1985) On the isothermal compressibility of silicate liquids at high pressure (Submitted).

Bottinga, Y., Richet, P., and Weill, D.F. (1983) Calculation of the density and thermal expansion coefficient of silicate liquids. *Bull. Mineral.*, 106, 129–138.

Bottinga, Y., Weill, D.F., and Richet, P. (1982) Density calculations for silicate liquids I. Revised method for aluminosilicate compositions. *Geochim. Cosmochim. Acta*, **46**, 909–919.

Bottinga, Y., and Weill, D.F. (1970) Densities of liquid silicate systems calculated from partial molar volumes of oxide components, *Amer. J. Sci.*, **269**, 169–182.

Bowen, N. L., and Schairer, J.F. (1932) The system $FeO-SiO_2$. *Amer. J. Sci.*, **24**, 5, 177–213.

Boyd, F.R., England, J.L., and Davis, B.T.C. (1964) Effects of pressure on the melting and polymorphism of enstatite, $MgSiO_3$. *J. Geophys. Res.*, **69**, 2101–2110.

Bucaro, J.A., and Dardy, H.D. (1976) Equilibrium compressibility of glassy SiO_2 between the transformation and melting temperature. *J. Noncrystall. Solids*, **20**, 149–151.

Brückner, R. (1970) Properties and structure of vitreous silica. *J. Noncrystall. Solids*, **5**, 123–175.

Carmichael, I.S.E., Nichols, J., Spera, F.J., Wood, B.J., and Nelson S.A. (1977) High temperature properties of silicate liquids: application to the equilibrium and ascent of basic magma, *Philos. Trans. Roy. Soc. Lond.* **A286**, 373–421.

Clark, S.P. (1959) Effect of pressure on the melting points of eight alkali halides. *J. Chem. Phys.*, **31**, 1526–1531.

Chen, C.H., and Presnall, D.C. (1975) The system $Mg_2SiO_4-SiO_2$ at pressures up to 25 kbars. *Amer. Mineral.*, **60**, 398–406.

Davis, B.T.C. (1964) The system diopside-forsterite-pyrope at 40 kbar. *Carnegie Inst. Washington Year Book*, **63**, 165–171.

Davis, B.T.C., and England, J.L. (1964) The melting of forsterite up to 50 kbars. *J. Geophys. Res.*, **64**, 1113–1116.

Frisillo, A.L., and Barsch, G.R. (1972) Measurement of single crystal elastic constants of bronzite as a function of pressure and temperature. *J. Geophys. Res.*, **77**, 6360–6384.

Ghiorso, M.S., Carmichael, I.S.E., Rivers, M.L., and Sack, R.O. (1984) The Gibbs free energy of mixing of natural silicate liquids: an expanded regular solution approximation for the calculation of magmatic intensive variable. *Contrib. Mineral. Petrol.*, **84**, 107–145.

Haselton, H.T., and Westrum, E.F. (1982) Low temperature heat capacities of synthetic pyrope, grossular, and $pyrope_{60}$-$grossular_{40}$. *Geochim. Cosmochim. Acta*, **46**, 909–919.

Herzberg, C.T. (1984) Chemical stratification in the silicate earth. *Earth Planet Sci. Lett.*, **67**, 249–260.

Holland, T.J.B. (1981), Thermodynamic analysis of simple mineral systems. *Adv. Phys. Geochem.* **1**, 19–34.

Jackson, I. (1976) Melting of the silica isotypes SiO_2, BeF_2 and GeO_2 at elevated pressures. *Phys. Earth Planet. Int.*, **13**, 218–231.

Kim, K.T., Vaidya S.N., and Kennedy, G.C. (1972) The effect of pressure on the temperature of eutectic minimums in two binary systems: NaF–NaCl and CsCl–NaCl. *J. Geophys. Res.*, **77**, 6984–6989.

Kumazawa, M., and Anderson, O.L. (1969) Elastic moduli, pressure derivatives of single-crystal olivine and single-crystal forsterite. *J. Geophys. Res.*, **74**, 5961–5972.

Levien, L., Prewitt C.T., and Weidner, D.J. (1979) Compression of pyrope. *Am. Mineral.*, **64**, 805–808.

Mo, X., Carmichael, I.S.E., Rivers, M., and Stebbins, J.F. (1982) The partial molar volume of Fe_2O_3 in multicomponent silicate liquids and the pressure dependence of oxygen fugacity in magmas. *Mineral. Mag.*, **45**, 237–245.

Mysen, B., and Kushiro, I. (1977) Compositional variations of coexisting phases with degree of melting of peridotite in the upper mantle. *Amer. Mineral.*, **62**, 843–865.

Navrotsky, A., and Akaogi, M. (1984), α, β, γ Phase relations in Fe_2SiO_4–Mg_2SiO_4 and Co_2SiO_4–Mg_2SiO_4: calculation from thermodynamic data and geophysical applications. *J. Geophys. Res.*, **89**, 10135–10140.

Navrotsky, A., Capobianco, C., and Stebbins, J.F. (1982) Some thermodynamic and experimental constraints on the melting of albite at atmospheric pressure, 1982. *J. Geol.*, **96**, 679–698.

Ohtani, E. (1983) Melting temperature distribution and fractionation in the lower mantle. *Phys. Earth Planet. Int.*, **33**, 12–25.

Ohtani, E. (1979) Melting relations of Fe_2SiO_4 up to about 200 kbar. *J. Phys. Earth*, **27**, 189–208.

Ohtani, E., Kumazawa, M., Kato, T., and Irfune, T. (1982) Melting of various silicates at elevated pressures. *Adv. Earth Planet. Sci.*, **12**, 259–270.

Ohtani, E., Irfune, T., and Fujino, F. (1981) Fusion of pyrope at high pressures and rapid crystal growth from the pyrope melt. *Nature*, **294**, 62–64.

Ohtani, E., and Kumazawa, M. (1981), Melting of forsterite Mg_2SiO_4 up to 15 GPa. *Phys. Earth Planet. Int.*, **27**, 32–38.

Orr, R.L. (1953) High temperature heat contents of magnesium orthosilicate and ferrous orthosilicate. *J. Am. Chem. Soc.*, **75**, 528–529.

Prigogine, I., and Defay, R. (1962) *Chemical Thermodynamics*. John Wiley & Sons, New York.

Richet, P., and Bottinga, Y. (1985) Heat capacity of aluminum free liquid silicates. *Geochim. Cosmochim. Acta*, **49**, 471–486.

Richet, P., and Bottinga, Y. (1984) Anorthite, andesine, wollastonite, diopside, cordierite and pyrope: thermodynamics of melting, glass transitions, and properties of the amorphous phases. *Earth Planet. Sci. Lett.*, **67**, 415–432.

Richet, P., Bottinga, Y., Denielou, L., Petitet, J.P. and Tequi, C. (1982) Thermodynamic properties of quartz, cristobalite and amorphous SiO_2: drop calorimetry measurements between 1000 and 1800 K and review from 0 to 2000 K. *Geochim. Cosmochim. Acta*, **46**, 2639–2658.

Rigden, S.M., Ahrens, T.J., and Stolper, E.M. (1984) Density of liquid silicates at high pressures. *Science*, **226**, 1071–1074.

Robie, R.A., Hemmingway, B.S., and Fisher, J.R. (1978) Thermodynamic properties of minerals and related substances at 298. 15 K and 1 bar (10^5 pascals) pressure and at high temperature. U.S. Geol. Survey Bull., 1452.

Robie, R.A., Bethke, P.M., Toulmin, M.S., and Edwards, J.L. (1966) X-ray crystallographic data, densities, and molar volumes of minerals. In *Handbook of Physical Constants*, pp. 27–73, *Geol. Soc. Am. Mem.* **97**.

Roeder, P.L., and Emslie, R.F. (1970) Olivine-liquid equilibrium. *Contrib. Mineral. Petrol.*, **29**, 275–289.

Skinner, B.J. (1966) Thermal expansion. In: *Handbook of Physical Constants*, pp. 78–96. *Geol. Soc. Amer. Mem.*, **96**.

Stebbins, J.F., and Carmichale, I.S.E. (1984) The heat of fusion of fayalite. *Amer. Mineral.*, **69**, 292–297.

Stebbins, J.F., Carmichale, I.S.E. and Moret, L.K. (1984), Heat capacities of silicate liquids and glasses: approximations and interpretations. *Contrib. Min. Petrol.*, **86**, 131–148.

Stolpter, E.M., Walker, D., Hager, B.A., and Hays, J.F. (1981) Melt segregation from partially molten source region: the importance of melt density and source region size. *J. Geophys. Res.*, **86**, 6261–6271.

Sumino, Y. (1979) The elastic constants of Mn_2SiO_4, Fe_2SiO_4, and Co_2SiO_4 and the elastic properties of olivine group minerals at high temperature. *J. Phys. Earth*, **27**, 209–238.

Sumino, Y., Nishizawa, O., Goto, T., Ohno, I., Ozima, M. (1977) Temperature variation of elastic constants of single crystal forsterite between $-190°$ and 400°C. *J. Phys. Earth*, **25**, 377–392.

Suzuki, I., Seya, K., Takei, H., and Sumino, Y. (1981) Thermal expansion of fayalite. *Phys. Chem. Mineral.*, **7**, 60–63.

Takahashi, E. (1978) Partitioning of Ni^{2+}, Co^{2+}, Fe^{2+}, Mn^{2+} and Mg^{2+} between olivine and silicate melts: compositional dependence of the partition coefficient. *Geochim. Cosmochim. Acta*, **42**, 1829–1844.

Takahashi, E., and Kushiro, I. (1983) Melting of dry peridotite at high pressures and basalt magma genesis. *Amer. Mineral.*, **68**, 859–879.

Takéuchi, Y., Yamanaka, T., Haga, N., and Hirano, M. (1984), High temperature crystallography of olivine and spinel. In: *Material Science of the Earth's Interior*. Terra Scientific Publishing, Tokyo.

Walker, D., Stolper, E.M., and Hays, J.F. (1978) A numerical treatment of melt/solid segregation: size of the eucrite parent body and stability of the terrestrial low-velocity zone. *J. Geophys. Res.*, **84**, 6005–6013.

Watanabe, H. (1982) Thermodynamic properties of synthetic high pressure compounds relevant to the earth's mantle. *Adv. Earth Planet. Sci.*, **12**, 441–464.

Weidner, D.J., Wang, H., and Ito, J. (1978) Elasticity of orthoenstatite. *Phys. Earth Planet. Int.*, **17**, 7–13.

Williams, D.W., and Kennedy, G.C. (1969) Melting curve of diopside to 50 kbars. *J. Geophys. Res.*, **74**, 4359–4366.

Wolf, G.H., and Jeanloz, R. (1984) Lindemann melting law: anharmonic correction and test of validity for minerals. *J. Geophys. Res.*, **89**, 7821–7835.

Yagi, T., Ida, Y., Sato, Y., and Akimoto, S. (1975) Effect of hydrostatic pressure on the lattice parameters of olivine up to 70 kbar. *Phys. Earth Planet. Int.*, **10**, 348–354.

Chapter 8
The Relationship Between Activities of Divalent Cation Oxides and the Solution of Sulfide in Silicate and Aluminosilicate Liquids

Christopher D. Doyle

Introduction

At oxygen fugacities below those of approximately the nickel-nickel oxide buffer, sulfur dissolves in silicate and aluminosilicate liquids by the gas-melt anion exchange reaction,

$$O^{2-}{}_{\text{melt}} + 0.5S_{2\,\text{gas}} = S^{2-}{}_{\text{melt}} + 0.5O_{2\,\text{gas}} \tag{1}$$

(Fincham and Richardson, 1954; Richardson and Fincham, 1954; Katsura and Nagashima, 1974; Connolly and Haughton, 1972). The corresponding equilibrium constant, K_1 is given by

$$K_1 = (a_{S^{2-}}/a_{O^{2-}})(f_{O_2}/f_{S_2})^{0.5} = (\gamma_{S^{2-}}/a_{O^{2-}})C_S^m, \tag{2}$$

where C_S^m (the *molar* sulfide capacity) is the sulfide content of a silicate melt that is normalized (Doyle, 1983) to account for the effect of the composition of the coexisting gas, a_i and γ_i are activity and activity coefficient for corresponding component i in the melt, respectively; f_i-fugacity of component i in the equilibrium gas.

In their pioneering study, Fincham and Richardson (1954) demonstrated that C_S^m in $MO-SiO_2$ melts varied markedly with change of the divalent cation M at constant mole fraction of silica. Consequently, many later investigators have tried to relate changes of the activity of a divalent cation oxide to the compositional dependence of C_S^m in the same liquids (Kalyanram et al., 1960; Carter and Macfarlane, 1957a,b; Abraham et al., 1960a,b; Richardson, 1963; Sharma and Richardson, 1965, Haughton et al., 1974; Doyle, 1983).

This paper presents a simple model for the mixing of silicate and aluminosilicate melts that predicts the observed relationship between the solution behavior of divalent cation oxides and the corresponding variations of the molar sulfide capacity. The generated short equations are shown to account readily for the available data on simple silicate liquids in the system SiO_2–MnO–(Mg, Ca)O. These relationships are then extended to natural melts.

Theory—The Ideal Mixing Model

Activities of Divalent Cation Oxides

In this study, a magma is assumed to consist of a variously interconnected silicate and aluminosilicate polymeric network that encloses positions or sites on which the divalent cations are randomly distributed. The predictions of this simple ideal mixing model are best visualized using the hypothetical ternary system Matrix–MO–NO (Fig. 1). The component termed Matrix refers to everything in these melts other than the divalent cation oxides MO and NO. A series of liquids, such as A–t–B (Fig. 1), that have the same mole fraction of Matrix ($X_{\mathrm{Matrix}} = 1 - X_{\mathrm{MO}} - X_{\mathrm{NO}}$) are assumed to be structurally similar. Therefore, according to the above hypothesis, the activities of the end-member pseudobinary melt components (A and B) equal their mole fractions in the ternary liquid t. However, the activities of MO, NO and Matrix do *not* equal their mole fractions but instead must obey simple relations in the compositional range A–B.

In this model, the formation of ternary liquid t from the binary melts A and B occurs with no enthalpy or volume of mixing and the entropy of mixing is solely configurational arising only from the random distribution of the divalent cations. Therefore, the molar Gibbs function, G^t, for the ternary liquid is

$$\begin{aligned} G^t/RT &= (m/(m+n))G^A/RT + (n/(m+n))G^B/RT + G_{\mathrm{mix}}/RT, \\ &= (m/(m+n))[s \ln a^A_{\mathrm{Matrix}} + (m+n)\ln a^A_{\mathrm{MO}}] + \\ &\quad (n/(m+n))[s \ln a^B_{\mathrm{Matrix}} + (m+n)\ln a^B_{\mathrm{NO}}] + \\ &\quad [(m/(m+n))\ln(m/(m+n)) + (n/(m+n))\ln(n/(m+n))] \qquad (3) \end{aligned}$$

where the superscripts refer to the liquids A, B, and t, G_{mix} is the Gibbs function for the ideal mixing of these melts and *m*, *n*, and *s* are the mole fractions of MO,

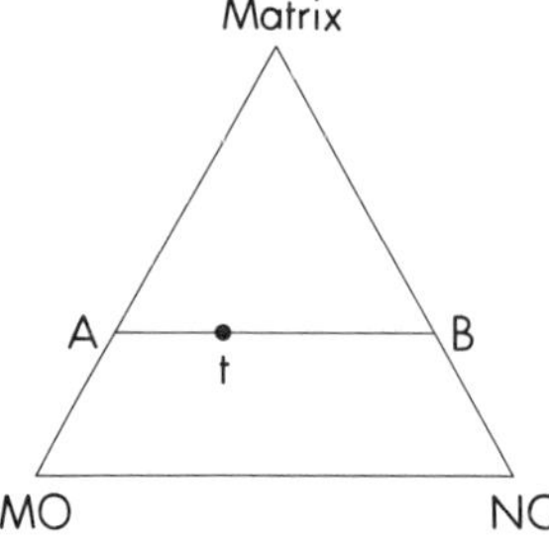

Fig. 1. Hypothetical ternary system Matrix–MO–NO. The component termed Matrix refers to everything in these melts other than the divalent cation oxides MO and NO. The composition 't' represents a melt lith the same mole fraction of Matrix as liquids A and B.

NO and Matrix respectively in ternary liquid t. Use of the well-known relation between G^t and a^t_{MO} at constant temperature and pressure,

$$\ln a^t_{MO} = G^t/RT + (1 - m)(\partial(G^t/RT)/\partial m)_{n/s} \quad (4)$$

yields the activity of the divalent cation oxide MO in ternary melt t,

$$a^t_{MO} = a^A_{MO}(m/(m + n))(a^A_{Matrix}/a^B_{Matrix})^{sn/(m+n)^2} \quad (5)$$

However, if the anionic natures of liquids A, B, and t are unaffected by the M–N substitution then a_{Matrix} is constant in the pseudobinary A–B. Using this simple assumption, the activity of MO in the melt t is

$$a^t_{MO} = a^A_{MO}(m/(m + n)) = a^A_{MO}(M/(M + N)) \quad (6)$$

where (M/(M + N)) is the fraction of the divalent cations that is M. Because N can be one or more divalent cations, this relation can reasonably be expected to hold for a variety of silicate and aluminosilicate systems.

The Solution of Sulfide

If the melt t (Fig. 1) is placed in a furnace and allowed to equilibrate with a sulfur- and oxygen-bearing gas then a small concentration of sulfide will dissolve in the liquid provided that the oxygen fugacity is below that of the nickel-nickel oxide buffer (Fincham and Richardson, 1954; Richardson and Fincham, 1954; Katsura and Nagashima, 1974; Connolly and Haughton, 1972). If the sulfur content remains low, the S^{2-} ions can reasonably be assumed to be randomly distributed within the anionic network of the liquid. If these sulfide anions are affected by a 'regular-solution' type interaction with the ideally-mixed divalent cations (M^{2+} and N^{2+}) then the molar excess Gibbs function (G^{XS}) (Hildebrand et al., 1970) of the ternary melt t is

$$G^{xs} = H_{mix} = W_{MS}(M/(M + N))A_S + W_{NS}(N/(M + N))A_S \quad (7)$$

In this equation, W_{MS} and W_{NS} are the molar coefficients for the interactions of the S^{2-} ions with the M^{2+} or N^{2+} cations respectively and A_S is the anion fraction of sulfide (i.e. the fraction of the network that is sulfide). The corresponding activity coefficient of sulfide ions ($\gamma_{S^{2-}}$) at constant temperature and pressure can be calculated from G^{XS} using the relation,

$$\ln \gamma_{S^{2-}} = G^{xs}/RT + (1 - A_S)(\partial(G^{xs}/RT)/\partial A_S)_{\text{see below}} \quad (8)$$

The partial differentiation takes place with no compositional change on the divalent cations sites and, with the exception of sulfide, the ratios of all anionic species in the melt remaining constant. Combination of the two previous equations yields

$$\ln \gamma_{S^{2-}} = (1/RT)[W_{MS}(M/(M + N)) + W_{NS}(N/(M + N))] \quad (9)$$

Therefore, the solution of sulfide is predicted to obey Henry's law because the activity coefficient $\gamma_{S^{2-}}$ is independent of A_S. However, $\gamma_{S^{2-}}$ is predicted to vary exponentially with the divalent cation fractions. Analogous equations for the activity coefficients of M^{2+} and N^{2+} indicate that the divalent cations are hardly affected by the interactions if the sulfide concentration is small.

Combination of Eq. (1) and (9) yields the molar sulfide capacity in melt t ($C_S^{m:t}$), given by

$$\log C_S^{m:t} = \log[A_S(f_{O_2}/f_{S_2})^{0.5}]$$
$$= (1/2.303RT)[W_{MS}(M/(M+N)) + W_{NS}(N/(M+N))] + \log(K_1 a^t_{O^{2-}}) \quad (10)$$

where $a^t_{O^{2-}}$ is the activity of oxide ions in the melt t. However, $a_{O^{2-}}$ is buffered by the condensation reactions between the silicate and aluminosilicate polymers in the network and is expected to be constant in the structurally similar melts A, B and t (Fig. 1) (Toop and Samis, 1962a, b; Masson, 1965, 1977; Hess, 1971, 1977). Therefore, writing similar equations for log C_S^m in the liquids A and B it is readily shown that,

$$\log C_S^{m:t} = (M/(M+N))\log C_S^{m:A} + (N/(M+N))\log C_S^{m:B} \quad (11)$$

This relation (known as the Flood-Grjotheim equation (Flood and Muan, 1950; Flood and Grjotheim, 1952, 1953)) predicts that log C_S^m varies linearly in the pseudobinary A–B (Fig. 1).

Simple Silicate Liquids

The purpose of this section is to show that the available data on activities of divalent cation oxides and molar sulfide capacities in geologically interesting simple silicate liquids are consistent with the predictions of Eq. (6) and (11) respectively. In these melts the component termed Matrix (Fig. 1) is SiO_2.

Activities of Divalent Cation Oxides

The available measurements of the activity of MnO in melts with constant mole fraction of silica in the systems SiO_2–MnO–MgO at 1923 K and SiO_2–MnO–CaO at 1773 and 1923 K are shown in Fig. 2 (Abraham, 1959; Abraham et al., 1960a; Mehta and Richardson, 1965; Gaskell, 1974). These pseudobinaries are chosen because accurate data on a_{MnO} are available over most of the compositional range. Though small variations do occur, these measurements are in good agreement with the predictions of Eq. (6). This agreement strongly suggests that the divalent cations mix essentially ideally on the sites that they occupy in the liquids of these simple silicate systems.

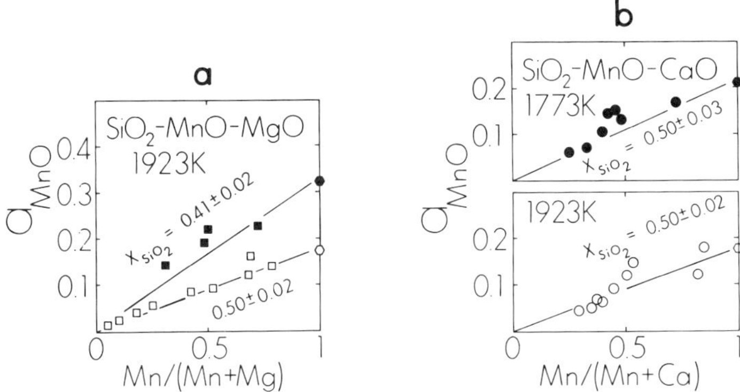

Fig. 2. Variation of the activity of MnO with change of the divalent cation fraction of manganese at constant mole fraction of silica in the ternary systems SiO_2–MnO–MgO (a) at 1923 K and SiO_2–MnO–CaO (b) at 1773 and 1923 K. The standard state is solid 'MnO'. The data are from Abraham (1959), Abraham et al. (1960a), Mehta and Richardson (1965) and Gaskell (1974). The hexagons represent interpolated data from the respective silicate binary systems.

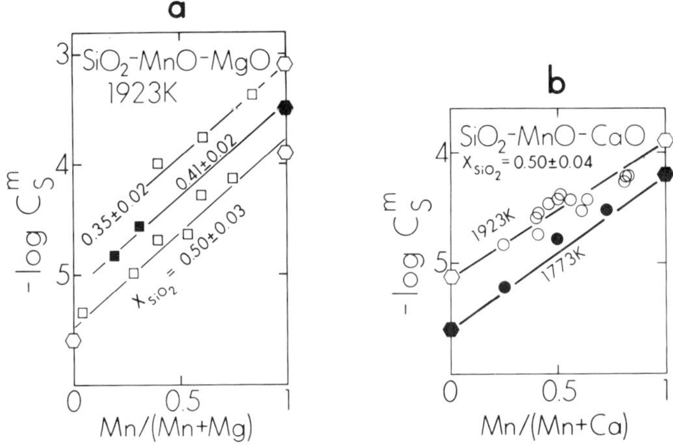

Fig. 3. Variation of the logarithm of the molar sulfide capacity (log C_S^m) with change of the divalent cation fraction of manganese at constant mole fraction of silica for approximately the same pseudobinary systems shown in Fig. 2. The data are from Fincham (1953), Davies (1955), Abraham (1959), Kalyanram (1959) and Sharma (1963). The hexagons represent interpolated data from the respective silicate binary systems.

The Solution of Sulfide

The corresponding determinations of C_S^m in approximately the same liquids are shown in Fig. 3 (Fincham, 1953; Davies, 1955; Abraham, 1959; Kalyanram, 1959; Sharma, 1963). Present knowledge of the constitution of the network of silicate liquids precludes determination of the anion fraction of sulfide (A_S). Therefore, C_S^m is calculated by assuming that A_S is given by the molar ratio

$$A_S = (S/SiO_2 + S)). \tag{12}$$

The data are clearly in good agreement with Eq. (11). Furthermore using Eq. (10), the molar interaction coefficients are essentially constant [$(W_{MnS} - W_{MgS}) = -57 \pm 5, -60 \pm 4$ and -64 ± 4 KJ/mol for the data in order of increasing silica content on Fig. 3a; $(W_{MnS} - W_{CaS}) = 47 \pm 3$ (1773 K) and -42 ± 4 (1923 K) KJ/mol for the data on Fig. 3b]. The demonstrated log-linearity and the similarity of the estimates for the interaction coefficients suggest that the proposed simple model is a good one.

Multicomponent Aluminosilicate Liquids

Activities of Divalent Cation Oxides

The ideal mixing model has recently been applied to the study of the compositional dependencies of the activities of divalent cation oxides in natural liquids. Doyle and Naldrett (1986) investigated variations of a_{FeO} at 1600 K in three ternary aluminosilicate systems (Fig. 4), of the type Matrix–FeO–MgO, that contain synthetic analogs of standard diabase W-1 (C1, Fig. 4A), MORB (C2, Fig. 4B) and high-TiO_2 mare basalt 74275 (C3, Fig. 4C) respectively. Using the ideal mixing model, the activity of FeO in a ternary liquid is given by

$$a_{FeO}^{ternary} = a_{FeO}^{Mg-free}(Fe/(Fe + Mg)), \tag{13}$$

or conversely,

$$a_{FeO}^{Mg-free} = a_{FeO}^{ternary}((Fe + Mg)/Fe). \tag{13a}$$

These relations are directly analogous to Eq. 6 and, therefore, their derivation requires a similar set of assumptions. Using Eq. 13a, the data from within each ternary are projected into the corresponding Mg-free binary in Fig. 5. Clearly, the binary (filed symbols in Fig. 5) and the projected ternary data from the different isoactivity curves (open symbols) are in excellent agreement with the predictions of the ideal mixing model at constant X_{Matrix}. These results strongly suggest that the divalent cations of ferrous iron and magnesium mix ideally on the sites that they occupy in the melts of the ternaries.

In a similar study, Doyle and Naldrett (1987) investigated the compositional dependence of a_{NiO} at 1673 K in a ternary system, of the type Matrix–NiO–MgO,

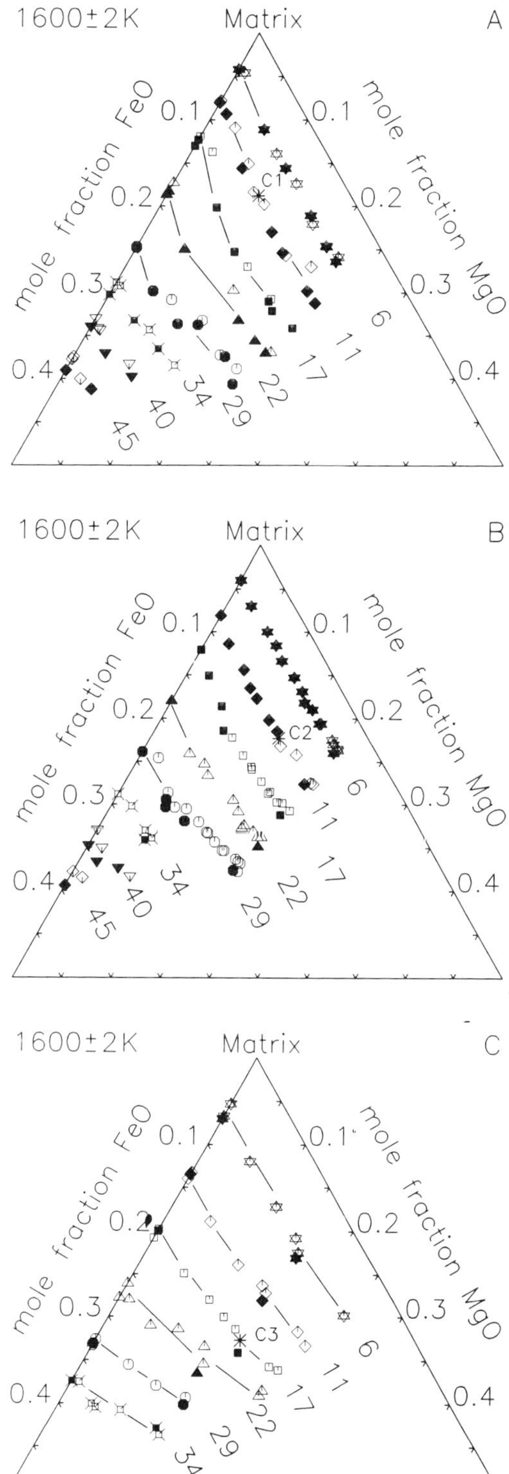

Fig. 4. Three Matrix–FeO–MgO aluminosilicate systems contoured for the activity of ferrous oxide (± 0.015) at 1600 ± 2 K. The standard state is liquid $Fe_{0.947}O$ (Doyle, 1988). Runs performed at the same a_{FeO} have the same symbol. Filled and open symbols represent runs from starting materials that were initially richer and poorer in FeO respectively. The Matrix components were prepared such that compositions C1, C2 and C3 (on Fig. 4A, 4B and 4C respectively) are chemically similar to standard diabase W-1, MORB and high-TiO_2 mare basalt 74275 respectively. The numbers next to each curve on these diagrams are $a_{FeO} \times 100$.

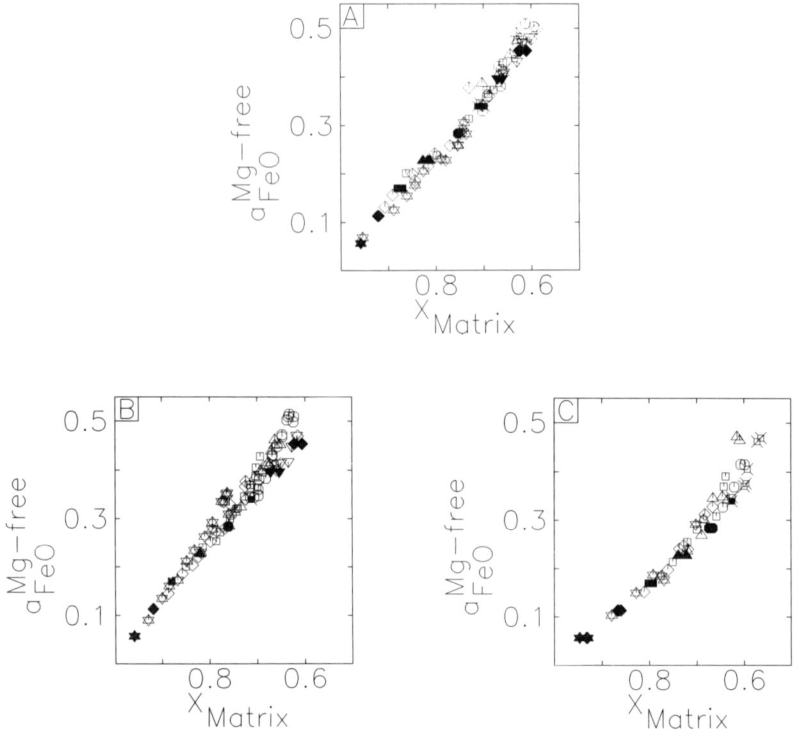

Fig. 5. Activity of ferrous oxide plotted against mole fraction of Matrix for the Mg-free binaries of the ternary systems shown in Fig. 4 (4A corresponds to 5A and so on). The standard state is liquid $Fe_{0.947}O$ (Doyle, 1988). The filled symbols are the data for the Mg-free binary liquids. The open symbols represent data projected to the binary from within the corresponding ternary using Eq. (13a). The symbols are the same as in Fig. 4.

that contains a synthetic analog of standard diabase W-1 (C4, Fig. 6). Using the relation,

$$a_{\text{NiO}}^{\text{Mg-free}} = a_{\text{NiO}}^{\text{ternary}}((\text{Ni} + \text{Mg})/\text{Ni}). \tag{14a}$$

the data from the three isoactivity curves in Fig. 6 are projected to the Mg-free binary in Fig. 7. Because they coincide to produce one line, this strongly suggests that the divalent cations of nickel and magnesium are randomly distributed on the sites that they occupy in the ternary melts.

In both the above studies, each of the investigated ternary systems (Fig. 4 and 6) contains a melt composition (C1–C4) that has a close chemical similarity with a geologically important mafic rock type. Because the ideal mixing hypothesis clearly holds for the data in Fig. 4 through 7, it was concluded that this model is also valid for the mafic magmas concerned.

Doyle and Naldrett (1986) and Doyle (1988) extended the ideal mixing hypothesis to account for all the available data on the activity of FeO in petrologically

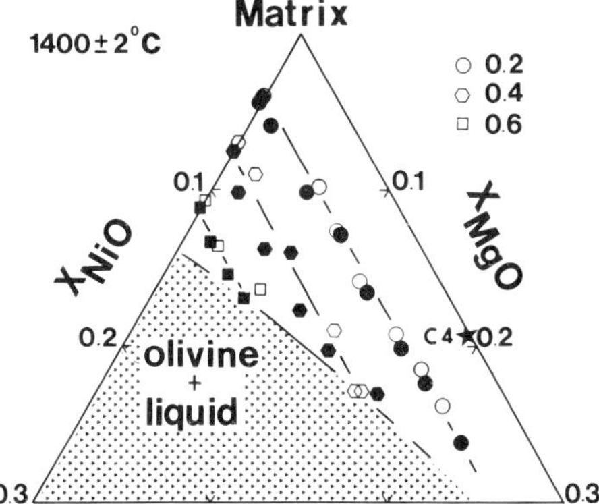

Fig. 6. Matrix–NiO–MgO aluminosilicate system contoured for the activity of nickel oxide ($\pm 5\%$) at 1673 $\pm$ 2 K. The standard state is solid NiO (Doyle and Naldrett, 1987). Runs performed at the same a_{NiO} have the same symbol. Filled and open symbols represent runs from starting materials that were initially richer and poorer in NiO respectively. The Matrix component was prepared such that composition C4 is chemically similar to standard diabase W-1.

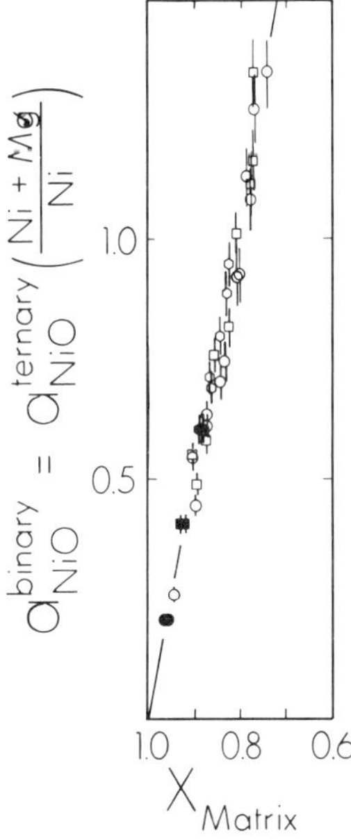

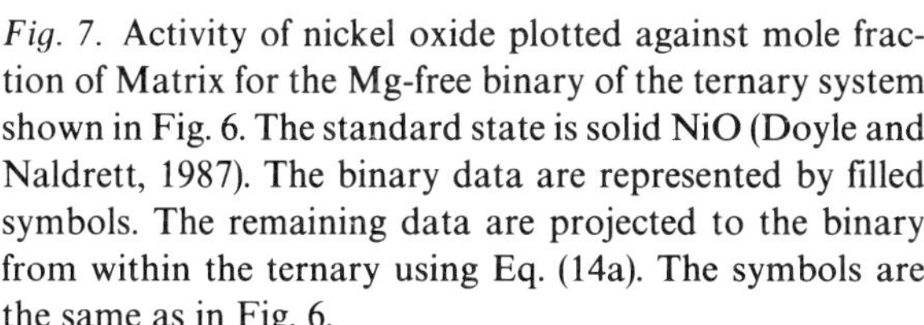
Fig. 7. Activity of nickel oxide plotted against mole fraction of Matrix for the Mg-free binary of the ternary system shown in Fig. 6. The standard state is solid NiO (Doyle and Naldrett, 1987). The binary data are represented by filled symbols. The remaining data are projected to the binary from within the ternary using Eq. (14a). The symbols are the same as in Fig. 6.

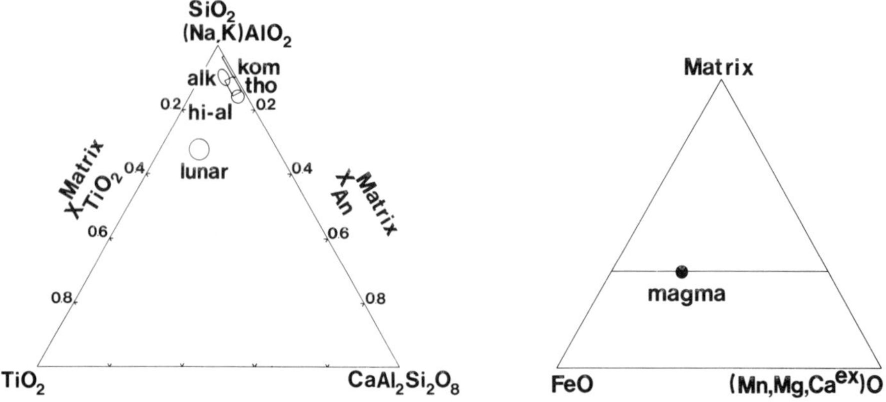

Fig. 8. Sexternary system $[SiO_2–(Na,K)AlO_2–CaAl_2Si_2O_8–TiO_2]–FeO–(Mn,Mg,Ca^{ex})O$ that contains the compositions of most metaluminous rocks. The system is drawn as two triangles combining all the Matrix components (a) to form a Matrix–FeO–$(Mn,Mg,Ca^{ex})O$ ternary system as shown in (b). The Matrix is drawn as a triangle by projecting the silica apex so that SiO_2 plots in the same position as $(Na,K)AlO_2$. In this way the ranges of X_{An}^{Matrix} and $X_{TiO_2}^{Matrix}$ of komatiites (kom), alkali basalts (alk), tholeiites (tho), high-alumina basalts (hi-al) and high-TiO_2 mare basalts (lunar) are clearly illustrated. No attempt has been made to distinguish these magma types according to their $(X_{(Na,K)AlO_2}^{Matrix}/X_{SiO_2}^{Matrix})$ though it is an important variable (Doyle, in 1983).

interesting melts. These authors considered that metaluminous liquids lie in ternary systems of the type Matrix–FeO–$(Mn,Mg,Ca^{ex})O$ (i.e. M = Fe and N = $(Mn + Mg + Ca^{ex})$ in Fig. 1). The Matrix component of a particular melt is made up of four constituents, namely SiO_2, $(Na,K)AlO_2$, $CaAl_2Si_2O_8$ (An) and TiO_2 (Fig. 8). The term $Ca^{ex}O$ (above) thus refers to the calcium remaining after a portion is allotted to form $CaAl_2Si_2O_8$ (An). A particular Matrix–FeO–$(Mn,Mg,Ca^{ex})O$ ternary is specified by the mole fractions of the constituents in its Matrix component (i.e. by specifying $X_{SiO_2}^{Matrix}$, $X_{(Na,K)AlO_2}^{Matrix}$, X_{An}^{Matrix} and $X_{TiO_2}^{Matrix}$).

The model is based on the well-known observations that $SiO_2–NaAlO_2–KAlO_2–CaAl_2Si_2O_8$ liquids are completely polymerized. (Seifert et al., 1982; Taylor and Brown, 1979a,b; Mcmillan et al., 1982; Henderson et al., 1985). Therefore, these four constituents make up the anionic network of a magma. Finally, TiO_2 is a unique Matrix constituent whose role in magma is poorly understood.

Recently, Doyle (1988) refined this model by adding two further simplifying assumptions. First, because TiO_2 contents of most terrestrial magmas are low ($X_{TiO_2}^{Matrix} \sim O$), he argued that data from the system $[SiO_2–(Na,K)AlO_2–CaAl_2Si_2O_8]–FeO–(Mn,Mg,Ca^{ex})O$ could be extrapolated to natural liquids (Fig. 8). Second, substitution of $NaAlO_2$ for $KAlO_2$ was assumed to have a negligible effect on the anionic nature of a magma (i.e. Na and K also mix ideally)

(Belton et al. 1974; Rammensee and Fraser, 1982; Fraser et al., 1985; Hervig and Navrotsky, 1984; Roy and Navrotsky, 1984). Using these assumptions, a_{FeO} in magma can be predicted from data for $[SiO_2–KAlO_2–CaAl_2Si_2O_8]–FeO$ (SKAnF) liquids using

$$a_{FeO}^{magma} = a_{FeO}^{SKAnF}(Fe/(Fe + Mn + Mg + Ca^{ex})), \quad (15)$$

This rearrangement of Eq. (6) is valid provided that the magma and the SKAnF liquid have the same mole fraction of Matrix ($X_{Matrix} = 1 - X_{FeO} - X_{MnO} - X_{MgO} - X_{Ca^{ex}O}$), $X_{(Na,K)AlO_2}^{Matrix}$ and X_{An}^{Matrix} (Fig. 8 with $X_{TiO_2}^{Matrix} = 0$).

Doyle (1988) performed experiments at 1600 K on melts in the system $[SiO_2–KAlO_2–CaAl_2Si_2O_8]–FeO$ and then used Eq. (15) to calculate activities of FeO in multicomponent magmas. Using this method, calculated values of a_{FeO} at 1600 K for the 218 multicomponent melts from Doyle and Naldrett's (1986) study are

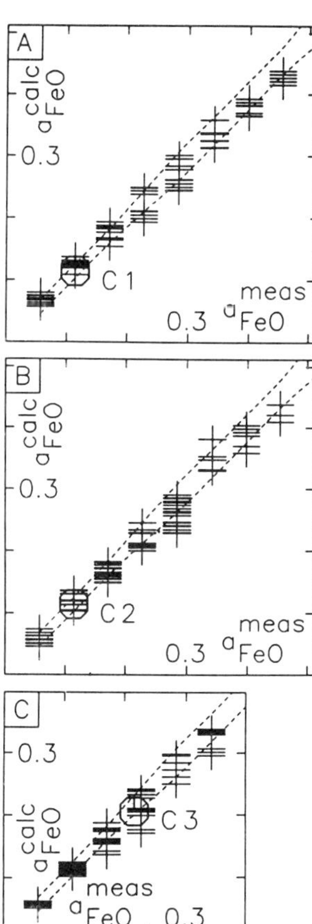

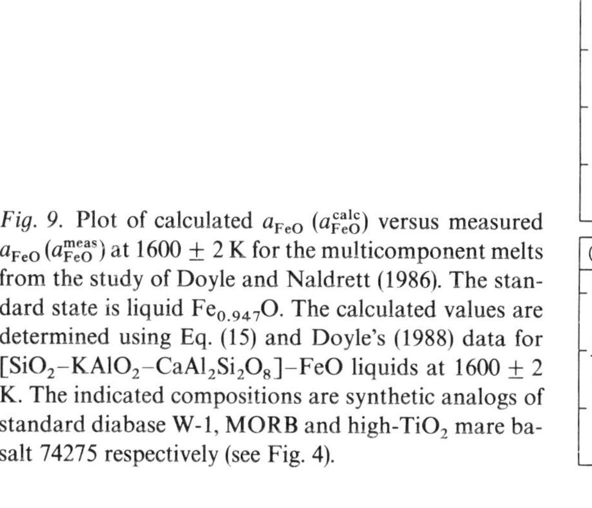

Fig. 9. Plot of calculated a_{FeO} (a_{FeO}^{calc}) versus measured a_{FeO} (a_{FeO}^{meas}) at 1600 ± 2 K for the multicomponent melts from the study of Doyle and Naldrett (1986). The standard state is liquid $Fe_{0.947}O$. The calculated values are determined using Eq. (15) and Doyle's (1988) data for $[SiO_2–KAlO_2–CaAl_2Si_2O_8]–FeO$ liquids at 1600 ± 2 K. The indicated compositions are synthetic analogs of standard diabase W-1, MORB and high-TiO_2 mare basalt 74275 respectively (see Fig. 4).

compared with the measured values from Fig. 4, also at 1600 K, in Fig. 9. The dashed curves indicate the error bar for the experimental determinations of a_{FeO}. The agreement of the calculated and measured activities is excellent. Therefore, the proposed simple model for the mixing of the components of molten magma is concluded to be a good one.

The Solution of Sulfide

Because the ideal mixing model has been so successful in predicting the solution behavior of divalent cation oxides in magmas, this model is also likely to provide a good base for studies of the solution of sulfide in petrologically interesting melts. Therefore, by direct analogy to the derivation of Eq. (11), the logarithm of the molar sulfide capacity ($\log C_S^m$) in a magma is given by

$$\begin{aligned}\log C_S^m = {} & (\mathrm{Fe}/(\mathrm{Fe} + \mathrm{Mn} + \mathrm{Mg} + \mathrm{Ca^{ex}})\log C_S^{m:\mathrm{Fe}} \\ & + (\mathrm{Mn}/(\mathrm{Fe} + \mathrm{Mn} + \mathrm{Mg} + \mathrm{Ca^{ex}})\log C_S^{m:\mathrm{Mn}} \\ & + (\mathrm{Mg}/(\mathrm{Fe} + \mathrm{Mn} + \mathrm{Mg} + \mathrm{Ca^{ex}})\log C_S^{m:\mathrm{Mg}} \\ & + (\mathrm{Ca^{ex}}/(\mathrm{Fe} + \mathrm{Mn} + \mathrm{Mg} + \mathrm{Ca^{ex}})\log C_S^{m:\mathrm{Ca^{ex}}} \end{aligned} \tag{16}$$

where terms of the type $C_S^{m:j}$ refer to the value of C_S^m in the liquid in which the only divalent cation is j. This relationship is strictly valid *only* for melts with the same Matrix composition, that are structurally similar (same X_{Matrix}) and thus have essentially constant $a_{O^{2-}}$ over the entire compositional range.

Unfortunately, no study has ever contoured for $\log C_S^m$ a ternary aluminosilicate system, of the type shown in Fig. 1, that contains a synthetic analog of a natural liquid. However, by assuming that (1) the ferrous iron term in Eq. 16 is dominant, (2) the activity of oxide ($a_{O^{2-}}$) is essentially constant in many natural liquids, and that (3) the anion fraction of sulfide (A_S) can be approximated by the molar ratio

$$A_S = \mathrm{S}/(\mathrm{SiO_2} + (\mathrm{Na,K})\mathrm{AlO_2} + \mathrm{CaAl_2Si_2O_8} + \mathrm{TiO_2} + \mathrm{S}), \tag{17}$$

all of the available data on $\log C_S^m$ for mafic and ultramafic ($X_{\mathrm{Matrix}} = 0.40 - 0.70$, $X_{\mathrm{SiO_2}}^{\mathrm{Matrix}} = 0.53 - 0.90$, $X_{(\mathrm{Na,K})\mathrm{AlO_2}}^{\mathrm{Matrix}} = 0.0 - 0.20$, $X_{\mathrm{An}}^{\mathrm{Matrix}} = 0.08 - 0.33$, $X_{\mathrm{TiO_2}}^{\mathrm{Matrix}} = 0.0 - 0.23$) aluminosilicate liquids at 1473 and 1523 K are plotted against Fe/(Fe + Mn + Mg + Ca^{ex}) in Fig. 10 (Haughton, 1970; Haughton et al., 1974 (series I and II); Buchanan and Nolan, 1979; Danckwerth et al., 1979). Clearly, some scatter occurs indicating that the above assumptions are not completely valid. However, the approximate linearity of the data at both temperatures suggests that the proposed model for the solution of sulfide holds in natural melts. Indeed, the similarity of the estimates for W_{FeS} (72 ± 7 KJ/mol at 1473 K and 85 ± 9 KJ/mol at 1523 K) calculated using the regression lines to the data for the widely varying compositions in Fig. 10 suggests that this simple approach warrants further testing.

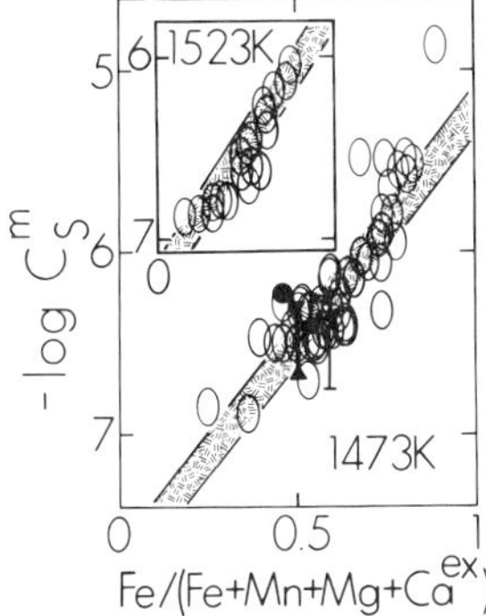

Fig. 10. Variation of the logarithm of the molar sulfide capacity (log C_S^m) with change of the divalent cation fraction of ferrous iron (Fe/(Fe + Mn + Mg + Ca^{ex}) for *all* the available data on mafic and ultramafic ($X_{Matrix} = 0.4 - 0.7$, $X_{SiO_2}^{Matrix} = 0.53 - 0.91$, $X_{(Na,K)AlO_2}^{Matrix} = 0.0 - 0.20$, $X_{An}^{Matrix} = 0.08 - 0.33$, $X_{TiO_2}^{Matrix} = 0.0 - 0.23$) aluminosilicate liquids at 1473 and 1523 K (inset). The data are from Haughton (1970), Haughton et al. (1974, Series I and II), Buchanan and Nolan (1979) and Danckwerth et al. (1979). The one standard deviation limits of Buchanan and Nolan's experiments on DB/3, DB/8 and PAL-685 at 1473 K are shown by the symbols (ⵐ), (•) and (I) respectively. The regression lines to the data are shown as zones consistent with the estimated uncertainty in the data.

Summary

In this study, a model for the ideal mixing of silicate and aluminosilicate melts has been proposed that accounts for the available data on the activities of divalent cation oxides in both simple silicate and multicomponent aluminosilicate liquids. The model has been extended to predict the compositional dependence of the solution of sulfide in geologically interesting melts. The developed simple relationships account for the observed variation of the molar sulfide capacity (C_S^m) in melts of the system SiO_2–MnO–(Mg, Ca)O and to a first approximation predict the variation of C_s in mafic and ultramafic aluminosilicate liquids.

Acknowledgments

This work is an extension of part of the author's Ph.D. dissertation from the University of Connecticut at Storrs. The author thanks Drs. N.H. Gray and A.R. Philpotts (Univ. of Conn.) and A.J. Naldrett (Univ. of Toronto) for their enthusiastic support. He further thanks Dr. C.J.B. Fincham, K.P. Abraham, M.W. Davies, C.B. Alcock and J.H.E. Jeffes for assistance in obtaining copies of the experimental data on simple silicate systems.

References

Abraham K.P. (1959) Thermodynamic aspects of some silicate melts. Ph.D. Dissertation, Univ. London. England.

Abraham K.P. Davies M.W. Richardson F.D. (1960a) Activities of manganese oxide in silicate melts. Iron Steel Inst. J. **196**: 8–87.

Abraham K.P. Davies M.W. Richardson F.D. (1960b) Sulphide capacities in silicate melts. Part I. Iron Steel Inst. J. **196**: 309–312.

Belton G.R., Choudary U.V. Gaskell (1974) Thermodynamics of mixing of sodium-potassium silicates. In Physical Chemistry of Process Metallurgy: The Richardson Conference (ed. J.H.E. Jeffes and R.J. Tait). Instn. Mining Metall., London.

Buchanan D.L. Nolan J. (1979) Solubility of sulfur and sulfide immiscibility in synthetic tholeiitic melts and their relevance to Bushveld-Complex rocks. Can. Mineral. **17**: 483–494.

Carter P.T. Macfarlane T.G. (1957a) Thermodynamics of slag systems. Part I—thermodynamic properties of CaO–Al_2O_3 slags. Iron Steel Inst. J. **185**: 54–62.

Carter P.T. Macfarlane T.G. (1957b) Thermodynamics of slag systems. Part II—thermodynamic properties of CaO–SiO_2 slags. Iron Steel Inst. J. **185**: 62–66.

Connolly J.W.D. Haughton D.R. (1972) The valence of sulfur in glass of basaltic composition formed under conditions of low oxidation potential. Amer. Mineral. **57**: 1515–1517.

Danckwerth P.A. Hess P.C. Rutherford M.J. (1979) The solubility of sulfur in high-TiO_2 mare basalts. Proc. Lunar Planet. Sci. Conf. **10**: 517–530.

Davies M.W. (1955) Thermodynamics of solutions of sulphides and oxides in silicate melts. Ph.D. Dissertation, Univ. London, England.

Doyle C.D. (1983) The solution of ferrous oxide in aluminosilicate melts and its effect on the solubility of sulphur. (U.M.I. Diss. Abs. **44**/04, p. 1033–b). Ph.D. Dissertation, Univ. of Connecticut, Storrs.

Doyle C.D. (1988) Prediction of the activity of FeO in multicomponent magma from known values in the system [SiO_2–$KAlO_2$–$CaAl_2Si_2O_8$]–FeO liquids. Geochim. Cosmochim. Acta **52**: 1827–1834.

Doyle C.D. Naldrett A.J. (1986) Ideal mixing of divalent cations in mafic magma and its effect on the solution of ferrous oxide. Geochim. Cosmochim. Acta **50**: 435–443.

Doyle C.D. Naldrett A.J. (1987) Ideal mixing of divalent cations in mafic magma. II. The solution of NiO and the partitioning of nickel between coexisting olivine and liquid. Geochim. Cosmochim. Acta **51**: 213–219.

Fincham C.J.B. (1953) Equilibria of sulfur and oxygen between gases and liquid silicates. Ph.D. Dissertation, Univ. London, England.

Fincham C.J.B. Richardson F.D. (1954) The behaviour of sulphur in silicate and aluminate melts. Roy. Soc. London Phil. Trans. **A223**: 40–62.

Flood H. Grjotheim K. (1952) Thermodynamic calculation of slag equilibria. Iron Steel Inst. J. **171**: 64–80.

Flood H. Grjotheim K. (1953) Discussion on the thermodynamics of steelmaking. Iron Steel Inst. J. **173**: 274–275.

Flood H. Muan A. (1950) The influence of the cation composition on anion—equilibria in molten salt mixtures. Acta Chem. Scand. **4**: 359–363.

Gaskell D.R. (1974) On the activity of MnO in MnO–SiO_2 melts. Metall. Trans. **5**: 776–778.

Fraser D.G. Rammensee W. Hardwick A. (1985) Determination of the mixing properties of molten silicates by Knudsen Cell mass spectrometry-II. The systems (Na–K)Al-

Si_4O_{10} and (Na–K)$AlSi_5O_{12}$. Geochim. Cosmochim. Acta **49**: 349–361.
Haughton D.R. (1970) The solubility of sulphur in basaltic melts. Ph.D. Dissertation, Queen's Univ., Kingston, Ontario.
Haughton D.R. Roeder P.L. Skinner B.J. (1974) Solubility of sulfur in mafic magmas. Econ. Geol. **69**: 451–467.
Henderson G.S. Bancroft G.M. Fleet M.E. (1985) Raman spectra of gallium and germanium substituted silicate glasses: variations in intermediate range order. Amer. Mineral. **70**: 946–960.
Hervig R.L. Navrotsky A. (1984) Thermochemical study of glasses in the system $NaAlSi_3O_8$–$KAlSi_3O_8$–Si_4O_8 and the join $Na_{1.6}Al_{1.6}Si_{2.4}O_8$–$K_{1.6}Al_{1.6}Si_{2.4}O_8$. Geochim. Cosmochim. Acta **48**: 513–522.
Hess P.C. (1971) Polymer model of silicate melts. Geochim. Cosmochim. Acta **35**: 289–306.
Hess P.C. (1977) Structure of silicate melts. Can. Mineral. **15**: 162–178.
Hildebrand J.H. Prausnitz J.M. Scott R.L. (1970) Regular and related solutions. Van Nostrand Reinhold, N.Y.
Kalyanram M.R. (1959) A study of activities of the constituents of molten slags. Ph.D. Dissertation, Univ. Glasgow, Scotland.
Kalyanram M.R. Macfarlane T.G. Bell H.B. (1960) The activity of calcium oxide in slags in the systems CaO–MgO–SiO_2, CaO–Al_2O_3–SiO_2, and CaO–MgO–Al_2O_3–SiO_2 at 1500°C. Iron Steel Inst. J. **195**: 58–64.
Katsura T. Nagashima S. (1974) Solubility of sulfur in some magmas at 1 atmosphere. Geochim. Cosmochim. Acta **38**, 517–531.
Masson C.R. (1965) An approach to the problem of ionic distribution in liquid silicate. Roy. Soc. London Proc. **A287**: 201–221.
Masson C.R. (1977) Anionic constitution of glass-forming melts. J. Noncryst. Solids **25**: 3–41.
McMillan P., Piriou B. Navrotsky A. (1982) A Raman spectroscopic study of glasses along the joins silica-calcium aluminate, silica-sodium aluminate, and silica-potassium aluminate. Geochim. Cosmochim. Acta **46**: 2021–2037.
Mehta S.R. Richardson F.D. (1965) Activities of manganese oxide and mixing relationships in silicate and aluminate melts. Iron Steel Inst. J. **203**: 524–528.
Rammensee W. Fraser D.G. (1982) Determination of activities in silicate melts by Knudsen Cell mass spectrometry-I. The system $NaAlSi_3O_8$–$KAlSi_3O_8$. Geochim. Cosmochim. Acta **46**: 2269–2278.
Richardson F.D. (1963) The properties and structure of phases and their relevance to process. In J.F. Elliot, ed., Steelmaking: the Chipman Conference. M.I.T. Press.
Richardson F.D. Fincham C.J.B. (1954) Sulphur in silicate and aluminate slags. Iron Steel Inst. J. **178**: 4–15.
Roy B.N. Navrotsky A. (1984) Thermochemistry of charge-coupled substitutions in silicate glasses: the systems $M_{1/nn^+}AlO_2$–SiO_2 (M = Li, Na, K, Rb, Cs, Mg, Ca, Sr, Ba, Pb). Amer. Ceram. Soc. J. **67**: 606–610.
Seifert F., Mysen B.O. Virgo D. (1982) Three-dimensional network structure of quenched melts (glass) in the systems SiO_2–$NaAlO_2$, SiO_2–$CaAl_2O_4$ and SiO_2–$MgAl_2O_4$. Amer. Mineral **67**: 696–717.
Sharma R.A. (1963) Activities in molten silicates. Ph.D. Dissertation, Univ. London, England.
Sharma R.A. Richardson F.D. (1965) Activities of manganese oxide, sulphide capacities, and activity coefficients in aluminate and silicate melts. A.I.M.E. Trans. **233**: 1586–1592.

Taylor M. Brown G.E. Jr. (1979a) Structure of mineral glasses—I. The feldspar glasses $NaAlSi_3O_8$, $KAlSi_3O_8$, $CaAl_2Si_2O_8$. Geochim. Cosmochim. Acta **43**: 61–75.

Taylor M. Brown G.E. Jr. (1979b) Structure of mineral glasses—II. The SiO_2–$NaAlSiO_4$ join. Geochim. Cosmochim. Acta **43**: 1467–1473.

Toop G.W. Samis C.S. (1962a) Activities of ions in silicate melts. A.I.M.E. Trans. **224**: 878–887.

Toop G.W. Samis C.S. (1962b) Some new ionic concepts of silicate slags. Can. Metal. Quart. **1**: 129–152.

Chapter 9
Thermodynamics of the Liquidus in the System Diopside–Water: A Review

Leonid L. Perchuk and Ikuo Kushiro

Existing Experimental Data

Boyd and England (1963) studied melting of diopside under dry conditions up to pressures of 50 kbar. Boettcher et al. (1982) corrected these data below 30 kbar. Eggler (1973) was the first to study the system diopside–water up to a temperature of 1430°C at a pressure of 20 kbar. Hodges (1974) published data at the same pressure up to 1500°C. Rosenhauer and Eggler (1975) repeated Hodges's runs at 20 kbar and discovered a large difference in composition of the system at the univariant point (see Table 1). These authors did not discard their own data three years later (Rosenhauer and Eggler, 1978). Eggler and Burnham (1984) reported data on the liquidus of the system diopside–water studied in a gas vessel at a pressure of 2 kbar. Table 1 compiles the present day data on the water-saturated liquidus.

Perchuk et al. (1988) reported new data on the diopside–water system. The isobaric sections of the system were thoroughly studied at pressures of 10, 15, 20, and 25 kbar (see Fig. 1) using a solid media piston cylinder apparatus with soft ceramic parts and specially designed graphite heater (Kushiro, 1976). The experimental methods are reported in more detail in the paper by Perchuk and Kushiro (1985). Diopside glass and crystals were used as starting materials. The accuracy of the water amount loaded in 2-mm diameter capsules was estimated to be ± 1 wt%. Two to five capsules with different water contents were loaded in one assembly, depending on its diameter, for each particular run. The most consistent data on the liquidus at given T and P in the course of a run were obtained using this approach. For example, the position of the liquidus at a pressure of 20 kbar was corrected: 10 datapoints of Rosenhauer and Eggler (1975) were supplemented by 15 new ones. As a result, the liquidus shows slight curvature in comparison with the previous line (see isobaric section 20 kb in Fig. 1). However, the univariant point appeared at the H_2O content of 18 ± 1 wt% and

Table 1. *T-P* parameters and water contents related to the wet liquidus for the system diopside–H_2O (experimental data)

P, kbar	*t*	H_2O, wt%	Source
2	1293	5.4 ± 0.5	Eggler and Burnham (1984)
5	1295	n.d.	Yoder (1975)
10	1270	10 ± 1	Perchuk, Kushiro, Kosyakov (1988)
15	1265	14.8 ± 1	
20	1245	17 ± 1	Eggler (1973); Eggler and Rosenhauer (1978)
20	1255	18 ± 1	Perchuk, Kushiro, Kosyakov (1988)
25	1250	20.5 ± 1	
30	1265	21.5 ± 1	Hodges (1984)
30	1245	30 ± 1	Eggler and Rosenhauer (1978), Rosenhauer and Eggler (1974)

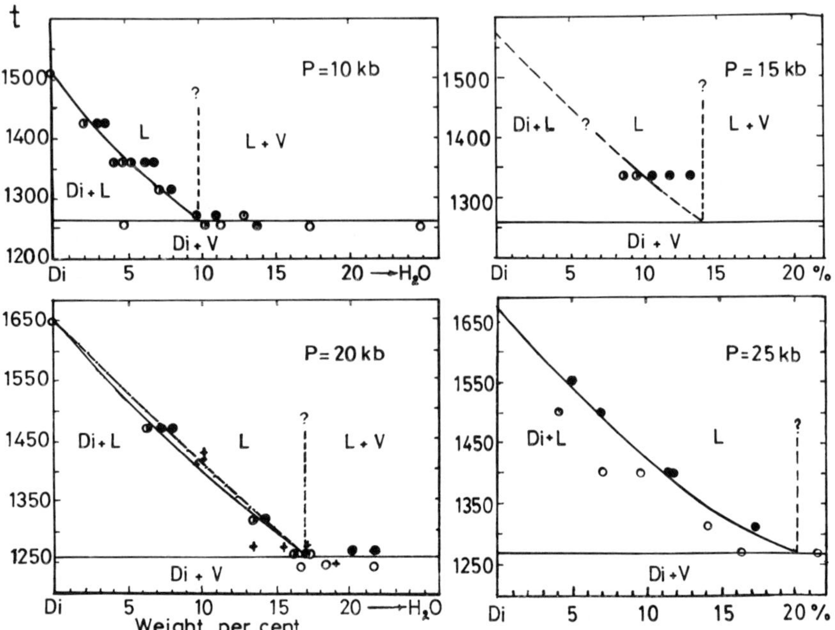

Fig. 1. Isobaric *T–X* sections for the water-undersaturated portion of the diopside–H_2O system at pressures of 10, 15, 20, and 25 kbar (Perchuk, Kushiro, and Kosyakov, 1988).

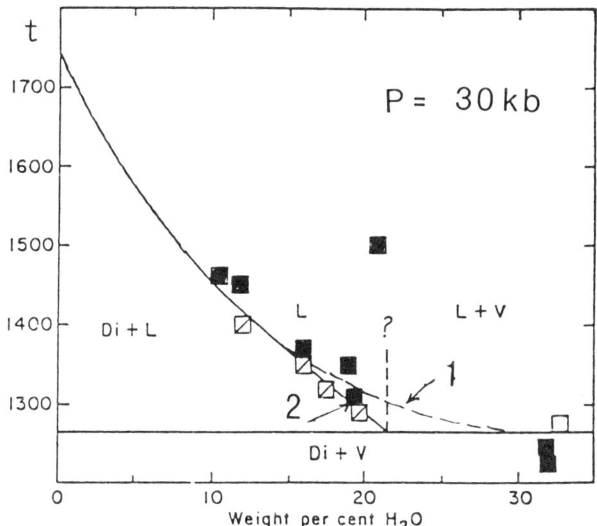

Fig. 2. Isobaric T–X section for the system diopside-water (depression of freezing point) at $P = 30$ kbar. 1, Data of Rosenhauer and Eggler (1975); 2, data of Hodges (1974)

1225°C, i.e. at the same composition, but the temperature estimated was 10°C higher than that reported by Rosenhauer and Eggler (1975).

Perchuk et al. (1988) showed there was inconsistency of the enthalpy and volume change at the fusion of minerals, with melting experiments calculated on the basis of the Schreder equation.

As mentioned above, Hodges (1974) and Eggler and Rosenhauer (1978) studied the diopside–water system at 30 kbar pressure and obtained different results for the water content at the univariant point—21.5 wt% at 1260°C and ~30 wt% at 1240°C, respectively (see Fig. 2). The correct values might be found by mean of thermodynamic treatment of the entire data set for the system concerned.

Perchuk (1983, 1985) and Perchuk and Kushiro (1985) studied the alkali basalt–water system with clinopyroxene at the liquidus over a wide range of pressure, temperature, and water content. The results were discussed from the viewpoint of the silicate melt structure.

A review of the papers published on the diopside–water equilibrium shows that the restricted and partly incompatible experimental data are the salient features of the system.

Data similar to Ferrier's (1968, 1971) measurements of the enthalpy of fusion of diopside were reported by Navrotsky et al. (1980) and Weil et al. (1980), who proposed the best fit for temperature dependence of the enthalpy:

$$\Delta H^{\circ}_{\mathrm{Di}} = 30.88T - 3.922 \cdot 10^{-3}T - 1.574 \cdot 10^{6}/T - 5518 \tag{1}$$

However, the data listed in Table 2 show that the enthalpy of fusion of diopside ranges from 18.5 to 34.1 kcal/mol.

Table 2. Enthalpy of fusion of diopside at a pressure of 1 bar

T, K	ΔH°_{Di}, cal	Method	Source
1665	30,700	High temperature calorimetry[a]	Ferrier (1968, 1971)
1665	34,085	High temperature calorimetry[a]	Navrotsky et al. (1980)
1665	34,085	High temperature calorimetry[a]	Weil et al. (1980)
1665	18,500	Method is not mentioned[b]	Robie et al. (1978)
1665	31,043	Melting experiments data	Carmichael et al. (1977)
1664	31,302	Melting experiments	Boettcher et al. (1982)

[a] ΔC_p for crystals-glass-melt transition

[b] ΔC_p for crystals-glass transition (reference to the method by Navrotsky et al. (1980)).

Navrotsky (1981) discussed the enthalpies of vitrification and fusion for diopside. She concluded that the value $\Delta H^\circ_{Di} = 18.5$ kcal/mol, given in Robie et al.'s (1978) handbook, reflects the enthalpy of vitrification of diopside rather than its fusion. Navrotsky (1981) also formulated the following rule: the more depolymerized the silicate liquid, the larger the change in heat capacity at the glass transition under dry conditions.

The enthalpy of fusion of diopside $\Delta H_{Di} = 18.5$ cal/mol given in Robie et al.'s (1978) handbook was also criticized by Boettcher et al. (1982) who revised the dry liquidi for diopside and albite and developed the thermodynamics of their melting. For diopside, the authors reported the following equation:

$$\Delta H_{Di} = 20.49T - 2795 + P(0.292 - 0.2299 \cdot 10^{-5}P + 0.447 \cdot 10^{-10}p^2) \text{ cal/mol,} \tag{2}$$

where the term in parentheses shows a pressure dependence of the volume change at the fusion of diopside. It is clearly seen that at a pressure of 1 bar Eq. (2) differs from Eq. (1) by ~2.7 kcal (see Table 2). According to Eq. (2), ΔH_{Di} 31.302 cal/mol at 1 bar and 1664 K [compare with other data in Table 2, data by Ferrier (1968, 1971), Navrotsky et al.(1980), Weil et al. (1980)].

Thermodynamic Analyses

Existing Models

Wasserburg (1957) applied a modified form of Flory's (1942) theory of linear high polymer solutions (Mayer and Lunderman, 1935) to the silicate melts in order to create a model for solubility of water in them. Wasserburg used the Schreder equation in the following form:

$$RT \ln X^L_{H_2O} = -\Delta S(T - T_0) + \Delta C_p(T_0 - T)^2/2T_0 - P\Delta V$$

where

$$\ln X^L_{H_2O} = \ln(1 - X^L_{sil}) = 1 - r\ln\left(\frac{rX^L_{sil}}{r + (r-1)X^L_{sil}}\right)$$

and the latter equation reflects a maximum of bridging oxygens in a silicate framework (for diopside $r = 6$). By using the handbook's thermodynamic data ($\Delta S^\circ_{Di} = 9.18$, $\Delta C_p = 3.755$, $\Delta V^\circ_{Di} = 0.2565$) for the albite–water system, Wasserburg (1957) provided his model with calculation of a phase diagram and compared the H_2O solubility in the melt with Goranson's (1936) experimental results. A similar model was proposed by Kurkjian and Russel (1958). Later Show (1964), Silver (1982), Silver and Stolper, (1985) developed Wasserburg's model and calculated phase diagrams for the silicate–water systems such as, for example, $Ab–H_2O$, $Di–H_2O$, $Ol–H_2O$.

Eggler and Burnham (1984) applied the well-known Burnham model to the system diopside–water and obtained good agreement with experimental data at low pressures. Silver and Stolper (1985) proposed a thermodynamic model for any hydrous silicate melt. This model will be discussed below.

Langmuir and Hanson (1981) noted that the majority of binary systems show a linear relationship between the partition coefficient and temperature along the liquidus.

According to Burnham and Davis (1970, 1971), the partial molar volumes of Ab and H_2O in the water-saturated silicate melt are not a function of its composition (H_2O content) within the experimental accuracy. The enthalpy of mixing for the water–albite melt varies within the first hundreds of cal/mole. This value is also within the range of experimental accuracy. Hence, to a first approximation, the water silicate melts can be regarded as an ideal mixture ($H^e = 0$ and $V^e = 0$; Wasserburg, 1957), and the $T–X$ diagrams might be calculated using the Schreder equation (Eq. (5)). In this event, data on the ΔH°_{Di} and dP/dT values for dry and water-saturated liquidi at $P_{H_2O} > 5$ kbar are needed, since the $H_2O^\bullet$ molecular species dominate in the melt (Stolper, 1982a,b; Epel'baum et al., 1984; Epel'baum, 1965; Silver and Stolper, 1985).

As mentioned above, high pressure data on the hydrous diopside melt are known from the publications of Rosenhauer and Eggler (1975), Boettcher et al. (1982), Perchuk et al. (1988) up to a pressure of 30 kbar.

On the basis of Eq. (1) Boettcher et al. (1982) obtained the following derivative:

$$\frac{dT}{dP} = \frac{\Delta H^\circ_{Di}}{T\Delta V^\circ_{Di}} = \frac{20.49T - 2795}{T(0.292 - 0.299810^{-5}P + 0.44710^{-10}P^2)} + \left(\frac{P}{T}\right)\text{bar/K} \quad (3)$$

which is valid for dry conditions. The value ΔV°_{Di} changes up to 20% with increases in pressure toward 30 kbar. According to Perchuk et al. (1988), the melting curve for diopside can be described with the following equation:

$$T_0 = 1664 + 0.0126124P(\text{bar}) - 0.485810^{-7}P^2 \quad (4)$$

but unrealistic results were obtained on the basis of the Schreder equation

$$\ln X_{Di}^{L} = \frac{\Delta H_{Di}}{1.987}\left(\frac{1}{T_0} - \frac{1}{T}\right) \tag{5}$$

coupled with Eq. (2) and the data in Fig. 1. For example, at $\Delta H_{Di}^{\circ} = 20.49T - 2795$ the values of water content (wt%) of 29, 37, 37.85, 47.78, and 57.26 were calculated at pressures 15, 20, 25, and 30 kbar, respectively.

Therefore, the values ΔH_{Di}° and ΔV_{Di}° from Eq. (2) can not be used for calculation with Eq. (5), which might be used to describe the existing experimental data. In using data of Perchuk et al. (1988), Eq. (4) and assuming temperature independence of ΔH_{Di}° the following values for Eq. (5) have been found:

$$\Delta H_{Di}^{\circ} = 18{,}500 \pm 1000 \text{ cal/mol} \tag{6}$$

$$\Delta V_{Di}^{\circ} = 0.14025 - 0.10955 \cdot 10^{-5} P - 25.498 \cdot 10^{-11} P^2 \text{ cal/bar} \tag{7}$$

Despite the low ΔV_{Di} value the enthalpy of fusion of diopside Eq. (6) is practically the same as that in Robie et al.'s (1978) handbook. Thus, values from Eq. (6) and (7) can be used for formal calculation of the water content along the liquidus in the system water–diopside with Eq. (5).

The above calculations were made with experimental data obtained at pressures of 10, 15, 20 (Fig. 1), and 30 kbar (Fig. 2). Several runs were then conducted at 25 kbar pressure to check values from Eq. (6) and (7). The results shown in Fig. 1 for $P = 25$ kbar are consistent with the values obtained at the above pressures, on the one hand, and support the data of Hodges (1974), on the other.

Thus is found a formal thermodynamic description of the liquidus surface in the system diopside–water. However, the problem with the difference between the enthalpy of fusion from Eq. (6) and that calculated on the basis of calorimetric and dry melting experimental data (see Table 2) has not yet been discussed in this paper. Solutions for the problem can be found on the basis of models for the water-silicate melts, like those proposed by Wasserburg (1957), Kurkjian and Russel (1958), Persikov (1975), Ryabchikov (1975), Eggler and Burnham (1984), Silver and Stolper (1985).

All the above authors consider the dissociation reaction of water in a melt as follows

$$H_2O^{\bullet} + O^{\bullet} = 2OH^{\bullet} \tag{8}$$

The only difference lies in the estimates of the $H_2O^{\bullet}/OH^{\bullet}$ ratio as a function of water solubility in a silicate melt. Kadik (1965) was the first to determine the ratio experimentally at high pressures and temperatures . Ryabchikov (1975) used equilibrium (Eq. (8)) for calculation of the PTX diagram for the granite-water system, assuming the total dissociation of water in the melt up to 10 kbar, in spite of the results of Persikov (1972), who explored the effect of pressure on the $H_2O^{\bullet}/OH^{\bullet}$ ratio. Later Stolper (1982a) and Epel'baum et al. (1984) supported Persikov's idea experimentally.

The most advanced thermodynamic model for hydrous silicate melts was developed later. Following Wasserburg (1957), Silver and Stolper (1985) considered the silicate melts as ideal mixtures of water molecules, hydroxyl groups, and oxygen atoms. The equilibrium of the species in the melt is described by the authors with the reaction of Eq. (8) and its constant

$$K_{(8)} = \frac{(a^2_{OH^\bullet})^L}{(a^L_{H_2O^\bullet})(a^L_{O^\bullet})} \cong \frac{(X^2_{OH^\bullet})^L}{(X^L_{H_2O^\bullet})(X^L_{H_2O})} \tag{9}$$

where

$$X^L_{H_2O^\bullet} = X_B - 0.5^L_{OH^\bullet}$$

$$X^L_{O^\bullet} = -2X_B - X^L_{H_2O^\bullet}$$

and

$$X_B = N^L_{H_2O^\bullet}/(N^L_{H_2O^\bullet} + rN^L_{sil}),$$

and where $N^L_{H_2O^\bullet}$ is the number of H_2O moles mixes with N^L_{sil} moles of silicate (sil) melt and r is the amount of oxygens in the formula of silicate ($r = 6$ for diopside).

Along the water-saturated liquidus, the solubility of water in a melt can be calculated on the basis of the equilibrium between melt and vapor

$$H_2O^v = H_2O^\bullet \tag{10}$$

with the reaction constant

$$K_{(10)} = \frac{X_{H_2O^\bullet}}{f_{H_2O}/f^0_{H_2O}} \tag{11}$$

Also, Silver and Stolper (1985) proposed an expression to calculate X_b using the equilibrium constants $K_{(8)}$ and $K_{(10)}$. For example,

$$X_B = 1 - X_{O^\bullet} + 0.25\{K_{(8)}X_{O^\bullet} - [(K_{(8)}X_{O^\bullet})^2 + 4K_{(8)}X_{O^\bullet} - 4K_{(8)}X^2_{O^\bullet}]^{1/2}\} \tag{12}$$

where

$$X_{O^\bullet} = \exp\left(\frac{1}{rRT}\int_{T^0}^{T_v} \Delta S_i \, dT\right)$$

For the systems diopside–water ($r = 6$) and albite–water ($r = 8$), the $\int_T^{T_v} \Delta S_{Di} \, dT$ values are written by the authors as integrated equations. However, using those equations a disagreement was found between X_B calculated with Eq. (12) and experimental data obtained for the diopside–water system at pressures of 15–25 kbar (Fig. 1). This disagreement may be the result of the incorrect value of $\Delta S_{Di} = f(T, P)$ used by Silver and Stolper (1985) from the literature because the model itself seems practically perfect.

According to Stolper (1982*b*), the ratio $H_2O^\bullet/OH^\bullet$ increases with the bulk water content X_B in the melt, i.e., molecules of $H_2O^\bullet$ dominate at high P_{H_2O}.

Models Proposed

The reaction of Eq (8) shows dissociation of water species in the melt, and the $K_{(8)}$ value reflects the degree of dissociation. From the statistical viewpoint, the $H_2O^{\bullet}$ molecules as well as the $OH^{\bullet}$ species in a melt are distributed among the energetically nonequivalent sites according to the available amount of the hydrated silicate species. In total hydration, this amount reaches the maximum value, whereas the statistical distribution of the species in a melt is controlled by $f^{v}_{H_2O}$ along the wet liquidus. The chemical potential of oxygen at the sites in hydrated and non-hydrated species must be equal in terms of the Korzhinskiy free energy thermodynamic potential. This equality can be achieved by subtracting a portion of the energy from the Gibbs free energy thermodynamic potential. The portion of energy concerned relates to those species which are in a position to equilibrate the system at given T and P. The sense of this equilibration can be described with the Korzhinskiy potential. On the other hand, from the viewpoint of thermodynamic formalism, this equilibration can be written as n interaction equations between water and all oxygens in the melt at the n sites:

$$\left.\begin{array}{l} H_2O^{\bullet}_{(1)} + O^{\bullet}_{(1)} = 2OH^{\bullet}_{(1)} \\ H_2O^{\bullet}_{(2)} + O^{\bullet}_{(2)} = 2OH^{\bullet}_{(2)} \\ \cdots\cdots\cdots\cdots\cdots\cdots \\ H_2O^{\bullet}_{(n)} + O^{\bullet}_{(n)} = 2OH^{\bullet}_{(n)} \end{array}\right\} \quad (13)$$

Then n parameter reflects an effective oxygen charge at the sites. For example, at $n = 1$ this charge is O^- and the system is close to an ideal mixture.

Along the water-saturated liquidus the equilibrium constant can be written as follows

$$K_{(13)} = \frac{a^{2n}_{OH^{\bullet}}}{a^{n}_{O^{\bullet}} \cdot a^{n}_{H_2O^{\bullet}}} = \exp\left(\frac{-\Delta G^{\circ}_{(13)}}{nRT}\right) \quad (14)$$

where $a^{n}_{H_2O^{\bullet}} = f_{H_2O}/f^{\circ}_{H_2O}$ and $a_{O^{\bullet}} = a^{L}_{Di}$. According to Eq. (2), for the dry diopside melting curve

$$\ln a^{L}_{Di} = \frac{\Delta G^{\circ}_{Di} + (P-1)\Delta V^{\circ}_{Di}}{RT} = 77.25 - 1406/T - 10.31 \ln T$$
$$+ P(0.147 - 1.5 \cdot 10^{-6} P + 2.25 \cdot 10^{-11} P^2)/T = \frac{\Delta G^{\circ}_{(13)}}{T} \quad (15)$$

In accordance with the data of Perchuk (1973), $\Delta G^{0}_{(13)}$ in Eq. (14) might be written as:

$$\Delta G^{\circ}_{(13)} = \Delta G^{\circ}_{Di} - (G^{\circ}_{T} - G^{\circ}_{0})_v = \Delta G_{Di} + 9904.09 + 45553.19(10^{-3} t°C)$$
$$+ 8693.45(10^{-3} t°C)^2 - 1585.63(10^{-3} t°C)^3 \text{ cal/mol} \quad (16)$$

and

$$2[\ln X_{OH^\bullet} + \ln \gamma_{OH^\bullet}] - [\ln X_{O^\bullet} + \ln \gamma_{O^\bullet}] = \ln f^\circ_{H_2O} - \Delta G^\circ_{(13)}/nRT \qquad (17)$$

Taking into account that $X_{O^\bullet} = X^L_{Di}$ and using the regular model for the water–diopside melt mixture (Ryabchikov et al., 1984), the following equations can be written:

$$RT \ln \gamma_{OH^\bullet} = W_1(1 - x_{O^\bullet})^2 = W_1(X^L_{Di})^2 \qquad (18)$$

$$RT \ln \gamma_{OH^\bullet} = W_1 X^2_{OH^\bullet} = W_1(1 - X^L_{Di})^2 \qquad (19)$$

Substituting (18) and (19) into (17) and comparing the result with (16), after rearranging the following can be obtained:

$$\ln X_{Di} + \ln f^\circ_{H_2O} - 2\ln(1 - X^L_{Di}) = \frac{W[2(X^L_{Di})^2 - (1 - X^L_{Di})^2] - \Delta G^\circ_{(13)}}{nRT}$$

where W is the interaction parameter for bridging oxygens in the diopside melt.

According to the regular model, the W_1/RT value does not depend on temperature, pressure and bulk composition of the system. For the diopside–water system, this value is practically constant over a very wide range of the above parameters (at $n = 1$):

$$W_1/RT = 2.358 \pm 0.003 \qquad (21)$$

The linear correlation between the molar fractions in both the left and right hand sides of Eq (18) and (19) is characterized by the correlation coefficient $r^2 = 0.993$. This result shows that the model reflects the entire dissociation of water in the diopside melt, but it disagrees with Stolper's (1982*a*) data.

At the univariant point of point depression and along the water-saturated liquidus equilibrium (10) is valid, and its constant is

$$K_{(10)} = a_{H_2O^\bullet}/f^\circ_{H_2O} = (X_{H_2O^\bullet}\gamma_{H_2O^\bullet})/f^\circ_{H_2O} \qquad (22)$$

For the regular model

$$RT \ln \gamma_{OH^\bullet} = W_2(X^L_{Di})^2/nRT \qquad (23)$$

By using the data from Fig. 2 and 3 and vapor properties from Kestlin et al.'s (1984) paper, the W_2/RT parameter was computed at n = 1 and different T, P, and X_b:

$$W_2/RT = 2.283 \pm 0.089 \qquad (24)$$

Eq. (24) gives the mean arithmetic value for the relationship between parameters $(X^L_{Di})^2$ and $[\ln(f^\circ_{H_2O}/\ln X_{O^\bullet}]$ taken from equation (25):

$$\ln(f^\circ_{H_2O}/X_{O^\bullet}) = W_2(X^L_{Di})^2 - \Delta G_{(10)}/RT \qquad (25)$$

The value of Eq. (24) differs from that of Eq. (21) by 0.089, i.e.

$$\frac{n_2}{n_1} = \frac{W_1}{W_2} = 1.032$$

and $W_1 \cong W_2$.

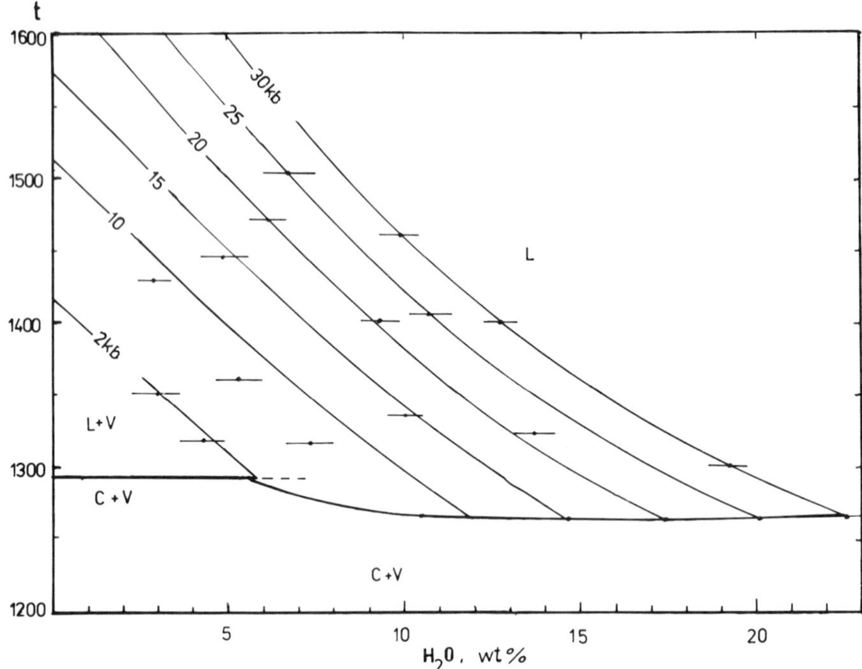

Fig. 3. The depressions of the freezing points of diopside as a function of H_2O content along the liquidi at pressures of 2, 10, 15, 20, and 25 kbar. The diagram is calculated using Eq. (30) with parameters from Eq. (6) and (7). Horizontal bars at the datapoints reflect the experimental accuracy (Eggler and Burnham, 1984; Perchuk, Kushiro and Kosyakov, 1988).

Hence anionic sites are practically absent and only $H_2O^\bullet$ and $OH^\bullet$ groups exist in the melt. In addition, it should be mentioned that the W_1 and W_2 values were computed with Eq. (20) and (23) by minimizing the n parameter that varies in the range 1.095–1.665 as a function of T and P. This result shows that the effective oxygen charge ranges from -1.1 to -1.7.

Eq. (5) is now rearranged with respect to a model for the diopside–water melt with the fixed oxygens:

$$T = \left(\frac{\Delta H^\circ_{Di} + \Delta V^\circ_{Di}(P-1)}{rRT} - \ln X_{O_2^\bullet}\right)^{-1} \left(\frac{\Delta H^\circ_{Di} + \Delta V^\circ_{Di}(P-1)}{rT}\right) \tag{26}$$

where $r = 6$ and r has the same meaning as that in Eq. (13), i.e. the amount of oxygens per one Di mol. It is helpful to define the first term in large brackets as α:

$$T = \alpha^{-1}[\Delta H^\circ_{Di} + (P-1)\Delta V^\circ_{Di}]/1.987r \tag{27}$$

where $\Delta H^\circ_{Di} \cong$ const at different temperatures. According to the Van Laar law, at $P =$ const, the liquidus line curvature at the point of dry melting of diopside is defined by the second derivative of T with respect to the Di molar fraction

$$\left(\frac{\partial^2 T}{\partial (X_{\mathrm{Di}}^L)^2}\right) = \frac{\Delta H_{\mathrm{Di}}^\circ + (P-1)\Delta V_{\mathrm{Di}}^\circ}{1.987 r \alpha^2 (X_{\mathrm{Di}}^L)^2}\left(\frac{2}{\alpha} - 1\right) \tag{28}$$

Fig. 2 and 3 show a linear relationship between temperature and composition *near the diopside side* of the diagrams. Such a relationship means that the second derivative (28) equals zero because (Moelwyn–Hughes, 1961)

$$\frac{\Delta H_{\mathrm{Di}}^\circ + (P-1)\Delta V_{\mathrm{Di}}^\circ}{1.987 \cdot r \cdot T_0} \cong 2 \tag{29}$$

For simplicity, Eq. (28) will now be defined as φ. Note that the linear relationship disappears with pressure at $P > 25$ kbar and $X_{\mathrm{H_2O}}^L$, i.e. when $\varphi > 0$, and

$$\frac{\Delta H_{\mathrm{Di}}^\circ + (P-1)\Delta V_{\mathrm{Di}}^\circ}{1.987 \cdot r \cdot T_0} < 2 \tag{30}$$

At $r =$ const, the curvature of the liquidus line in the freezing point depression ($P =$ const) is mainly determined by the enthalpy of fusion of a dry silicate. For water–silicate systems, a decrease in temperature along the liquidus leads to a change in the amount of the bridging oxygens $O^\bullet$ equilibrated with $H_2O^\bullet$ and $OH^\bullet$.

As mentioned above, for the dry diopside melt $r = 6$ which is, however, not constant, being a function of composition (Di°/H_2O), temperature and pressure:

$$r = Q - \mathbf{A}\psi(1-\psi)(P-3) \tag{31}$$

where P is in kbar, $\mathbf{A}$ is an empiric coefficient defining the value of derivative (28) for the molten mixture at $X_{\mathrm{H_2O}}^L > 0.1$, Q reflects a morphology of the liquidus at constant pressure and

$$\psi = \frac{T_0 - T}{T_0 - T_w} \tag{32}$$

From the data on the water-saturated liquidus (see Table 2) $\mathbf{A} = 0.33$. Hence according to the Schreder equation (5) and expression (32), the liquidus surface in the $P - T - X_{\mathrm{H_2O}}^L$ coordinates can be calculated with the following formula:

$$N_{\mathrm{Di}}^L = \frac{X_{\mathrm{Di}}^L \cdot 21655.2/6}{X_{\mathrm{Di}}^L \cdot 216.552/6 + X_{\mathrm{Di}}^L \cdot 18} \tag{33}$$

where

$$X_{\mathrm{Di}}^L = 1 - X_{\mathrm{H_2O}}^L = \exp\frac{\Delta H_{\mathrm{Di}}^\circ + P\Delta V_{\mathrm{Di}}^\circ}{Q \cdot 1.987}\left(\frac{1}{T_0} - \frac{1}{T}\right) \tag{34}$$

and where

$$Q = 6 + 0.33\psi \cdot (1-\psi)(P-3) \tag{35}$$

Results of the calculations using Eq. (30) show a good agreement with the experimental data except for the 10 kbar isobar: the liquidus is located near the experimental data points but away from the bars showing the accuracy. However,

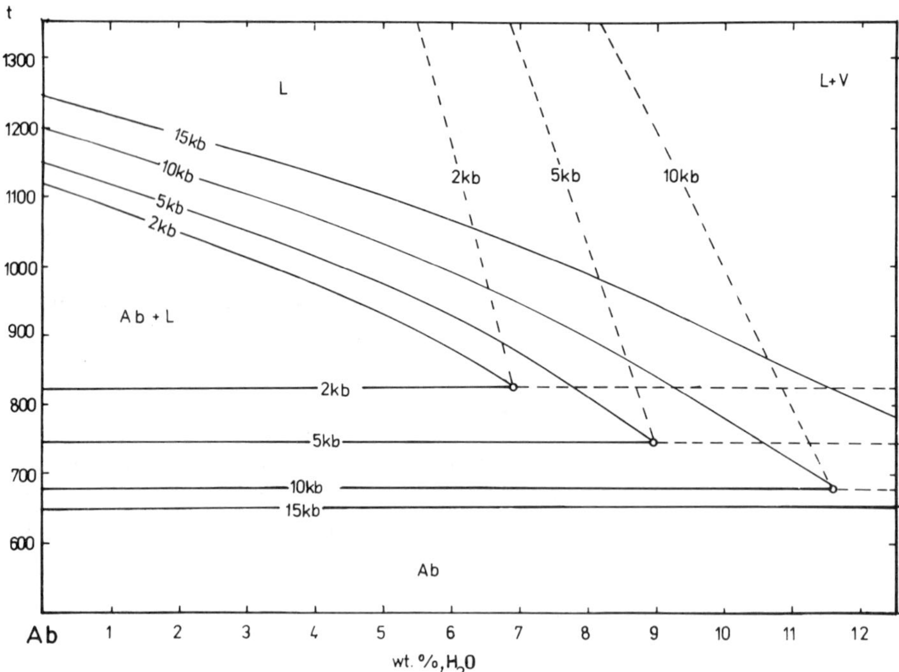

Fig. 4. The depressions of the freezing points of albite as a function of H_2O content along the liquidi at different pressures.

the 10 kbar isobaric freezing point depression might be easily described with the Schreder equation (5) by incorporating Eq. (3) and (4).

By using the same approach isobars have been calculated in the freezing point depression for the system albite-water (see Fig. 4).

Thus, satisfactory models for thermodynamic description of the system diopside–water were developed, but they do not solve the structural part of the problem. In other words, the enthalpy of fusion of diopside based on the results shown in Fig. 2 and 3 agrees with Boettcher et al.'s (1982) melting experiments, on the one hand, and with the calorimetric data (see Table 1), on the other.

According to Stolper's (1982a, b) model

$$X^L_{H_2O} = X_b = X_{H_2O^\bullet} + 0.5X_{OH^\bullet} = \frac{n_{H_2O^\bullet}}{n_{H_2O^\bullet} + 6n_{O^\bullet}} \tag{36}$$

where

$$n_{H_2O^\bullet} = H_2O(\text{wt\%})/18.015$$

$$n_{O^\bullet} = [100 - H_2O(\text{wt\%})]/216.552/6$$

The equilibrium constant of reaction (Eq. (8)) is connected with the bulk H_2O content in the system and molar fraction of $H_2O^\bullet$ as follows:

$$X_{OH^\bullet} = \frac{0.5 - \sqrt{[0.25 - (K_{(8)} - 4)]/K_{(6)} \cdot (X_b - X_b^2)}}{(K_{(8)} - 4)/K_{(8)}} \tag{37}$$

where $K_{(8)} \cong 0.2$ (Silver and Stolper, 1985)

The value X_b can be calculated with the equation

$$X_b = X^L_{H_2O} = \frac{N^L_{Di}/216.552/6}{N^L_{Di}/216.552/6 + N^L_{H_2O}/18.015} \tag{38}$$

and $K_{(8)}$—with Eq. (14) and (15) at given T and P. All the values in Eq. (37) are connected via the following equation

$$X_{H_2O^\bullet} = 1 - X_{O^\bullet} - 0.5X_{OH^\bullet} \tag{39}$$

In combining Eq. (39) with Eq. (36)–(38), the molar fractions of different species were calculated. The results of the calculations for the system diopside–water are shown in Fig. 5.

In the diagram, Fig. 5*b*, the highest value of $X_b = 0.385$ at $P = 30$ kbar is much less than that of 0.806 calculated from Eq. (5) with data from Eq. (6) and (7), or using Eq. (34) with data from Eq. (2). The difference between the values calculated reflects an energetic contribution of the speciation of water to the melt. In other words, for thermodynamic description of the liquidus surface in the water–silicate system both the enthalpy of fusion and mixing energy of $O^\bullet$ and $OH^\bullet$ should be taken into account. In this instance, the H_2O molar fraction for equilibrium (Eq. (8)) can be calculated with the formula

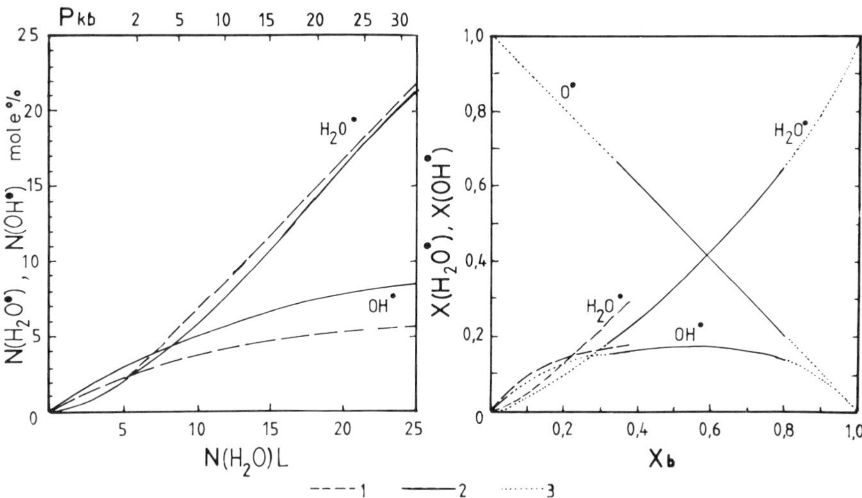

Fig. 5. Ratios of $H_2O^\bullet$, $OH^\bullet$ and $O^\bullet$ species along the water-saturated liquidus for the system diopside–water as a function of the bulk content of H_2O in the melt. 1, After Silver and Stolper (1985); 2, calculation with Eq. (40)–(43) and (37); 3, extrapolation.

$$X_b = X^L_{H_2O} = X_{H_2O\bullet} + 0.5X_{OH\bullet} = 1 - X^L_{Di} \quad (40)$$

Rearranging (40), gives

$$X_{OH\bullet} = 2(1 - X^L_{Di} - X_{H_2O\bullet}) \quad (41)$$

$$X_{H_2O\bullet} = 2X_b - 1 + X^L_{Di} \quad (42)$$

$$X^L_{H_2O} = 1 - X_b = 1 - 0.5X_{OH\bullet} - X_{H_2O\bullet} \quad (43)$$

where X^L_{Di} corresponds strongly to the value calculated using the Schreder equation (5) and the enthalpy of fusion of diopside taken from Eq. (2) and X_b—from Eq. (38). Eq. (41) through (43) define the molar fraction of $OH^\bullet$, $H_2O^\bullet$ and $O^\bullet$ (Di_L) with masses of 17, 18 and 216.552 g/mol, respectively. Using very simple Eq. (41–43) and parameters (6) and (7), or the values $X_b = X^L_{H_2O} = 1 - X^L_{Di}$ from Eq. (34), the variations of the above species in the melt have been calculated. The diagram in Fig. 5*a* shows variations of $X_{H_2O\bullet}$ and $X_{OH\bullet}$ with the bulk composition of the system along the liquidus. The results of this calculation are in agreement with Stolper's model. However, a noticeable difference may be seen

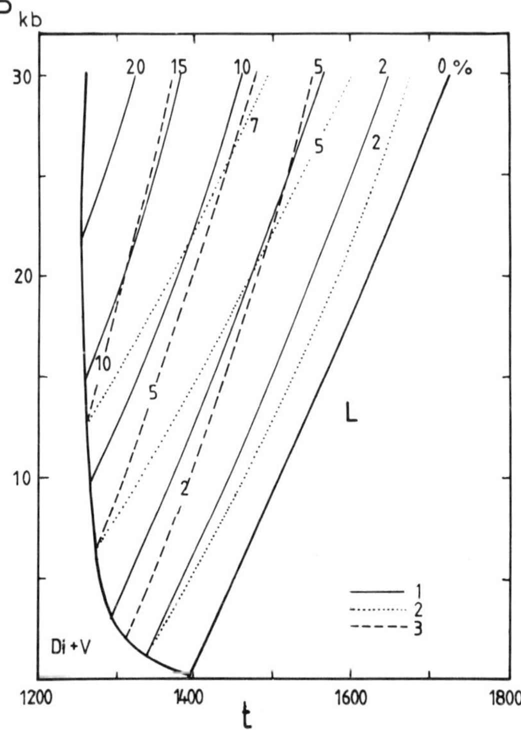

Fig. 6. Projection of the liquidus surface of the diopside–water system onto the *P–T* plane. Isopleths show the contents of different species of water (wt%): 1, H_2O in the melt calculated with Eq. (4)–(7), and (35) and (43); is 2, $OH^\bullet$ and 3, $H_2O^\bullet$ calculated with Eq. (42) and (35).

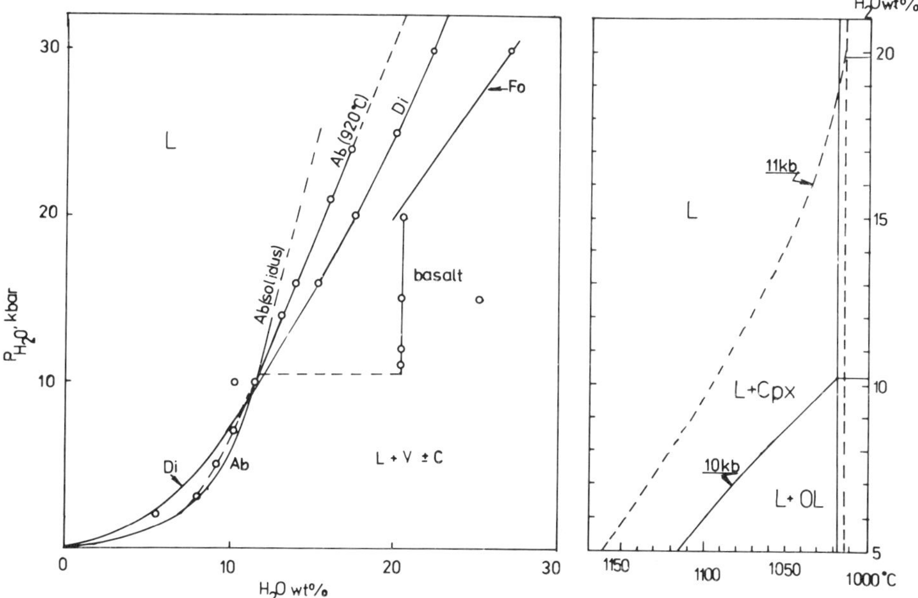

Fig. 7. Solubility of water in the silicate melts of different compositions as a function of pressure and temperature. Experimental data on the alkali basalt melt-water system after Perchuk and Kushiro (1985); the olivine-water system after Hodges (1973).

in the diagram of Fig. 5b, where X_b was calculated with molar weight of diopside of 216.552 g/mol. A symmetric line in the diagram for concentration of $OH^{\bullet}$ reflects an ideal mixing of $OH^{\bullet}$ and $H_2O^{\bullet}$ species. With the help of the diagrams in Fig. 5 it can be predicted that melting of diopside at low pressure starts from the almost total dissociation of water in the melt. The higher the pressure (along the water-saturated liquidus) the higher the concentration of $H_2O^{\bullet}$ in the melt.

Fig. 7b shows projections of the liquidus surface onto the $P-T$ plane in reference to speciation of water in the diopside melt. At relatively low content of water in the melt (up to 5 wt%) the $OH^{\bullet}$ species dominate. A decrease in temperature at constant pressure involves an increase of $H_2O^{\bullet}$ molecules in the melt. At high pressure these relations become inverted.

By using the models proposed the water solubility in the albite and olivine melts has been calculated as a function of pressure at a temperature near the wet solidus. Fig. 7 illustrates the results of this calculation, along with the experimental data: on the basis of reversals Perchuk and Kushiro (1985) obtained extraordinary results concerning the liquidus of the system alkali basalt–water. The diagram at the right in Fig. 7 shows a water solubility gap in the basalt melt between pressures 10 and 11 kilobars. These results are still unexplained and cannot be treated thermodynamically despite the appearance of Cpx crystals on the liquidus.

Conclusion

A review of experimental data concerning melting studies of the system diopside–water up to a pressure of 30 kbar reveals a formal calculation of the liquidus surface with Eq. (5)–(7). The value of ΔH°_{Di} does not correspond to calorimetric measurements of the heat of fusion of diopside. The models considered permit estimation of energy factors that are responsible for the presence of the $OH^{\bullet}$, $H_2O^{\bullet}$ and $O^{\bullet}$ ($CaMgSi_2O_6$) species in equilibrium in the melt and variations in the equilibrium constant of reaction (8) with T and P. The models proposed are in good agreement with Stolper's model and permit calculation of the liquidus in the water–diopside system regarding the speciation.

Acknowledgments

The authors have profited from collaboration with A. Kosyakov and G. Gonchar. Helpful reviews by A. Kadik and I. Ryabchikov are also acknowledged.

Abbreviations

Symbols for components and phases used:

$H_2O^{\bullet}$ water dissolved in a silicate melt in molecular form
$OH^{\bullet}$ hydroxyl groups attached to a silicate polymer
$O^{\bullet}$ bridging oxygens in a silicate melt
Di diopside
Ol olivine
Cpx clinopyroxene
C crystals of silicate (sil)
L liquid (melt)
v vapor
i component of the given system
Di° molecule of diopside in dry melt;

Thermodynamic parameters:

T temperature, K
t temperature, °C
R universal gas constant 1.987 cal/K/mol
P pressure, bar
T_0 melting temperature of diopside under dry conditions at a given pressure
T_v melting temperature of diopside at a given water pressure
P_{H_2O} melting pressure along the water-saturated liquidus

$K_{(n)}$	equilibrium constant for reaction *n*
ΔG°	Gibbs free energy of fusion of diopside under dry conditions, cal/mol
ΔC_p	heat capacity of fusion of diopside under dry conditions, cal/mol/deg
ΔS_{Di}	entropy of fusion of diopside under dry conditions, cal/mol/deg
ΔH°_{Di}	enthalpy of fusion of diopside under dry conditions, cal/mol
ΔV°_{Di}	volume change at the fusion of diopside, cal/bar/mol
X^L_i	molar fraction of component *i* in a melt
N_i	concentration of component *i* in a melt, wt%
α_i	activity of component *i* in a phase
γ_i	activity coefficient of component *i* in a phase
W	interaction parameter of components in a melt, cal/mol
$f^\circ_{H_2O}$	water fugacity at given temperature and pressure
$\Delta G^\circ_{(n)}$	Gibbs free energy of reaction *n*, cal
$(G^\circ_T - G^\circ_0)_{H_2O}$	Gibbs free energy change for vapor with increasing temperature from $T = 0$ to a given T at $P = 1$ bar, cal/mol
G^m_{Di}	partial Gibbs free energy of mixing for Di in the water-bearing melt, cal/mol
S^m_{Di}	partial entropy of mixing for Di in the water-bearing melt, cal/mol/K
$f^\circ_{H_2O}$	standard water (vapor) fugacity at given T and P

References

Boettcher, A.L., Burnham, C.W., Windom, K.E., and Bohlen, S.R. (1982) Liquids, glasses, and the melting of silicates at high pressures. *J. Geol.*, **90**, 127–138.

Boyd, R.F., and England, J.L. (1963) Effect of pressure on the melting of diopside, $CaMgSi_2O_6$, and albite, $NaAlSi_3O_8$, in the range up to 50 kilobars. *J. Geophys. Res.*, **68**, 311–323.

Burnham, C.W., and Davis, N.F. (1974) The role of H_2O in silicate melts: Part. II. Thermodynamics and phase relations in the system $NaAlSi_3O_8$–H_2O to 10 kb, 750°C to 1,100°C. *Amer. J. Sci.*, **274**, 902–940.

Burnham, C.W., and Davis, N.F. (1970) Thermodynamic properties of water-bearing magmas. *Phys. Earth Planet Int.*, **3**, 332–348.

Carmichael, I.S., Nicholls, J., Spera, F.J., Wood, B.J., and Nelson, S.A. (1977) High temperature properties of silicate liquids: applications to the equilibration and ascent of basic magma. *Philos. Trans. Roy. Soc. Lond. [A]*, **286**, 373–431.

Eggler, D.H., and Burnham, C.W. (1984) Solution of H_2O in diopside melts: a thermodynamic model. *Contrib. Mineral. Petrol.*, **85**, 58–66.

Eggler, D.H. (1973) Role of CO_2 in melting processes in the mantle. *Carnegie Inst. Wash.*, **72**, 457–467.

Eggler, D.H., and Rosenhauer, M. (1978) Carbon dioxide in silicate melts: II. Solubilities of CO_2 and H_2O in $CaMgSi_2O_6$ (diopside) liquids and vapors at pressures to 40 kb. *Amer. J. Sci.*, **278**, 64–94.

Epel'baum, M.B. (1985) The structures and properties of hydrous granitic melts. *Geol.*

Zbornik—Geol. Carpathica, **36**, 491–498.
Epel'baum, M.B., Persikov, E.S., and Zhigun, I.G. (1984) Relations of the different water species in the hydrous albite glass. *Contrib. Phys. Chem. Petrol.*, **12**, 72–78.
Ferrier, A. (1968) Chimie pyroxene—Mesure de l'enthalpe du diopside sinthetique entre 298 et i 885, K. *Comp. Rend. Acad. Sci. Paris. Ser. C.*, **267**, 101–106.
Ferrier, A. (1971) Etude experimentale de l'enthalpie de cristallisation du diopside et de l'anorthite sinthetique. *Rev. Int. Hautes Temp. Refract.* **8**, 31–36.
Flory, P.J. (1942) The thermodynamics of high polymers. *J. Chem. Phys.*, **10**, 51–61.
Goranson, (1936) The solubility of H_2O in albite melts. *Amer. Geoph. Union Trans.* **17**, 257–268.
Kadik, A.A. (1965) State of the water and silicate components in the melts (magmas) of acid composition at high pressure of H_2O vapor. In: *Geochemical Studies at High Temperatures and Pressures*, edited by N. I. Khitarov, pp. 5–15. Nauka, Moscow.
Kestin, J., Sengers, J.V., Kandar-Parsi, B., and Levelt Sengars, J. M. H. Thermophysical properties of fluid H_2O. *J. Phys. Chem. Ref. Data*, **3**, 175–183.
Kracek, F.C. (1953) Contributions of thermochemical and X-ray data to the problem of mineral stability. *Carnegie Inst. Wash.*, **52**, 69–75.
Kushiro, I. (1976) Changes in viscosity and structure of melts of $NaAlSi_3O_8$ composition at high pressure. *J. Geophys. Res.*, **81**, 6347–6350.
Kurkjian, C.R., and Russel, L.E. (1958) Solubility of water in molten alkali silicates. *J. Soc. Glass. Technol.*, **42**, 130–144.
Langmuir, C.H., and Hanson, G.N. (1981) Calculating mineral-melt equilibria with stoichiometry, mass balance, and single-component distribution coefficients. In: *Advances in Physical Geochemistry, Vol. 1*, edited by S.K. Saxena, pp. 247–271. Springer-Verlag, New York.
Hodges, F.N. (1973) Solubility of H_2O in forsterite melt at 20 kb. *Carnegie Inst. Wash.*, **72**, 495–497.
Hodges, F.N. (1974) The solubility of H_2O in silicate melts. *Carnegie Inst. Wash.*, **73**, 251–255.
Moelwyn-Hughes, E.A. (1961) *Physical Chemistry*. Pergamon Press, London and New York.
Meyer, K.N., and Lunderman, R. (1935) Über das Verhalten hoher molekularer Verbindung in Lösung. *Helvet. Chim. Acta*, **18**, 307–325.
Navrotsky, A. (1980) Thermodynamics of mixing in silicate glasses and melts. In *Advances in Physical Geochemistry, Vol. 1*, edited by S.K. Saxena, pp. 189–206. Springer-Verlag, New York.
Navrotsky, A., and Coons, W.E. (1976) Thermochemistry of some pyroxenes and related compounds. *Geochim. Cosmochim. Acta*, **40**, 1281–1288.
Navrotsky, A., Hon, R., Weil, D.F., and Henry, D. (1980) Thermochemistry of glasses and liquids in the systems $CaMgSi_2O_6$–$CaAl_2Si_2O_8$–$NaAlSi_3O_8$, SiO_2–$CaAl_2Si_2O_8$–$NaAlSi_3O_8$ and SiO_2–Al_2O_3–CaO–Na_2O. *Geochim. Cosmochim. Acta*, **44**, 1409–1423.
Persikov, E.C. Experimental study of H_2O solubility in granitic melt and kinetics of equilibria of granitic melt-H_2O at high pressures. *Geology Geophys.*, **9**, 3–9.
Perchuk, L.L. (1973) *Thermodynamic Regime of Depth Petrogenesis*. Nauka Press, Moscow, 316 pp.
Perchuk, L.L. (1983) System alkali basalt—water: I. Analyses of run products near the liquidus at pressure 15 kbar. *Contrib. Phys.-Chem. Petrol.* **11**, 103–120.
Perchuk, L.L. (1985) System alkaline basalt—water. II. Liquidus surface at pressures 1–20,000 bar. *Contrib. Phys.-Chem. Petrol.*, **13**, 66–80.

Perchuk, L.L., and Vaganov, V.I. (1978) Temperature regime of crystallization and differentiation of basic and ultrabasic magmas. *Contrib. Phys.-Chem. Petrol.*, **6**, 142–174.
Perchuk, L.L., and Kushiro, I. (1985) Experimental study of the system alkali basalt-water up to pressure 20 kbar in respect of estimation of H_2O content in the original magmas beneath the island arcs. *Geol. Zbornik.—Geol. Carpathica*, **36**, 3, 359–368.
Perchuk L.L., Kushiro, I., and Kosyakov, A.V. (1988). Experimental determination of the liquidus surface in the system diopside–water. *Geokhimia*, **7**, 942–955.
Robie, R.A., Hemingway, B.S., and Fisher, J.R. (1978) *Thermodynamic properties of minerals and related substances at 298, 15, K and 1 bar pressure and at higher temperatures.* US Government Printing Office, 1452, 456.
Rosenhauer, M., and Eggler, D.H. (1975) Solution of H_2O and CO_2 in diopside melt. *Carnegie Inst. Wash.*, **74**, 474–479.
Ryabchikov, I.D. (1975) *Thermodynamics of Fluid Phase of Granitic Magmas.* Nauka, Moscow, 320 pp.
Ryabchikov, I.D., Solovova, I.P., Dmitriyev, Yu.I., and Muravitskaya, G.N. (1984) Water in parental magma of oceanic Fe-basalts. *Geokhimia*, **2**, 209–216.
Show, H.R. (1964) Theoretical solubility of H_2O in silicate melts: quasi-crystalline models. *J. Geol.*, **72**, 601–617.
Silver, L., and Stolper, E.A. (1985) Thermodynamic model for hydrous silicate melts. *J. Geol.*, **93**, 161–177.
Stolper, E. (1982,*a*) Water in silicate melts: an infrared spectroscopic study. *Contrib. Mineral. Petrol.*, **81**, 1–17.
Stolper, E. (1982,*b*) The specification of water in silicate melts. *Geochim. Cosmochim. Acta*, **46**, 2609–2620.
Yoder, H.S. (1965) Diopside-anorthite-water at 5–10 kbar and its bearing on explosive volcanism. *Carnegie Inst. Wash.*, **64**, 82–89.
Yoder, H.S. (1975) Heat of melting of simple systems related to basalts and eclogites. *Carnegie Inst. Wash.*, **74**, 515–519.
Wasserburg, G.J. (1957) Effect H_2O in silicate systems. *J. Geol.*, **65**, 15–23.
Weil, D.F., Hon, R., and Navrotsky, A. (1980) The igneous system $CaMgSi_2O_6$–$CaAl_2Si_2O_8$–$NaAlSi_3O_8$: variations on a classic theme by Bowen. In *Physics of Magmatic Processes*, edited by R.B. Hargraves, pp. 49–92. Princeton University Press, Princeton, NJ.

Chapter 10
Origin of Subduction Zone Magmas Based on Experimental Petrology

Yoshiyuki Tatsumi

Introduction

More than 70% of the subaerial volcanoes of the Earth are distributed at convergent plate boundaries such as in the Andes, Cascades, Aleutian Islands, Japan, and New Zealand. It is widely accepted that formation of magmas erupted by those volcanoes is closely related to subduction of the oceanic lithosphere beneath volcanic arcs. For example, the following four phenomena make a strong impression that the existence of the downgoing slab is a necessary condition to produce arc magmas: (1) segmented subduction of the plate is closely related to segmented distribution of volcanoes, as is clearly observed in the Andes area (Carr et al., 1974); (2) change of the mode of plate boundaries (from transform to subduction) corresponds to the appearance of volcanoes in the Aleutian area (McKenzie and Parker, 1967); (3) initiation of subduction corresponds to that of volcanism in Tertiary Southwest Japan and Bonin areas (Tatsumi and Ishizaka, 1982; Tatsumi, 1983); and (4) ^{10}Be, which is accumulated in sediments on the ocean floor, is detected only in subduction zone rocks (Brown et al., 1982).

Chemical compositions of subduction-related arc magmas erupting on the surface are essentially governed by chemical processes at the following three stages;

1. Formation of initial magmas. At a temperature just above the solidus of source materials, initial partial melts are formed as small droplets at a junction of constituent minerals. Although the major composition of a magma changes through later processes, some geochemical characteristics (e.g., isotope and incompatible element ratios) of the magma are determined at this initial stage.
2. Progressive melting and segregation of magmas. Along with an increasing degree of partial melting, the magma migrates upward within the mantle wedge, probably in the form of a partially molten diapir (Green and Ringwood, 1968; Ringwood, 1975; Sakuyama, 1983; Tatsumi et al., 1983). Through

this process, the diapir can not be kept in a closed system and the chemical compositions of the magma in a diapir are changed by the reaction with the surrounding mantle wedge material. Magmas are then segregated from the ascending diapir and at this stage, the constituent minerals of the diapir will cease to control magma chemistry. P–T–H_2O conditions at this stage critically determine the liquid composition. A magma at this stage is called the primary magma. Four types of primary magmas generally appear in subduction zones: olivine tholeiites, high-alumina basalts, alkali olivine basalts, and high-magnesian andesites. They are, however, by no means discrete magma types.

3. Differentiation and mixing of magmas. An ascending magma may ultimately form a magma chamber within the crust. Crystallization differentiation occurs in the magma chamber and mixing of magmas probably in the conduit between a magma chamber and the surface. Through these processes the composition of a magma changes drastically.

In attempts to understand the above processes, many models were proposed up to the end of the 1970s on the basis of experimental data combined with petrographical and geochemical information (e.g., Kuno, 1959; Green et al., 1967; Green and Ringwood, 1967; McBirney, 1969; Kushiro, 1974; Ringwood, 1974; Yoder, 1976; Wyllie, 1979). This review discusses the origin of subduction zone magmas on the basis of experimental and geochemical data mainly obtained in the early 1980s.

This review chapter was written in 1985. Since then, many papers have been published and have greatly improved the concept of magma genesis in subduction zones. Although this paper does not take into account the papers published in the second half of the 1980's, readers are requested to refer to the following papers from our laboratory to learn of recent developments in the author's concept of the topic: Tatsumi and Nakamura (1986); Tatsumi et al. (1988); Tatsumi and Isoyama (1988); Tatsumi (1989); Tatsumi et al. (1989); Sudo and Tatsumi (1990); Goto and Tatsumi (1990).

Across-Arc Compositional Variation of Arc Magmas

Several authors have pointed out the across-arc variation of compositions in subduction zone lavas since Kuno (1959). Although criticisms of general or easygoing application of the concept of across-arc lateral variations in subduction zones have appeared (e.g., Johnson, 1976; Arculus and Johnson, (1978), a clear relationship between compositions of volcanic rocks and distance from the trench is established beyond question in several arc-trench systems. Several kinds of across-arc compositional variations are recognized; examples are systematic change of magma types, contents and ratios of incompatible elements, FeO*/MgO ratios, and isotope ratios (Gill, 1981).

Magma Types and Contents of Incompatible Elements

Early experimental data on both simple and natural systems (e.g., Green and Ringwood, 1967; Green et al., 1967; O'Hara, 1968; Kushiro, 1969) indicate that compositions of liquids formed by partial melting of peridotites or crystal fractionation become poor in SiO_2 with increasing pressure, which suggests that the depth of formation of magma becomes deeper with increasing distance from the trench.

Direct Partial Melting of Peridotites

Numbers of experiments have been done in attempts to determine compositions of liquids produced by partial melting of peridotites under various temperature-pressure conditions (e.g., Kushiro et al., 1972; Kushiro, 1973; Mysen and Boettcher, 1975; Mysen and Kushiro, 1977). Jaques and Green (1979), on the other hand, criticized the methods of determination of liquid compositions previously used in direct partial melting experiments on peridotite compositions. Jaques and Green (1980) conducted anhydrous melting experiments on two synthetic peridotite compositions at temperatures ranging from the solidus to about 200°C above the solidus within the pressure range 0–15 kbar. These investigators also estimated liquid compositions in equilibrium with peridotites by mass balance calculations on the basis of chemical and modal compositions of residual phases. Their results clearly indicate that partial melts change compositions from tholeiitic to alkalic with increasing pressure or decreasing degree of partial melting. Takahashi and Kushiro (1983) confirmed the above tendency (Fig. 1) using a special experimental technique in which a thin layer of basalt powder is sand-

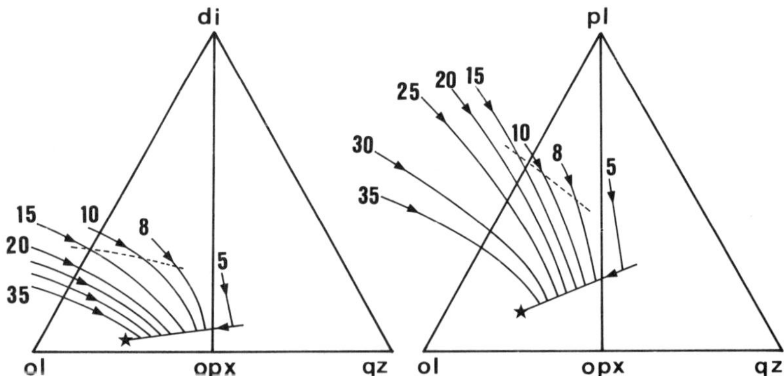

Fig. 1. Isobaric compositional trends of partial melts from a peridotite (Takahashi and Kushiro, 1983). Normative compositions are calculated after Walker et al. (1979). Numbers, stars, and broken lines indicate pressures, starting compositions, and iso-degree of partial melting lines, respectively. It is suggested that partial melts change compositions from tholeiitic to alkalic with increasing pressures or decreasing degree of partial melting. Abbreviations: di, diopside; ol, olivine; opx, orthopyroxene; qz, quartz; pl, plagioclase.

wiched between compressed peridotite minerals and is equilibrated with its host at melting temperatures. The above experimental data may enable the depth of segregation of a basalt magma from its chemical composition to be estimated by fitting the obtained isobaric liquid trends, if the magma was produced under anhydrous conditions.

Multiple Saturation Approaches

A multiple saturation approach at high pressures for natural rock compositions (e.g., Thompson, 1974) can give essential information about the $P-T$ condition of segregation of the magma, assuming that it has compositions of a primary magma; that is, the $P-T$ condition of a multiple saturation point with lherzolitic or harzburgitic minerals indicates the conditions of segregation of the starting magma from residues of mantle source rock, if these phases crystallizing on the liquidus have compositions close to those of mantle minerals. In general, however, most subduction zone lavas, even basalts, are derivatives after crystallization differentiation, and do not preserve compositions of primary magmas in equilibrium with upper mantle minerals (e.g., O'Hara, 1965; Nicholls and Whitford, 1976; Perfit et al., 1980). Furthermore, subduction zone magmas are generally H_2O-bearing, which is supported by several lines of evidence including; explosive eruptions of arc volcanoes, existence of hydrous minerals as phenocrysts, existence of Ca-rich plagioclase phenocrysts, absence of quartz and plagioclase phenocrysts in some felsic rocks. (Yoder, 1969; Tatsumi and Ishizaka, 1982; Sakuyama, 1983). As many experimental results have demonstrated, water strongly affects both phase relations and the stability field of crystallizing phases in magmas (e.g., Yoder and Tilley, 1962; Kushiro, 1972, 1974; Eggler, 1972; Nicholls and Ringwood, 1972; Tatsumi, 1981). For example, Eggler (1972) determined phase relations for an andesite at pressures up to 10 kbar for various H_2O contents in the melt. Eggler indicated that the liquidus phase changes from plagioclase to orthopyroxene with the increase of H_2O content in liquid and demonstrated that the slope of the liquidus for a given phase becomes less positive as water content is raised.

Kushiro (1972) indicated experimentally that the olivine liquidus field expands under H_2O-saturated conditions in the system diopside-forsterite-silica, and that similar observations applied to natural basaltic systems (Nicholls and Ringwood, 1972). Tatsumi (1981) conducted multiple saturation researches for a Mg-rich arc basalt and showed that the point of multiple saturation shifts toward lower pressures and higher temperatures under H_2O-undersaturated conditions.

Therefore, in order to understand the $P-T$ conditions for segregation of primary magmas by multiple saturation experiments, it is necessary to choose the starting material carefully and estimate the H_2O content in the magma.

Experimental simulation of the phenocryst assemblage and the crystallization sequence of natural rocks (Eggler, 1972; Eggler and Burnham, 1973; Maaløe and Wyllie, 1975; Sekine et al., 1979) contributed to the estimation of H_2O content in arc magmas. For example, Eggler (1972) compared the geometry of H_2O-undersaturated liquidi and experimental phase compositions for an andesite and concluded that phenocrysts in the starting andesite may be interpreted as having crystallized from a magma with 2.2 ± 0.5 wt.% H_2O at a temperature of 1110 ± 40°C. Also, fluorine geochemistry (Ishikawa et al., 1980), petrographic considerations (Sakuyama, 1979) for Northeastern Japanese volcanic rocks and estimation of H_2O content in a fluid inclusion in phenocrysts (e.g., Anderson, 1979), suggest that arc basalt magmas contain about 3 wt% H_2O as a maximum.

On the basis of these data, together with the across-arc lateral variation of H_2O content in arc magmas (Sakuyama, 1979), it was proposed that primary alkali olivine basalt and olivine tholeiite magmas contain about 3 and 0.6–0.7 wt% of water, respectively, assuming that K_2O and H_2O are coherent in magmatic processes (Sakuyama, 1983). Such a small amount of water in an olivine tholeiite magma would have little effect on the melting phase relations, suggesting that the olivine tholeiite magmas are produced under anhydrous

Table 1. Chemical and CIPW normative compositions of starting primary basalt magmas estimated on the basis of olivine maximum fractionation model.

	AOB	HAB	OTB
SiO_2	49.11	49.39	49.71
TiO_2	1.01	0.85	0.74
Al_2O_3	15.45	15.70	14.97
FeO*	9.42	9.76	10.57
MnO	0.13	0.15	0.14
MgO	11.59	12.05	13.03
CaO	9.66	9.43	9.00
Na_2O	2.54	2.33	1.56
K_2O	1.09	0.34	0.28
or	6.44	2.01	1.65
ab	20.26	19.71	13.20
an	27.53	31.38	33.02
di wo	8.51	6.43	4.85
di en	5.18	3.90	2.93
di fs	2.85	2.18	1.66
hy en	—	7.54	17.83
hy fs	—	4.21	10.13
ol fo	16.59	13.01	8.19
ol fa	10.06	8.02	5.13
ne	0.66	—	—
il	1.92	1.61	1.41

AOB, alkali olivine basalt; HAB, high-alumina basalt; OTB, olivine tholeiite; FeO*, total iron as FeO.

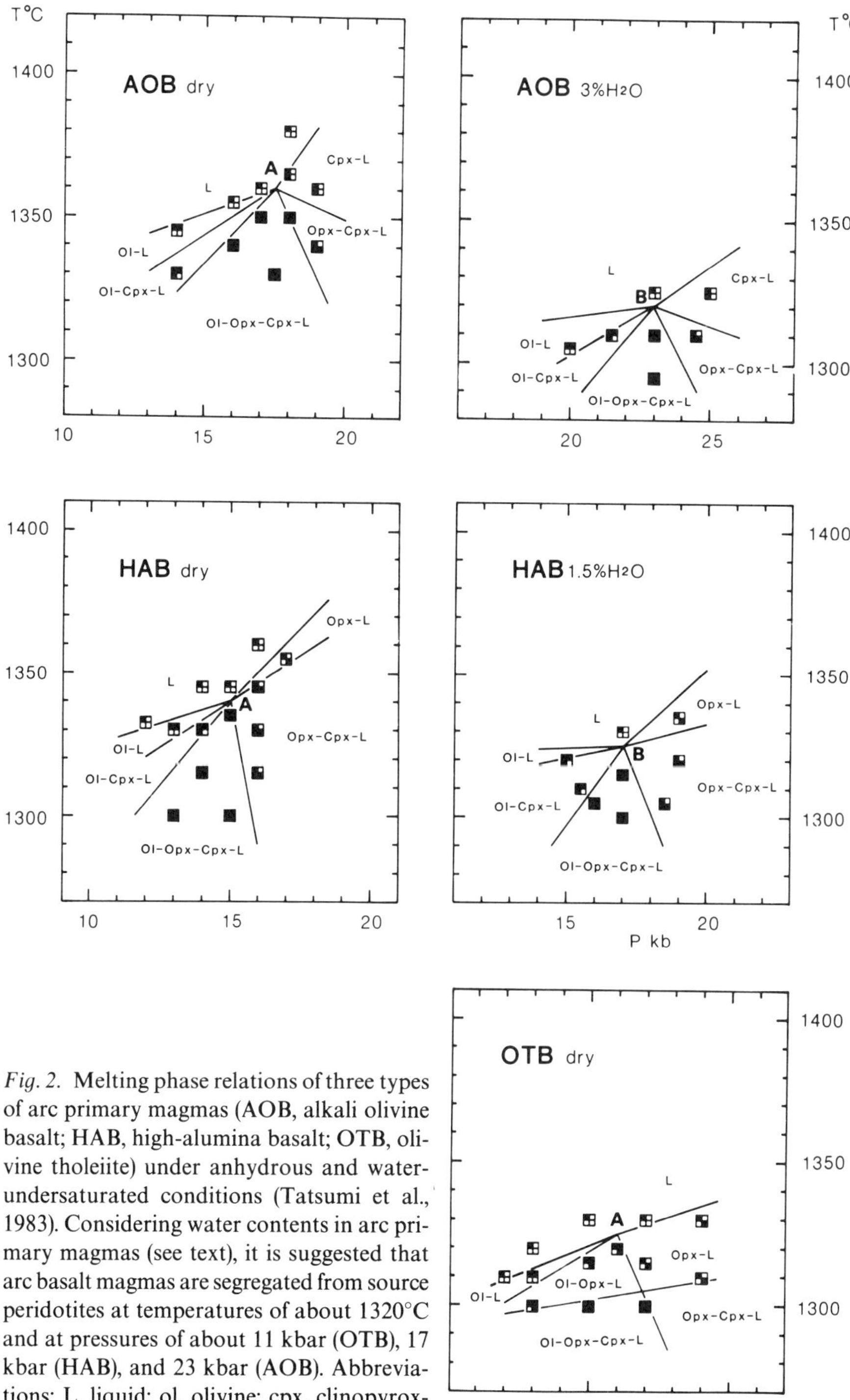

Fig. 2. Melting phase relations of three types of arc primary magmas (AOB, alkali olivine basalt; HAB, high-alumina basalt; OTB, olivine tholeiite) under anhydrous and water-undersaturated conditions (Tatsumi et al., 1983). Considering water contents in arc primary magmas (see text), it is suggested that arc basalt magmas are segregated from source peridotites at temperatures of about 1320°C and at pressures of about 11 kbar (OTB), 17 kbar (HAB), and 23 kbar (AOB). Abbreviations: L, liquid; ol, olivine; cpx, clinopyroxene; opx, orthopyroxene.

conditions. The primary high-alumina basalt magma were estimated to contain a medium amount of H_2O (=1.5wt%), because chemical characteristics of high-alumina basalts are transitional between those of another two.

Tatsumi et al. (1983) conducted multiple saturation experiments on three types of arc basalt compositions and estimated $P-T$ conditions of segregation of those magmas. The chemical compositions of primary magmas of olivine tholeiite, high-alumina basalt and alkali olivine basalt in Table 1 were obtained using the olivine maximum fractionation model (Nicholls and Whitford, 1976; Tatsumi and Ishizaka, 1982) for magnesian basalts from Japanese Quaternary volcanoes. These primary magma compositions were examined from various viewpoints such as the effect of subtraction or accumulation of plagioclase, pyroxene, and olivine including multiple fractionation and the effect of assumed mantle olivine compositions. It is concluded that the estimated compositions of the primary magmas are not strongly dependent on these variables.

Those synthetic primary magmas were studied at high pressures and temperatures (Fig. 2). The estimated primary alkali olivine basalt magma coexists with lherzolitic minerals at 17 kbar and 1360°C under anhydrous conditions and at 23 kbar and 1320°C in the presence of 3 wt% water. The high-alumina basalt magma also coexists with lherzolitic minerals at 15 kbar and 1340°C under anhydrous conditions and at 17 kbar and 1325°C in the presence of 1.5 wt% water. The olivine tholeiite magma, on the other hand, coexists with harzburgitic minerals at 11 kbar and 1320°C under anhydrous conditions. These data support the previous suggestion that arc basalt magmas are segregated from their source mantles at deeper levels away from the trench. The difference in residual phases for olivine tholeiite and the other two gives evidence that the degree of partial melting decreases toward the back-arc side, which is in harmony with the systematic increase of contents of incompatible elements from the volcanic front.

FeO*/MgO Ratio,

Kushiro (1983) examined FeO*/MgO ratios of quaternary basalts in the Japanese Islands and indicated that the maximum FeO*/MgO ratios decrease from the volcanic front toward the back-arc side. As Gill (1981) mentioned, on the other hand, this across-arc variation can be recognized only in the Japanese Islands and the northern Kuriles but not in other active volcanic arcs.

In order to interpret the across-arc lateral variation in FeO*/MgO ratios, Kushiro (1983) measured densities of representative olivine tholeiite and alkali olivine basalt magmas at high pressures with the falling-sphere method developed by Kushiro et al. (1976) and Fujii and Kushiro (1977). The results of the measurements are shown in Fig. 3, in which a possible density profile of the crust in the NE Japan arc is also indicated. Fig. 3 shows that the primary tholeiite melts are denser than the average granites at pressures greater than about 2 kbar, whereas the alkali olivine basalt melts are less dense than granitic rocks. As mentioned above, olivine tholeiite magmas are nearly anhydrous and the density

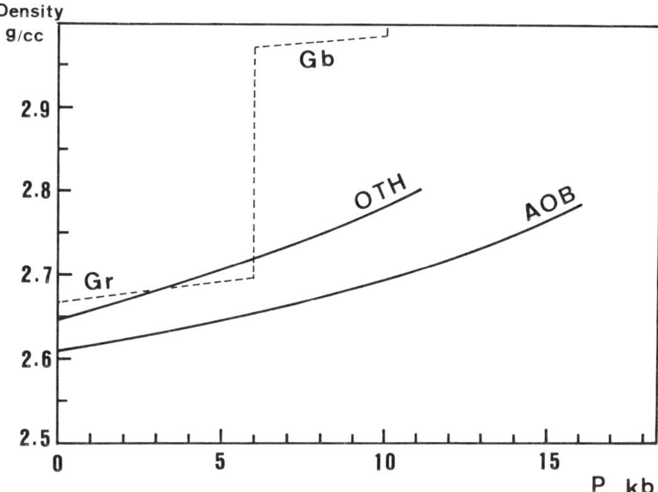

Fig. 3. Densities of magnesian olivine tholeiite (OTH) and alkali olivine basalt (AOB) determined by the falling sphere method (Kushiro, 1983). The density profile in the crust is based on densities of average granite (Gr) and gabbro (Gb) given by Daly et al. (1966) and the compressibility data of those rocks given by Birch (1966). Olivine tholeiite magmas are denser than granitic crust at pressures around 5 kbar, suggesting that the olivine tholeiite magma cannot ascend beyond the upper crust by buoyancy alone.

data can be applied for natural magmas. These data support the suggestion that the primary olivine tholeiite magmas cannot ascend beyond the upper crust by buoyancy alone, and the magmas would fractionate to produce less dense magmas. On the other hand, alkali olivine basalt magmas are likely to contain water and should be less dense than the experimental results. The magmas can easily ascend within the crust. The across-arc variation in FeO*/MgO ratios may thus be due partially to the difference in density of magma.

The Role of Subducted Lithosphere

The existence of subducted lithosphere is a necessary condition for generation of arc magmas. However, the role of subducted slab has been a matter of much debate. Two petrogenetically distinct hypotheses have been proposed: the slab produces (1) partial melts that are primary magmas for calc-alkaline andesites or partial melts that metasomatize the overlying mantle wedge to form solid sources for arc magmas, and (2) H_2O, which migrates upward to cause partial melting of the mantle wedge. The decisive factor controlling the above two mechanisms is the temperature distribution within the subducted lithosphere and the mantle wedge. There remains, however, considerable uncertainty about the precise distribution of isotherms, and petrologists must study which mechanism will enable the characteristics of arc magmatism to be better understood.

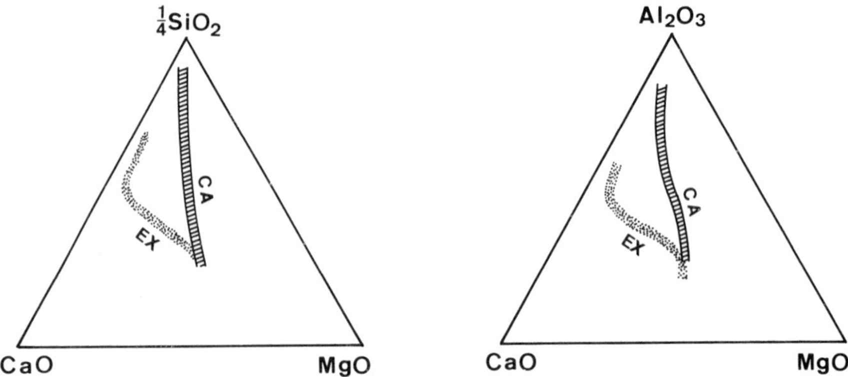

Fig. 4. Equilibrium liquid path (EX) for quartz eclogite with 7.5% H_2O at 30 kbar projected from Cao–MgO–Al_2O_3–SiO_2 onto two surfaces (Sekine et al., 1981). The hatched curve (CA) corresponds to the average chemical variation of calc-alkaline rocks. It is indicated that partial melting of the subducted crust produces liquids with a range of intermediate SiO_2 content but with Ca/(Ca + Mg) ratios higher than calc-alkaline rocks.

Partial Melting of the Subducted Lithosphere

Several experimental petrologists claimed that the basaltic oceanic crust, converted to quartz eclogite, could partially melt to yield residual quartz-free eclogite and andesitic magmas belonging to the calc-alkaline affinity (e.g., Green and Ringwood, 1968; Yoder, 1969; Holloway and Burnham, 1972; Allen et al., 1975; Allen and Boettcher, 1978). In order to test these early experimental predictions, phase relations of both natural and simple systems of basaltic compositions at high pressures were determined under hydrous conditions (Stern and Wyllie, 1973; Stern, 1974; Stern and Wyllie, 1978; Sekine et al., 1981). For example, Sekine et al. (1981) studied a synthetic oceanic tholeiite in the system CaO–Al_2O_3–MgO–SiO_2 at 30 kbar with 3.5 to 32.5 wt.% H_2O. They confirmed the conclusion of Stern and Wyllie (1978) that partial melting of quartz eclogite in the subducted oceanic crust at 100 km depth produces liquids with a range of intermediate SiO_2 contents, but with Ca/(Ca + Mg) ratios higher than calc-alkaline andesites (Fig. 4). At the present stage, then, the hypothesis that subduction zone andesites are primary magmas from the subducted oceanic crust has few adherents.

Nicholls and Ringwood (1973) and Ringwood (1974) proposed a model in which hydrous silicious melts produced by partial melting of the subducted oceanic crust react with the overlying mantle, converting olivine into pyroxenes. It was further suggested that diapirs of wet pyroxenite then rise and undergo partial melting to produce calc-alkaline andesite magmas. Series of works were conducted in order to support the above idea on both synthetic and natural rock systems (Sekine and Wyllie, 1982a,b,c). Sekine and Wyllie (1982a) constructed the H_2O-saturated liquidus surface for the system $KAlSiO_4$–Mg_2SiO_4–SiO_2–

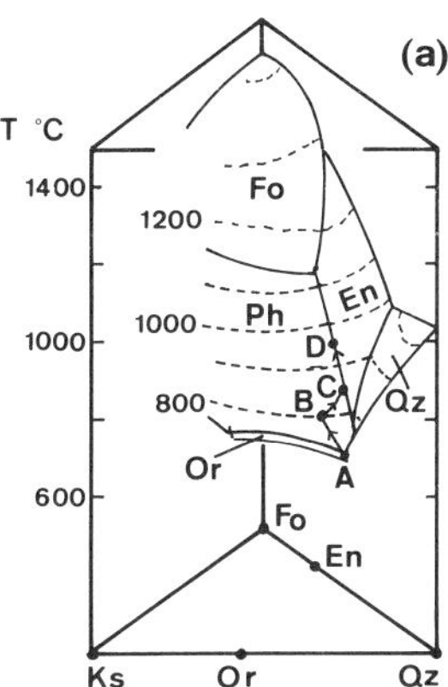

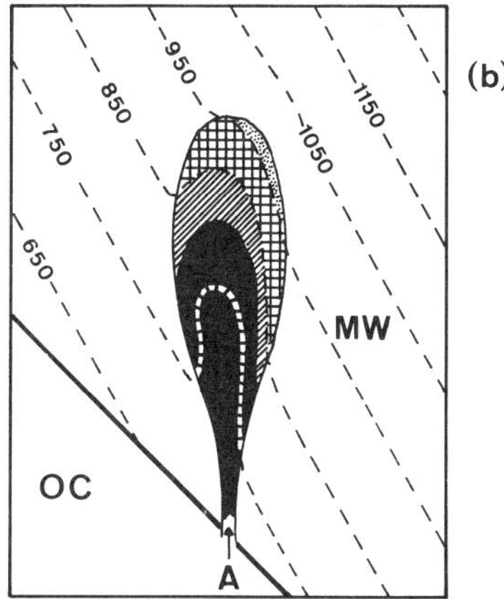

Fig. 5. (a) H_2O-saturated liquidus surface of the system $KAlSiO_4$–Mg_2SiO_4–SiO_2–H_2O at 20 kbar (Sekine and Wyllie, 1982a). The path A–B–C–D indicates the compositional change of a partial melt of the oceanic crust with compositions of A during the hybridization with wedge peridotite. Abbreviations: Fo, forsterite; Ks, kalsilite; Or, sanidine; Qz, quartz; En, enstatite. (b) Schematic representation of a silicious magma body rising from the oceanic crust (OC) after Wyllie and Sekine (1982). During the process of hybridization between silicious melts (A) and peridotite in the mantle wedge (MW), the melt change assemblage from liquid only (black area) through liquid + phlogopite (hatched area) and liquid + phlogopite + enstatite (crossed area) to phlogopite + enstatite (dotted area). The mineral precipitates are modelled from the liquid path in Fig. 5(a). After solidification, the body releases water that migrates upward to cause partial melting of peridotite. Numbers and broken lines indicate temperatures (°C) and possible isotherms, respectively.

H_2O at 20 kbar based on various experimental data (Fig. 5a), which includes model representatives of hydrous silicious magma from the subducted crust and the overlying mantle wedge. Using the obtained liquidus surface, Wyllie and Sekine (1982) discussed the process of hybridization between the slab-derived, hydrous silicious melts and mantle wedge peridotites (Fig. 5b), and proposed a model for generation of subduction zone magmas, reviewed in a following section.

Dehydration of the Subducted Lithosphere

Water held in hydrous minerals that have crystallized on the ocean floor and in the subducted lithosphere goes down with the slab. This water is released when the thermal or pressure stability limit of the hydrous minerals is exceeded. Several petrologists have related the dehydration in a downgoing slab with production of magmas in subduction zones since it was first suggested by Coats (1962), because water can drastically lower the solidus temperature of mantle materials (e.g., Kushiro et al., 1968). When the significance of the stability of hydrous minerals in

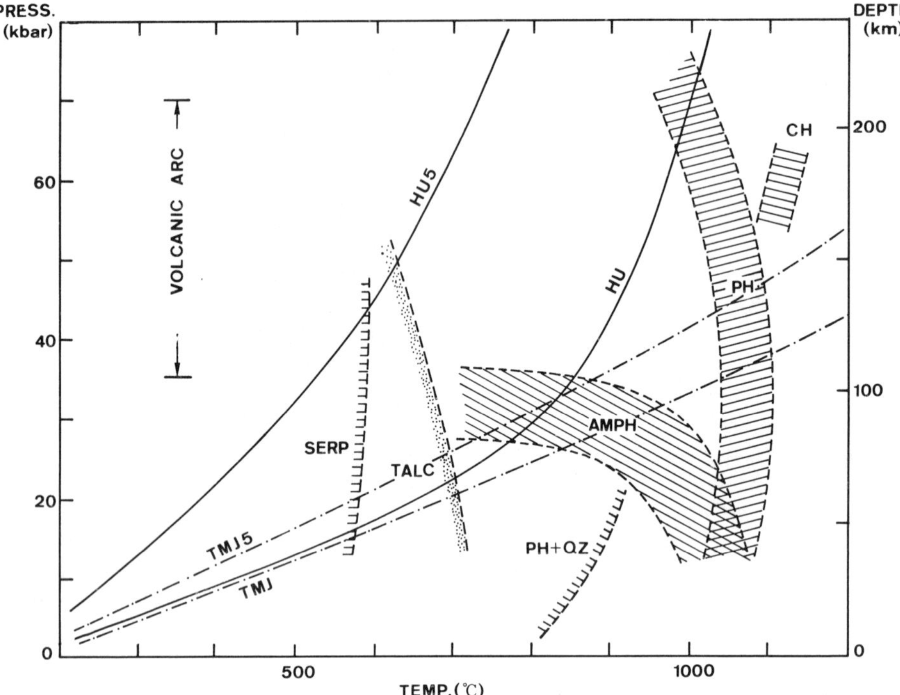

Fig. 6. Stability fields of hydrous phases, compared with the estimated range of temperatures within 5km-thick subducted oceanic crust, according to two calculations: TMJ and TMJ5, top and 5 km below the surface of the slab by Toksoz et al. (1971); HU and HU5, those by Honda and Uyeda (1983). Abbreviations: CH, clinohumite; PH, phlogopite; AMPH, amphiboles; QZ, quartz; TALC, talc; SERP, serpentine.

producing water is discussed, the choice of thermal model is essential. For published thermal models of subduction zones, however, the absolute value of temperature is widely variable; for example, the temperature at the slab/wedge interface at a depth of 100 km is estimated to be 400°C by Andrews and Sleep (1974) and 1400°C by Griggs. In the following, geothermal models by Tokzos et al. (1971) and Honda and Uyeda (1983) are tentatively used for considering the stability of hydrous phases in the slab; these models provide intermediate values of temperature among a number of geothermal models. Amphiboles, magnesium hydrosilicates and phlogopite are the main candidates of hydrous minerals in the subducted lithosphere. Stability limits of these minerals are shown in Fig. 6 together with two thermal models.

Amphiboles

Amphiboles are likely to exist in the downgoing lithosphere and often occur in arc lavas. Thus numbers of experiments have been conducted on the stability of amphiboles under various conditions of total pressures, temperatures and H_2O-fugacity. Although the experimental results may be debated among the experts, there is a consensus that amphiboles in basaltic systems (i.e., in the oceanic crust) decompose at a pressure lower than 30 kbar (e.g., Lambert and Wyllie, 1972; Allen et al., 1975; Allen and Boettcher, 1978, 1983); that is, amphibole is not a likely mineral to release fluid phases beneath volcanic arcs. On the other hand, amphiboles in peridotitic systems may be stable to pressures in excess of 30 kbar, probably to pressures around 40 kbar (e.g., Millhollen et al., 1974). If a thermal model that gives higher temperatures within a slab is chosen, amphibolites converted from basalt should melt. This process may not represent the existence of a volcanic front very well, as discussed in a following section.

Magnesium Hydrosilicates

Hydrous minerals in the system $MgO-SiO_2-H_2O$, could be distributed to a certain extent, especially in cumulate layers of the oceanic lithosphere. Several experiments on serpentine (e.g., Kitahara et al., 1966; Tatsumi et al., 1985) indicated that the stability limit of serpentine does not depend on pressure at pressures below 40 kbar (Fig. 6). If a thermal model with extremely low temperature distribution is not used, therefore, serpentine may decompose at pressures below 40 kbar. Serpentine contains about 15 wt.% of H_2O and significant element migration of the serpentine-derived fluid may occur. This migration should affect the compositions of the mantle wedge and its derivative, subduction zone magmas. This important geochemical problem will be discussed in a later section. Clinohumite and chondrodite whose stability fields were determined by Yamamoto and Akimoto (1977) could be shown to decompose in the slab beneath volcanic arcs if geothermal models that provide relatively high temperatures in the slab are used (Fig. 6). However, amounts of those phases would be limited in the slab and may not play a major role as the source of water.

Phlogopite

Since the pioneer work by Kushiro et al. (1967) a large amount of experimental data has been accumulated on the stability of phlogopite (e.g., Modreski and Boettcher, 1972). This work shows that phlogopite in a peridotite system is stable up to about 70 kbar under temperature conditions proposed for the subducted lithosphere (Fig. 6). On the other hand, phlogopite mainly exists in a pelitic layer of the oceanic crust which is saturated with SiO_2 and its stability in peridotite systems could not be directly applied to the dehydration process in subduction zones. According to the experiments of Bohlen et al. (1983), the thermal stability limit of phlogopite coexisting with quartz is more than 100°C lower than that in peridotite systems at pressures below 20 kbar. Furthermore, phlogopite in a pelitic layer should be Fe-rich biotite, and the stability range of biotite in basic to intermediate composition is more restricted than that of amphibole (Stern et al., 1975). Thus, it is questionable that phlogopite is a source of H_2O beneath volcanic arcs. Within the mantle wedge, on the other hand, phlogopite is still a strong candidate among hydrous minerals which persist beyond the breakdown pressure of amphiboles (i.e., a depth greater than 40 kbar).

Volcanic Fronts as a Clue to the Role of Subducted Lithospheres

At convergent plate boundaries, in general, the trenchward limit of the distribution of volcanoes defines a sharp line, the volcanic front of Sugimura (1960). The most noticeable fact on the volcanic front may be that the depth of the Wadati-Benioff zone beneath it is quite constant (124 ± 38 km; Gill, 1981) in each subduction zone. Two interpretations may be applied to this tectonic feature. One is that the ascent of magma to the surface is possible only behind the volcanic front. In other words, magmas are produced on both sides of the volcanic front; near-trench volcanism existing in some fore-arc regions supports this idea.

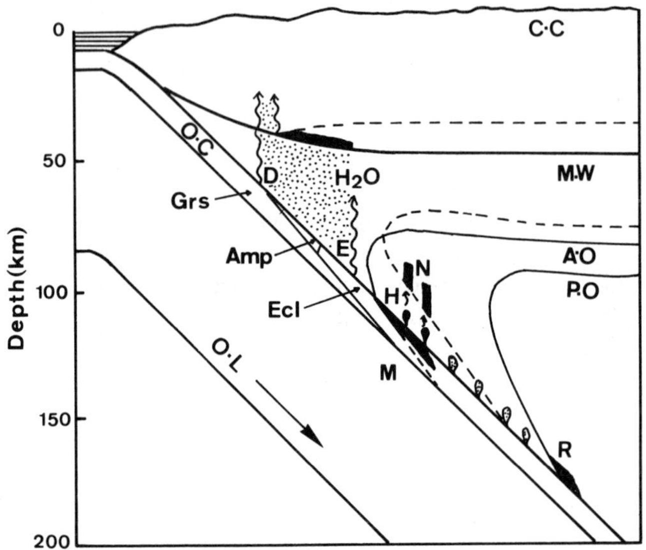

Fig. 7. A possible cross-section through the subducted oceanic lithosphere (O.L) proposed by Wyllie and Sekine (1982), in which amphibole-out (A.O), phlogopite-out (P.O), hydrous solidus (broken lines) of the mantle wedge (M.W) and the continental crust (C.C) are shown. Black areas indicate partial melting zones. The subducted oceanic crust (O.C) changes facies from greenschist (Grs) through amphibolite (Amp) to eclogite (Ecl) in which process H_2O is released between D and E to form hornblende peridotite in the mantle wedge. Eclogite with water may partially melt at M to produce hydrous silicious melts that react with mantle peridotite as shown in Fig. 5. Water released from the hybridized body at H causes partial melting of peridotite at N. The residual bodies composed of phlogopite and enstatite (dotted diapirs) are transported downward along the slab by the induced convection to form alkalic initial liquid at R through phlogopite-out reaction.

According to this interpretation, the volcanic front must be a boundary of physical properties such as structures and stress fields in the crust. However, no evidence indicating such a boundary is found in subduction zones. Therefore, the other interpretation, that magmas are formed in the slab or the mantle wedge only behind the volcanic front, is more likely; that is, the geotherm cuts the solidus temperature of the wedge mantle or the slab beneath the volcanic front.

Wyllie and Sekine (1982) discussed that the subducted slab partially melts under hydrous conditions produced by the amphibole-out reaction within the slab to form metasomatizing hydrous silicious magmas (Fig. 7). This mechanism could govern the formation of the volcanic front in some subduction zones. If this is true in general, however, the hydrous solidus temperature of an amphibole-bearing basaltic composition must be distributed within the slab just beneath a volcanic front: this result seems to be rather accidental because an oceanic lithosphere with variable age and temperature distribution subducts into the mantle wedge. Furthermore, the depth of breakdown of amphiboles in the slab may be shallower than 100 km which is a smaller value than that of the slab beneath volcanic fronts (about 120 km). Metamorphic petrologists (e.g., Reinsch, 1979) have discussed much shallower decomposition of amphibolite to eclogite assemblage by a reaction between epidote and amphibole.

It is suggested by Tatsumi (1986) that the site of the volcanic front is controlled by decomposition of amphibole in mantle wedge peridotites at 35 kbar (Fig. 8): the amphibole-bearing peridotite can be formed by reaction between wedge peridotites and slab-derived water from intra-slab amphiboles at shallower levels (see the previous section) to be transported downward along a slab by the induced convection in the mantle wedge. An advantageous point in this mechanism is that it does not restrict the temperature beneath volcanic fronts in the slab or the mantle wedge at 35 kbar to the solidus temperature but only demands the hydrous solidus temperature at any level between the top of the slab and the bottom of the crust.

Geochemical Role of the Subducted Lithosphere

It has been well established that subduction zone lavas are distinct from mid-oceanic ridge basalts (MORBs) by the following points; the former are enriched in large ion lithosphere elements (LILE) and depleted in high field strength elements (HFSE) (Wood et al., 1979; Perfit et al., 1980; Sounders et al., 1980; Arculus and Johnson, 1981; Tatsumi and Ishizaka, 1982); the former also have higher $^{87}Sr/^{86}Sr$ and lower $^{143}Nd/^{144}Nd$ ratios (e.g., DePaolo and Wasserburg, 1977; Hawkesworth et al., 1977; 1979; Nohda and Wasserburg, 1981; Ishizaka and Carlson, 1983). The slab-derived fluid phase should play an important role in giving rise to the above geochemical characteristics of subduction zone magmas. Recently some experimental studies have been done to try to understand the geochemical role of the subducted lithosphere and those data provide qualitative interpretation for LILE-enrichments and HFSE-depletions in subduction zone magmas.

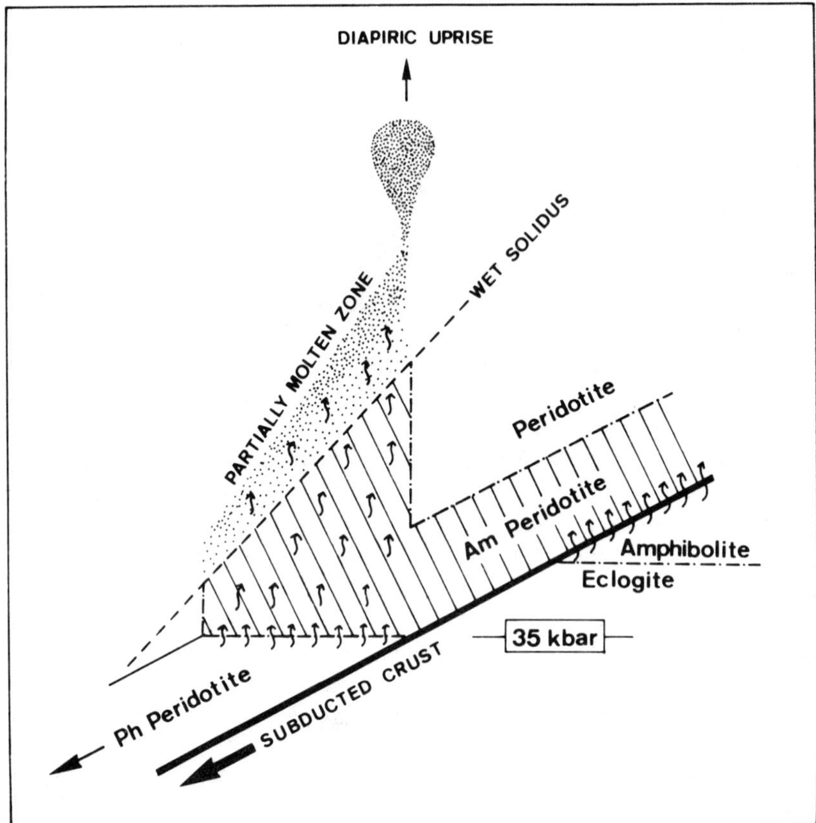

Fig. 8. Schematic process governing the formation of volcanic front (Tatsumi, 1986). The subducted oceanic crust releases water (arrows) through the phase transition from amphibolite to eclogite at a pressure less than 30 kbar to form amphibole (Am) peridotite in the mantle wedge; the hydrous peridotites are transported downward along the slab by the induced convection in the mantle wedge to release water at 35 kbar which is the upper stability limit of amphibole in peridotites. The water migrates upward with forming hornblende peridotite and cause partial melting of peridotites at a temperature of wet solidus. With increasing temperature, amounts of liquid increase and partially molten peridotites uprise as a form of diapir to produce olivine tholeiite magma distributed at the volcanic front.

Enrichment of Large Ion Lithophile Elements

Dissolution of K_2O in an aqueous vapor can be estimated from Yoder and Kushiro's experiments (1969) in which the subsolidus assemblage is phlogopite plus forsterite for the join phlogopite-H_2O. Ryabchikov and Boettcher (1980) determined the solubility of potassium in an aqueous fluid phase in equilibrium with phlogopite at 1100°C up to 30 kbar. The concentration of K_2O is 4 gr/100 gr H_2O at 11 kbar and increases to 25 gr/100 gr H_2O at 30 kbar. These experimental results are in harmony with petrographic evidence that mantle peridotites often contain amphibole-phlogopite veins, probably formed by K_2O–H_2O metasomatism.

Mysen (1979) determined partition coefficients of rare-earth elements (REEs) involving an aqueous vapor and the constituent minerals of garnet peridotites up to 30 kbar using the beta-track mapping technique. He indicated that light REEs (rare earth elements) are more strongly partitioned into H_2O vapor than medium or heavy REEs, especially for the vapor-garnet system. It has therefore been suggested that the mantle wedge overlying the subducted oceanic crust with an eclogite mineral assemblage would attain considerable light-REE enrichment as a result of the addition of fluid phases equilibrated in the slab with garnet. On the other hand it is questionable whether an aqueous fluid phase equilibrates with constituent minerals of the subducted slab because an H_2O-rich fluid infiltrates crystalline mantle minerals at a high rate of more than several millimeters per hour at pressures above 15 kbar (Mysen et al., 1977).

Some geochemical studies indicate that LILE in the upper mantle and the subducted lithosphere could be distributed along grain boundaries; Frey and Green (1974) found, in anhydrous lherzolites, grain boundaries with high concentrations of light REEs and LILEs; Basu and Murthy (1977) also indicated the high concentration of K, Rb, and Sr along the grain boundary of lherzolite;

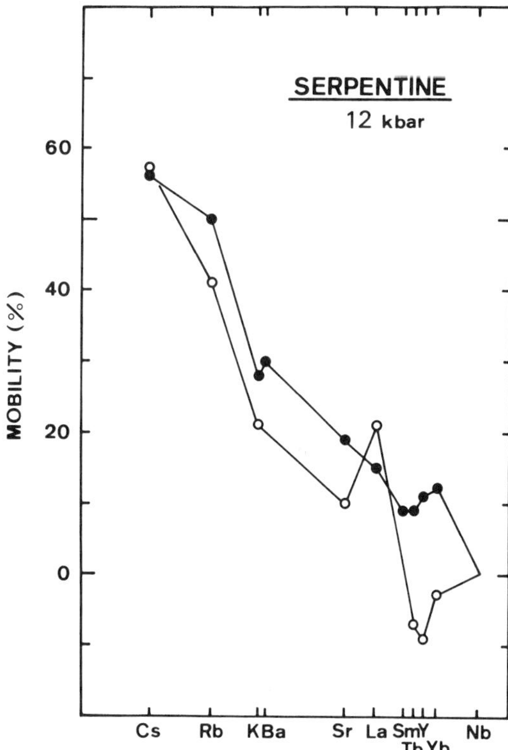

Fig. 9. Mobility of elements by the fluid phase released through dehydration of element-doped synthetic serpentine at 12 kbar (Tatsumi et al., 1985). Mobility is defined as amounts lost through the process divided by amounts initially present in the starting serpentine. Open circles indicate results using serpentine washed by HCl in order to remove the more loosely bound elements, and closed circles results of unwashed serpentine. It is suggested that elements with larger ionic radii are more readily transported by the H_2O gas phase.

Suzuki (1984) suggested that those elements are not trapped in any of the crystal structures. It is thus necessary to determine the degree of transportation of such elements distributed along grain boundaries by an aqueous fluid phase at high pressures.

Tatsumi et al. (1985) conducted dehydration experiments on a synthetic serpentine containing 11 trace elements (Cs, Rb, K, Ba, Sr, La, Sm, Tb, Y, and Nb). Chemical analyses of both the starting serpentine and the run products consisting of forsterite and enstatite indicated that an element with a larger ionic radius is more readily transported by an aqueous fluid phase released from serpentine (Fig. 9). Drastic reduction of doped-element concentration was observed after washing the starting serpentine with HCl, confirming that most spiked elements in the starting serpentine are distributed along grain boundaries, not in the crystal structure. Thus, the experimental results may be realistic under the descending slab environments. The pressure at which the element migration was measured was 12 kbar, which is about 3 times smaller than the actual pressure for the dehydration of serpentine in the subducted lithosphere. The results obtained in the experiments thus are not directly applicable to element transportation with slab-derived fluid phases. However, the general conclusion of mobility against ionic radii shown in Fig. 9 may hold good at higher pressures, and those elements should be more soluble under such conditions.

Depletion of High Field Strength Elements

HFSE behave as incompatible elements in the process of partial melting of "normal peridotites" consisting of olivine, orthopyroxene, clinopyroxene, and spinel or garnet. Although concentration of another group of incompatible elements (LILE) in arc magmas increases systematically toward the back-arc side, there is no difference in contents of HFSE in fore- and back-arc region magmas produced by different degrees of partial melting (Tatsumi and Nakano, 1984). This geochemical evidence indicates that the bulk distribution coefficient for HFSE should be near unity between melts and peridotite sources, assuming a constant peridotite composition. However, the distribution coefficients of HFSE, D_{HFSE}, are smaller than unity for major constituent minerals of peridotites (e.g., Gill, 1981). Therefore, minor minerals with D_{HFSE} much higher than unity would have to be present as residual phases after arc magma production. Candidates with such distribution coefficients are zircon, sphene, perovskite, and rutile.

Expansion of the stability field of the above phases, expected only in the mantle wedge environment, should also be understood in connection with subduction of the oceanic lithosphere. A series of melting experiments by Hellman and Green (1979) and Green (1981) provide important data showing that the stability field of rutile expands under hydrous conditions in an above-solidus basalt system (Fig. 10). A possible interpretation arising from the above geochemical evidence is that such HFSE-rich phases become stable under hydrous conditions induced by addition of the slab-derived fluid phases in a source peridotite for arc magmas. These experiments include some problems of which

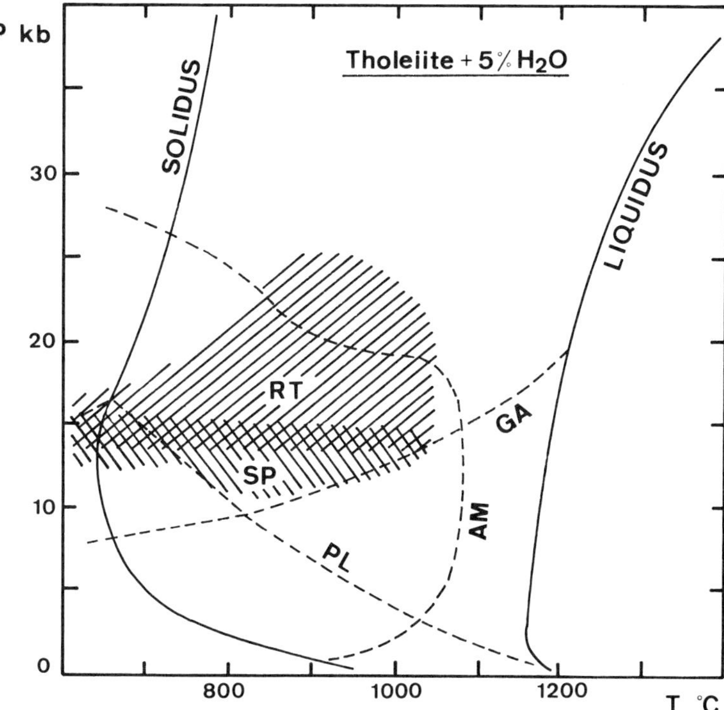

Fig. 10. Simplified phase relations of olivine tholeiite +5% H_2O after Green (1981). Stability fields of rutile (RT) and sphene (SP) rich in HFSE expand under hydrous conditions, providing a possible interpretation for HFSE-depletion in arc magmas formed by the addition of slab-derived water. Broken lines indicate phase-out lines of garnet (GA), amphibole (AM) and plagioclase (PL).

the most important is that such phases are not crystallized on the liquidus; in other words, there is no confidence that such phases are stable as residual phases at the time of magma segregation even under hydrous conditions.

Material Movement in the Mantle Wedge

The mantle wedge is by no means in a static condition: initial magmas produced at rather deeper levels in the mantle wedge should migrate upward to be segregated from their residues; the subduction of oceanic lithospheres may induce convective currents in the mantle wedge. Compositions of subduction zone magmas and magma sources should be influenced significantly through such processes of material movements. Diapiric uprise and convective transportation of mantle material are highlighted in the following section.

Mantle Diapir

The existence of mantle diapir which is high temperature mantle material ascending from the deeper level of the upper mantle has been suggested by several petrologists (e.g., Green and Ringwood, 1967; Wyllie, 1971; Ramberg, 1972; Cawthorn, 1975; Sakuyama, 1983). The following observations may reinforce understanding of the production of subduction zone magmas through diapiric uprise of mantle materials; First, the temperature of primary magma at the stage of magma segregation from its source may be much higher than that in the uppermost mantle wedge. Tatsumi et al (1983) suggested that primary olivine tholeiite magmas occurring on the volcanic front are segregated from the residual mantle at 11 kbar and 1320°C. Such segregation means that a high temperature region of magma segregation must be located at a depth of only about 35 km, which is just beneath Moho as estimated by the explosion seismology at the Northeastern Japan arc (Hashizume et al., 1968). On the other hand, the solidus temperature of an anhydrous lherzolite and a gabbro with a small amount of water is between 1200 and 1000°C at 10 kbar (Takahashi and Kushiro, 1983; Lambert and Wyllie, 1972), This condition means that the uppermost mantle and the lowermost crust would be melted to a certain extent if the temperature of 1320°C is maintained in a steady-state regional geotherm at a depth of about 35 km. However, there is no geophysical evidence of large-scale melting in the region. Secondly, a column with low velocity can be observed in the mantle wedge beneath a volcanic region by three-dimensional P-wave velocity analyses (Hasemi et al., 1984). This observation supports the theory of ascent of high-temperature mantle materials from a deeper part of the wedge. Thirdly, chemical compositions of volcanic rocks, especially those with content of incompatible elements, are different in different volcanoes (Kawano et al., 1961; Katsui et al., 1978; Onuma et al., 1981; Sakuyama, 1981). This effect of the difference in chemical composition implies that an individual diapir produces only a single volcano.

The temperature of a mantle diapir at a deeper level must be higher than that at the level of magma segregation, because a diapir consumes heat for the latent heat of fusion (Green and Ringwood, 1967; Cawthorn, 1975; Fukuyama, 1984). Fukuyama (1984) experimentally determined the heat of formation of basaltic liquid (150 cal per gram on average) in the system anorthite-diopside, forsterite-diopside-SiO_2 forsterite-diopside-anorthite and natural peridotite. Using the results, Fukuyama also calculated the heat content of partially-molten diapir and gave a set of the equal heat content contour on a $P-T$ diagram for peridotite (Fig. 11). Tatsumi et al. (1983) combined their data for $P-T$ conditions of segregation of arc basalt magma with the experimental data of Fukuyama, and suggested the existence of high temperature region of more than 1400°C within the mantle wedge.

Introducing the concept of diapiric uprise of mantle materials and associated magma production might explain the across-arc variation in volume of volcanic rocks. Sugimura et al. (1963) indicated that the volume of volcanic materials

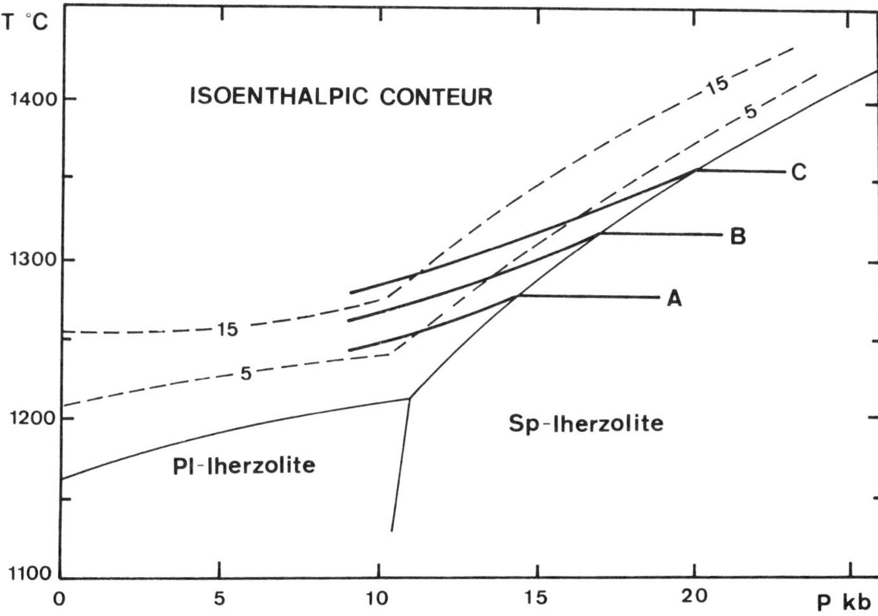

Fig. 11. The isoenthalpic contour lines (A, B and C) compared with the solidus temperatures of a peridotite (Fukuyama, 1984). Broken lines indicate contour of degree of partial melting

decreases toward the back-arc side from the volcanic front. To understand this observation it should be noted that there is no across-arc lateral variation in volume of one stratovolcano; in other words, the lateral variation in volume of volcanic materials partially corresponds to that in a number of volcanoes; more mantle diapirs ascend from the deeper part of the mantle wedge beneath the volcanic front than beneath the bach-arc region. Kushiro (1983) pointed out that the above phenomenon is well reproduced by the experiments of Marsh (1979) in which diapiric uprise is most frequently observed from the top of an inclined layer whose density and viscosity is lower than the overlying material. This model demands the existence of an inclined, partially-molten, layer, probably parallel to the subducted slab (Kushiro, 1983). However, it is highly questionable that partial melts are continuously formed along the slab.

Induced Convection in the Mantle Wedge

A hypothesis that vortex in the low-Q, low-V, layer is induced by the subducted slab has been proposed by many geophysists (e.g., McKenzie, 1969; Tokzos, 1971; Andrews and Sleep, 1974; Tokzos and Bird, 1977; Tokzos and Hsui, 1978, 1979; Hsui and Tokzos, 1981; Ito et al., 1985). The induced convection is considered to be a possible driving force for back-arc spreading (e.g., Hsui and Tokzos, 1981). The material migration in the mantle wedge associated with the induced convection must be taken into account in order to discuss the source material for arc magmas.

Wyllie and Sekine (1982) proposed the along-slab migration of phlogopite pyroxenite bodies by induced convection (Fig. 7). The phlogopite pyroxenite body is formed by hybridization between hydrous silicious melts from the slab and mantle wedge peridotites as reviewed in a previous section, and the body would begin to partially melt when it crossed the solidus phase boundary to yield alkalic liquids, initial liquids for alkalic magmas in the back-arc side of volcanic arcs.

The existence of induced convection in the mantle wedge is petrologically predicted (Sakuyama and Nesbitt, 1985; Tatsumi et al., 1985) on the basis of the following considerations. First, the subducted lithosphere is essentially anhydrous beneath volcanic arcs as mentioned before, and it cannot supply water, which causes partial melting of the mantle wedge. Therefore, polluted or metasomatized mantle materials (amphibole peridotite) formed beneath the fore-arc region must be transported to deeper levels. Secondly, across-arc compositional variations in isotopic ratios (lower $^{87}Sr/^{86}Sr$ ratios in back-arc side lavas) can be well understood by introducing the existence of convective current in the mantle wedge along the subducted slab. The "island-arc source" which is probably amphibole peridotite formed beneath the fore-arc region by reaction between wedge peridotite and slab-derived fluid phase may have higher ratios of $^{87}Sr : {}^{86}Sr$ than normal wedge peridotite as mentioned in a previous section. It is likely that the normal peridotite has chemical compositions similar to depleted N-type MORB source (Sakuyama and Nesbitt, 1985; Tatsumi et al., 1985; Nohda and Tatsumi, 1985). Thus the isotopic lateral variation may be formed by mixing between the "island-arc source" and the MORB source which is associated with along-slab migration of the "island-arc source" induced by the convection in the mantle wedge.

Origin of High-Magnesian Andesites

Voluminous eruption of calc-alkaline andesites, well-defined on an iron-magnesium-alkali diagram, characterizes subduction zones, and several mechanisms have been proposed to interpret their origin:

1. fractional crystallization of basalt magmas (e.g., Osborn, 1959; Kuno, 1968; Cawthorn and O'Hara, 1976; Irvine, 1976); for example, Irvine (1976) indicated that the liquidus field of olivine becomes wider relative to silica minerals with increasing the $KAlSi_3O_8$ component in the system Mg_2SiO_4–Fe_2SiO_4–$CaAl_2Si_2O_8$–$KAlSi_3O_8$–SiO_2 and suggested that calc-alkaline andesites are produced from a parent basalt magma with higher $KAlSi_3O_8$ component than for a tholeiitic andesite,

2. crustal anatexis (e.g., Pichler and Zeil, 1969; 1972),
3. crustal assimilation (e.g., Kuno, 1950; Wilcox, 1954; Ewart and Stipp, 1968),
4. magma mixing (e.g., Anderson, 1976; Eichelberger, 1978; Sakuyama, 1979; Luhr and Carmichael, 1980),
5. melting of the subducted crust (e.g., Green and Ringwood, 1968; Holloway and Burnham, 1972) which was reviewed in a previous section,
6. melting of hydrous peridotites (e.g., O'Hara, 1965; Yoder, 1969; Kushiro, 1974; Mysen and Boettcher, 1975).

Among the above, fractional crystallization and magma mixing seem to essentially determine the calc-alkaline trend as recently pointed out by Sakuyama (1983). On the other hand, rare but not insignificant characteristic andesites which form a typical calc-alkaline trend appear in subduction zones. These andesites are characteristically rich in MgO and thus are called "high-magnesian andesites" (HMAs; Tatsumi and Ishizaka, 1981), and they cannot be produced by the above two mechanisms (Sato, 1977; Tatsumi and Ishizaka, 1981; 1982). Recent experimental works provide clear interpretation for the origin of HMAs, and are reviewed in the following.

There are two essential problems to be addressed concerning the origin of HMA magmas; one is water content and the other is the origin of two distinct types of phenocryst assemblages, olivine plus orthopyroxene (opx-HMAs) and olivine plus clinopyroxene (cpx-HMAs). To clarify the above, two problems of melting-phase relations at high pressures were investigated for the two types of natural HMAs (Tatsumi, 1981; 1982). These HMAs characterize a tertiary volcanic belt extending about 1000 km in the southwestern Japan arc. (Tatsumi and Ishizaka, 1981).

Chemical compositions of starting HMAs are listed in Table 2, and experimental results are shown in Fig. 12; the cpx-HMA is in equilibrium with lherzolitic minerals at 15 kbar and 1030°C under H_2O-saturated conditions, and at 10 kbar and 1070°C in the presence of 7wt% H_2O; on the other hand, the opx-HMA is in equilibrium with harzburgitic minerals at 15.5 kbar and 1080°C under H_2O-saturated conditions, and at 11.5 kbar and 1120°C in the presence of 8 wt% H_2O. These experimental results indicate that both HMAs are direct partial melting products of upper mantle peridotites under hydrous conditions.

Water Contents in HMA Magmas

It has been believed that HMA magmas are formed under water-saturated conditions. More than 15 wt% water can be dissolved in an andesite melt under P–T conditions of the upper mantle (Hamilton et al., 1972; Sakuyama and Kushiro, 1978; Tatsumi, 1981). It is, however, questionable that such a large amount of H_2O exists in mantle peridotites.

The water contents in the experiments on two HMAs (7–8 wt% in the melt) correspond to water-undersaturated conditions based on the solubility limit of water in an HMA melt (15wt% at 12 kbar; Tatsumi, 1981). Thus, the above

Table 2. Chemical and C.I.P.W. normative compositions of starting high-magnesian andesites.

	TGI	SD-261
SiO_2	59.44	57.00
TiO_2	0.44	0.73
Al_2O_3	13.51	15.79
Fe_2O_3	2.54	2.32
FeO	4.01	4.22
MnO	0.12	0.13
MgO	9.62	7.37
CaO	6.23	7.15
Na_2O	2.65	2.88
K_2O	1.30	2.26
P_2O_5	0.13	0.14
FeO*/MgO	0.65	0.86
Q	11.51	5.88
or	7.67	13.35
ab	22.41	24.35
an	21.13	23.48
di wo	3.72	4.62
di en	2.79	3.32
di fs	0.55	0.88
hy en	21.15	15.02
hy fs	4.20	3.98
mt	3.68	3.36
il	0.83	1.38
ap	0.30	0.33

TGI, bronzite olivine andesite
SD-261, augite olivine andesite

experimental results indicate that HMA magmas can be produced by H_2O-undersaturated partial melting of mantle peridotites. However, there still remains an unsolved problem whether or not magmas with larger amounts of H_2O can uprise to the surface without solidification.

Origin of Two Types of Phenocryst Assemblages

The melting experiments indicate that residual mantle material for the cpx- and opx-HMA are lherzolite and harzburgite, respectively. It is established that clinopyroxene is the first pyroxene phase to disappear among lherzolite minerals (Green, 1973; Mysen and Kushiro, 1977; Jaques and Green, 1980). Therefore, the difference of residual phases observed in the experiments depends on the degree of partial melting; that is opx-HMA magmas are formed by higher degrees of

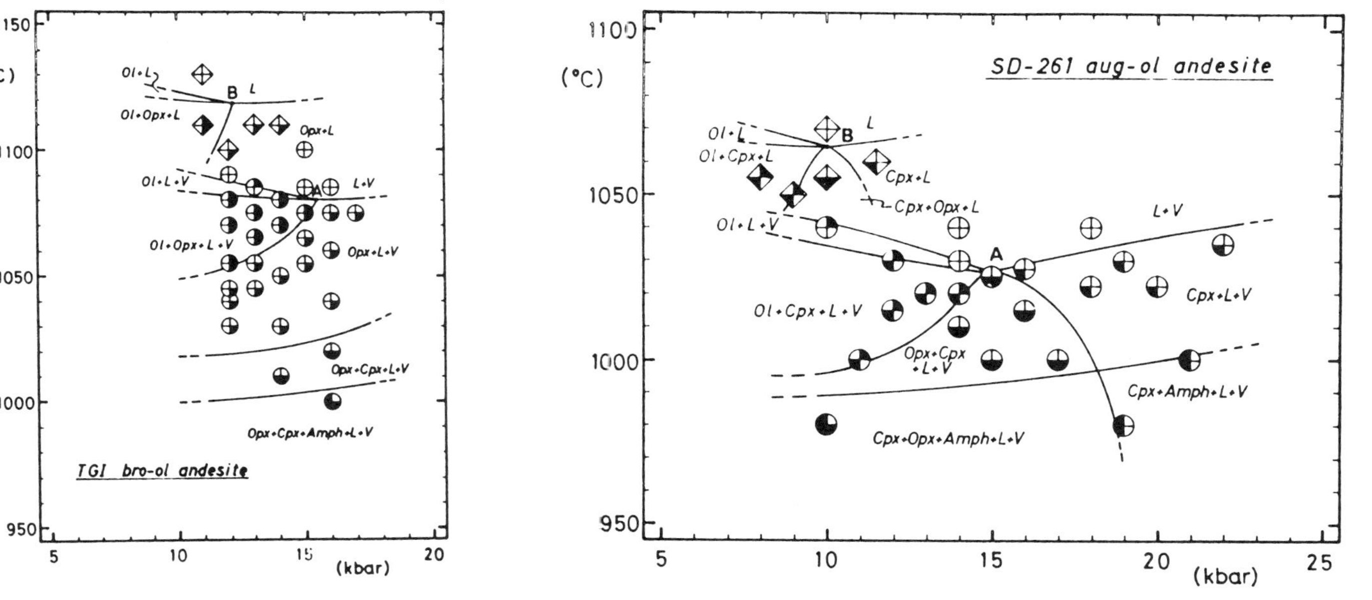

Fig. 12. Melting phase relations of two types of HMAs. Circles indicate phase relations under water-saturated conditions, and diamonds under water-undersaturated conditions with 8 and 7% water for TGI and SD-261, respectively. The results indicate that HMA magmas can be formed by partial melting of mantle peridotites under water-undersaturated conditions. The HMA TGI can coexist with harzburgitic minerals, and the HMA SD-261 with lherzolitic minerals, suggesting that the former are produced by higher degrees of partial melting.

partial melting than cpx-HMA magmas. This observation is supported by the fact that the temperature of multiple saturation point for the opx-HMA is about 50°C higher than that for the cpx-HMA. Chemical compositions of the two types of HMAs also reinforce the above idea; opx-HMAs contain less incompatible elements, and have lower FeO*/MgO ratios than cpx-HMAs.

A Model for Generation of Subduction Zone Magmas

Based on experimental results together with geochemical and seismological data, a plausible cross-section of an arc-trench system can be drawn (Fig. 13), which could provide reasonable solutions for problems on arc magma generation mentioned above.

As the oceanic plate descends, H_2O held in the slab is released and migrates upward into the mantle wedge. Combining data for stability limits of hydrous

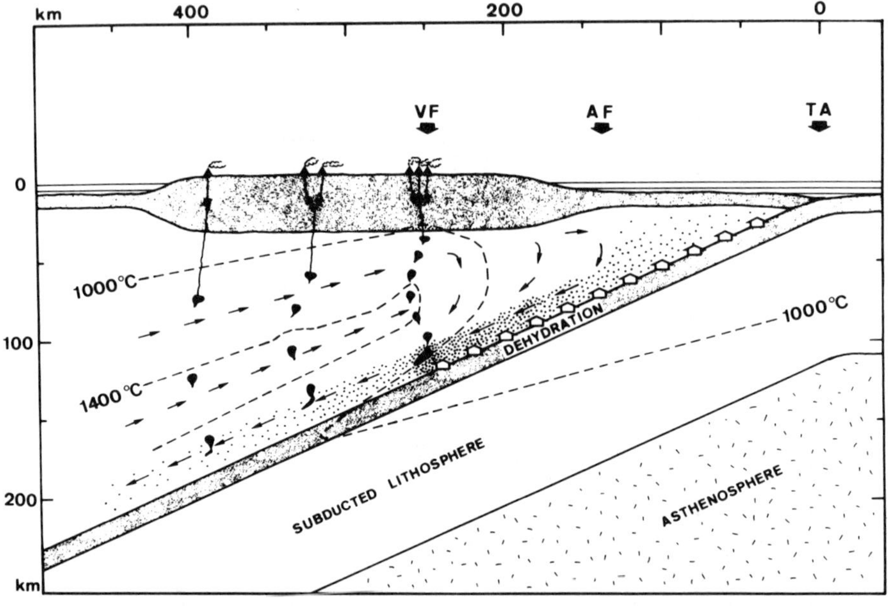

Fig. 13. A plausible cross-section of a subduction zone (Tatsumi et al., 1985). The dotted area indicates the mantle peridotite polluted by the slab-derived fluid phase, the "island-arc source" (see text). The polluted peridotites are transported by the convective current represented by arrows. The aseismic front (AF) corresponds to the trenchward limit of convective current. Voluminous felsic rocks distributed mainly along the volcanic front (VF) are formed by partial melting of the lowermost crust. Diapirs raised through the region with temperatures higher than 1400°C stop to release primary magmas at levels shown on the figure.

minerals with thermal structure in the slab shows that the downgoing slab is anhydrous at a depth greater than 120 km, which corresponds to that of the slab beneath the volcanic front. Through this dehydration process, LILE are added to the mantle wedge beneath fore-arc regions with the slab-derived fluid phases. As a result, the "island-arc source" is formed, and is distinct from the "MORB source" in that the former is more enriched in incompatible elements with larger ionic radii than the latter. The slab-derived fluid containing such elements should react with mantle peridotites to form amphiboles, because stability fields of amphiboles in a peridotite system expand to higher $P-T$ regions than those in a basalt system representing the slab.

Production of initial melts beneath the volcanic front is controlled by the decomposition of amphiboles in wedge peridotites (Figs. 6 and 8). Released H_2O migrates upward to produce water-saturated amphibole peridotites in the mantle wedge. At a level where the solidus temperature of hydrous peridotite is distributed, the amphibole peridotite begins melting to produced initial liquids of olivine tholeiite magmas (Fig 8). Although magmas at this initial stage are saturated with water, H_2O-content in the magma decreases relatively through the process of diapiric uprise of partially molten materials along with an increasing degree of partial melting.

On the other hand, initial melts beneath regions behind the volcanic front are produced by partial melting of phlogopite-peridotites formed after decomposition of amphibole-peridotites that are to be transported to deeper levels along the slab by the induced convection in the mantle wedge.

From the partially molten layer just above the slab formed through the above-mentioned processes, the initial melts migrate upward probably as a form of mantle diapir. The largest number of diapirs are produced beneath the volcanic front. The ascending path of mantle diapirs could be affected by convection in the mantle wedge. At the earlier stage of ascent, especially, the amount of melts in the diapir is smaller, that is, the speed of diapiric uprise is smaller, and consequently the path is bent toward the back-arc side because of the effect of convection current. This observation is reinforced by seismological evidence that a high-velocity region exists at the middle part of the mantle wedge beneath the volcanic front.

The mantle diapir ascends more slowly at the early stage because of the small fraction of partial melts in the diapir. Thus the diapir ascends non-adiabatically, and it can get heat from the surrounding mantle to attain temperatures of more than 1400°C. The temperature of the ascending heated diapir becomes higher than that of the anhydrous solidus at a depth of about 70 to 80 km. After this stage, the degree of partial melting increases abruptly, and the density of a diapir becomes much less than that of surrounding materials. Consequently, the mantle diapir could ascend faster than before, nearly adiabatically. At depths indicated in Fig. 13, the diapir stops rising and releases primary magmas. The uprising diapir may stop because of the increase of viscosity of uppermost mantle materials, which is controlled by the thermal structure of the upper mantle.

Beneath the volcanic front numbers of diapirs with higher temperatures come from deeper levels, and temperatures around Moho could exceed the solidus of the lowermost crustal materials. This condition is consistent with the fact that voluminous felsic pyroclastic flows are observed only along the volcanic front.

Production of HMA magmas can be also interpreted in the above scheme; in order to form HMA magmas, hydrous partial melting of wedge materials must have taken place at shallower levels; the isotherm thus must expand more widely toward the fore-arc side, and this tectonic setting could be obtained at an initial stage of slab subduction (Tatsumi, 1983); HMA magmas are direct partial melting products of the "island-arc source".

Acknowledgments

It will be recognized that many of the thoughts expressed here, however imperfect, were inspired by the remarkable genius of Dr. M. Sakuyama and H. Fukuyama. The author would like to thank Prof. I. Kushiro and S. Banno, and an anonymous reviewer for their critical and helpful comments on the manuscript. Financial support was provided by a Grant-in-Aid for Special Project Research (No. 60221011) from the Ministry of Education, Science and Culture of Japan.

References

Allen, J.C., and Boettcher, A.L. 1978, Amphiboles in andesite and basalt: II. Stability as a function of P–T–f_{H_2O}–f_{O2}, *Amer. Mineral.* **63**, 1074–1087.

Allen, J.C. and Boettcher, A.L. 1983, The stability of amphibole in andesite and basalt at high pressures, Am. Mineral. **68** 307–314.

Allen, J.C., Boettcher A.L. and Marland, G. 1975, Amphiboles in andesite and basalt: I. Stability as a function of P–T–f_{O2}, Am. Mineral. **60** 1069–1085.

Anderson, A.T., 1976, Magma mixing: petrological process and volcanological tool, J. Volcanol. Geotherm. Res. **1**, 3–33.

Anderson, A.T., 1979, Water in some hypersthenic magmas, J. Geol. **87** 509–531.

Andrews, D.J. and Sleep, N.H. 1974, Numerical modelling of tectonic flow behind island arcs, Geophys. J. R. Astron. Soc. **38** 237–251.

Arculus, R.J. and Johnson, R.W. 1981, Island-arc magma sources: a geochemical assessment of the roles of slab derived components and crustal contamination, Geochem. J. **15**, 109–133.

Basu, A.R. and Murthy, V.R. 1977, Ancient lithospheric lherzolite xenolith in alkali basalt from Baja California, Earth Planet. Sci. Lett. **35**, 239–246.

Birch, F., 1966, Compressibility; elastic constants, *Handbook of Physical Constants* (Clark, S.P. ed.), Geol. Soc. Am. Mem., **97**–173.

Bohlen, S.R., Boettcher, A.L., Wall V.J. and Clements, J.D. 1983, Stability of phlogopite-quartz and sanidine-quartz: a model for melting in the lower crust, Contrib. Mineral. Petrol. **83** 270–277.

Brown, L., Klein, J., Middleton, R., Sacks I.S. and Tera, F. 1982, ^{10}Be in island-arc volcanoes and implications for subduction, Nature **299**, 718–720.

Carr, M.J., Stoiber R.E. and Drake, C.L. 1974, The segmental nature of some continental

margins, *The Geology of Continental Margins* (*C.A. Burke and C.L. Drake eds.*), Springer-Verlag, Berlin.
Cawthorn, R.G., 1975, Degrees of melting in mantle diapirs and the origin of ultrabasic liquids, Earth Planet. Sci. Lett. **27** 113–120.
Cawthorn, R.G. and O'Hara, M.J. 1976, Amphibole fractionation in calcalkaline magma genesis, Am. J. Sci. **276** 309–329.
Coats, R.R., 1962, Magma types and crustal structure in the Aleutian arc, *Crust of the Pacific Basin*, Geophys. Monogr. Am. Geophys. Union **6** 92–109.
Daly, R.A., Manger G.E. and Clark, S.P. 1966, Density of rocks, *Handbook of Physical Constant* (Clark, S.P. ed.) Geol. Soc. Am. Mem., 19–26.
DePaolo, D.J. and G.J. Wasserburg, 1977, The sources of island arc as indicated by Nd and Sr isotopic studies, Geophys. Res. Lett. **4** 465–468.
Egaler, D.H., 1972, Water-saturated and undersaturated melting relations in a Paricutin andesite and an estimate of water content in the natural magma, Contrib. Mineral. Petrol. **34** 261–271.
Eggler, D.H., and Burnham, C.W. 1973, Crystallization and fractionation trends in the system andesite–H_2O–CO_2–O_2 at pressures to 10 kbar, Geol. Soc. Am. Bull. **84** 2512–2532.
Eicherberger, J.C., 1975, Origin of andesite and dacite: evidence of mixing at Glass Mountain in California and at other circum-Pacific volcanoes, Geol. Soc. Am. Bull. **86** 1381–1391.
Ewart, A., and Stipp, J.J. 1968, Petrogenesis of the volcanic rocks of the Central North Island, New Zealand, as indicated by a study of $Sr^{87/86}$ ratios, and Sr, Rb, K, U, and Th abundances, Geochim. Cosmochim. Acta. **32** 699–736.
Frey, F.A., and Green, D.H. 1974, The mineralogy, geochemistry and origin of lherzolite inclusions in Victoria basanites, Geochim. Cosmochim. Acta. **38** 10123–1059.
Fujii, T., and Kushiro, I. 1977, Density, viscosity, and compressibility of basaltic liquid at high pressures, Carnegie Inst. Wash. Yearb. **76** 419–424.
Gill, J.B., 1981, *Orogenic andesites and plate tectonics*, Springer Verlag, Berlin.
Goto, A., and Tatsumi, R. Stability of chlorize in the upper mantle, Am. Mineral., **75**, 105–108.
Green, D.H., 1973, Experimental melting studies on a model upper mantle composition at high pressure under water-saturated and water-undersaturated conditions, Earth Planet. Sci. Lett. **19** 37–53.
Green, D.H., and Ringwood, A.E. 1967, The genesis of basaltic magmas, Contrib. Mineral. Petrol. **15** 103–190.
Green, T.H., 1980, Island arc and continent-building magmatism—a review of petrologic models based on experimental petrology and geochemistry, Tectonophys. **63** 367–385.
Green, T.H., 1981, Experimental evidence for the role of accessory phases in magma genesis, J. Volcanol. Geotherm. Res. **10** 405–422.
Green, T.H., Green, D.H., and Ringwood, A.E. 1967, The origin of high-alumina basalts and their relationships to quarts tholeiites and alkali basalts, Earth Planet. Sci. Lett. **2** 41–51.
Green, T.H., and Ringwood, A.E. 1968, Genesis of calc-alkaline igneous rock suite, Contrib. Mineral. Petrol. **18** 105–162.
Hamilton, D.L., Burnham, C.W., and Osborn, E.F. 1972, The solubility of water and effect of oxygen fugasity and water content on crystallization of mafic magmas, J. Petrol. **5** 21–39.
Hasemi, A.H., Ishii, H., and Takagi, A. 1981, Fine structure beneath the Tohoku district,

northeastern Japan arc, as derived by an inversion of P-wave arrival times from local earthquakes,

Hashizume, M., Oike, K., Asano, S., Hamaguchi, H., Okada, A., Murauchi, S., Shima, E., and Noguchi, M. 1968, Crustal structure in profile across the northeastern part of Honshu, Japan, as derived from explosion seismic observations, 2, crustal structure, Bull. Earthq. Res. Inst. Univ. Tokyo **46** 607–630.

Hawkesworth, C.J., Norry, M.J., Roddick, J.C., Baker, P.E., Francis P.W., and Thorpe, R.S. 1979, $^{143}Nd/^{144}Nd$, $^{87}Sr/^{86}Sr$, and incompatible element variations in calc-alkaline andesites and plateau lavas from south America, Earth Planet. Sci. Lett. **42** 45–57.

Hawkesworth, C.J., O'Nion, R.K., Punkhurst, R.J., Hamilton P.J., and Evensen, N.M. 1977, A geochemical study of island-arc and back-arc tholeiites from the Scotia Sea, Earth Planet. Sci. Lett. **36** 253–262.

Hellman, P.L., and Green, T.H. 1979, The role of sphene as an accessory phase in the high pressure partial melting of hydrous mafic compositions, Earth Planet. Sci. Lett. **42** 1981–201.

Holloway, J.R., and Burnham, C.W. 1972, Melting relations of basalt with equilibrium water pressure less than total pressure, J. Petrol. **13** 1–29.

Honda, S., and Uyeda, S. 1983, Thermal process in subduction zones—a review and preliminary approach on the origin of arc volcanism—, *Arc Volcanism: Physics and Tectonics* (D. Shimozuru and I. Yokoyama eds.), Terra Scientific Publishing Company, Tokyo.

Hsui, A.T., and Toksoz, M.N. 1981, Back-arc spreading: trench migration, continental pull or induced Convection ?, Tectonophys. **74** 89–98.

Irvine, T.N., 1976, Metastable liquid immiscibility and MgO–FeO–SiO_2 fractionation patterns in the system Mg_2SiO_4–Fe_2SiO_4–$CaAl_2Si_2O_8$–$KAlSi_3O_8$–SiO_2, Carnegie Inst. Wash. Yearb **75** 597–611.

Ito, H., Masuda, Y., and Kinoshita, O. 1983, Mantle vortex induced by downgoing slab: experimental simulation and its application to trench-arc system, Bull. Univ. Osaka Pref. **A-32** 47–63.

Ishikawa, K., Kanisawa, S., and Aoki, K., 1980, Content and behavior of fluorine in Japanese Quaternary volcanic rocks and petrogenetic application, J. Volcanol. geotherm. Res. **8** 161–175.

Ishizaka, K., and Carlson, R.W., 1983, Nd–Sr systematics of the Setouchi volcanic rocks, southwest Japan: a clue to the origin of orogenic andesite, Earth Planet. Sci. Lett. **64** 327–340.

Jaques, A.L., and Green, D.H., 1973, Determination of liquid compositions in experimental high pressure melting of peridotite, Am. Mineral. **64** 1312–1321.

Jaques, A.L., and Green, D.H., 1980, Anhydrous melting of peridotite at 0–15 kbar pressure and the genesis of tholeiite basalt, Contrib. Mineral. Petrol. **73** 287–310.

Johnson, R.W., 1976, Pottasium variation across the New Britain volcanic arc, Earth Planet. Sci. Lett. **31** 189–191.

Katsui, Y., Oba, Y., Ando, S., Nishimura, S., Masuda, Y., Kurasawa, H., and Fujimaki, H. 1978, Petrochemistry of the Quaternary volcanic rocks of Jokkaido, north Japan, J. Fac. Sci., Hokkaido Univ., **IV–18** 449–484.

Kawano, Y., Yagi, K., and Aoki, K. 1961, Petrography and petrochemistry of the volcanic rocks of Quaternary volcanoes of northeastern Japan, Sci. Rep., Tohoku Univ. **III–7** 1–46.

Kitahara, S., Takenouchi, S., and Kennedy, G.C. 1966, Phase relations in the system

$MgO-SiO_2-H_2O$ at high temperatures and pressures, Am. J. Sci. **264** 223–233.
Kuno, H., 1950, Petrology of Hakone volcano and the adjacent areas, Japan, Geol. Soc. Am. Bull. **61** 957–1019.
Kuno, H., 1959, Origin of Cenozoic petrographic provinces of Japan and surrounding areas, Bull. Volcanol. **20** 37–76.
Kuno, H., 1966, Lateral variation of basalt magma type across continental margins and island arcs, Bull. Volcanol. **29** 195–222.
Kuno, H., 1968, Differentiation of basaltic magmas, *Basalts*, Hess, H.H. and Poldervaart, A. eds., John Wiley and Sons, Inc., New York.
Kushiro, I., 1969, The system forsterite-diopside-silica with and without water at high pressures, Am. J. Sci. **267A** 269–291.
Kushiro, I., 1970, Stability of amphibole and phlogopite in the upper mantle, Carnegie Inst. Wash. Yearb. **68** 245–247.
Kushiro, I., 1972, Effect of water on the composition of magmas formed at high pressures, J. Petrol. **13** 311–334.
Kushiro, I., 1973, Origin of some magmas in oceanic and circum-oceanic regions, Tectonophys. **17** 211–222.
Kushiro, I., 1974, Melting of hydrous upper mantle and possible generation of andesitic magma: an approach from synthetic systems, Earth Planet. Sci. Lett. **2** 294–299.
Kushiro, I., 1983, On the lateral variations in chemical composition and volume of Quaternary volcanic rocks across Japanese arcs, J. Volcanol. Geotherm. Res. **18** 435–447.
Kushiro, I., Shimizu, N., Nakamura, Y., and Akimoto, S-I. 1972, Compositions of coexisting liquid and solid phases formed upon melting of natural garnet abd spinel lherzolite at high pressures: A preliminary report, Earth Planet. Sci. Lett. **14** 19–25.
Kushiro, I., Shono, Y., and Akimoto, S-I. 1967, Stability of phlogopite at high pressures and possible presence of phlogopite in the Earth's upper mantle, Earth Planet. Sci. Lett. **3** 197–203.
Kushiro, I., Shono Y., and Akimoto, S-I. 1968, Melting of a peridotite nodule at high pressures and high water pressures, J. Geophys. Res. **73** 6023–6029.
Kushiro, I., Yoder, H.S., and Mysen, B.O. 1976, Viscosities of basalt and andesite melts at high pressures, J. Geophys. Res. **81** 6351–6356.
Lambert, I.B., and Wyllie, P.J. 1972, Melting of gabbro (quartz eclogite) with excess water to 35 kilobars with geological applications, J. Geol. **80** 693–720.
Luhr, J.F., and Carmichael, I.S.E. 1980, The Colima volcanic complex, Mexico, I, post-caldera andesite from Volcan Colima, Contrib. Mineral. Petrol. **71** 343–372.
Maaløe, S., and Wyllie, P.J., 1975, Water content of a granite magma deduced from the sequence of crystallization determined experimentally with water-undersaturated conditions, Contrib. Mineral. Petrol. **52** 175–191.
Marsh, B.D., 1979, Island arc development: some observations, experiments and speculations, J. Geol. **87** 687–713.
McBirney, A.R., 1969, Andesitic and rhyolitic volcanism of orogenic belts, Am. Geophys. Union Monogr. **13** 501–506.
McKenzie, C.P., 1969, Speculations on the consequences and causes of plate motions. Geophys. J. Roy. Actron. Soc. **18** 1–32.
McKenzie, D.P., 1972, Active tectonics of the Mediterranean region, Geophys. J. R. Astro. Soc. **30** 173–190.
McKenzie, D.P., and Parker, R.L. 1967, The north Pacific, and example of tectonics on a sphere, Nature **216** 1276–1280.

Merrill, R.B., and Wyllie, P.J. 1975, Kaersuitite and kaersuitite eclogite from Kakanui, New Zealand—water-excess and water-deficient melting to 30 kbars, Geol. Soc. Am. Bull. **86** 555–570.

Millhollen, G.L., Irving, A.J., and Wyllie, P.J. 1974, Melting interval of peridotite with 5.7 per cent water to 30 kbars, J. Geol. **82** 575–587.

Millhollen, G.L., and Wyllie, P.J. 1974, Melting relations of brown-hornbrende mylonite from St. Paul's rocks under water-saturated and water-undersaturated conditions to 30 kilobars, J. Geol. **82** 589–606.

Modreski, P.J., and Boettcher, A.L. 1972, The stability of phlogopite + enstatite at high pressures: a model for micas in the interior of the Earth, Am. J. Sci. **272** 852–869.

Modreski, P.J., and Boettcher, A.L. 1975, Phase relationships of phlogopite in the system K_2O–MgO–CaO–Al_2O_3–SiO_2–H_2O to 35 kilobars: a better model for micas in the interior of the Earth, Am. J. Sci. **273** 385–414.

Mysen, B.O., 1979, Trace element partitioning between garnet peridotite minerals and water-rich vapor: experimental data from 5 to 30 kbar, Am. Mineral. **64** 274–287.

Mysen, B.O., and Boettcher, A.L. 1975, Melting of a hydrous mantle. II. geochemistry of crystals and liquids formed by anatexis of mantle peridotite at high pressures and high temperatures as a function of controlled activities of water, hydrogen and carbon dioxide, J. Petrol. **16** 549–590.

Mysen, B.O., and Boettcher, A.L. 1976, Melting of a hydrous mantle. III. phase relations of garnet websterite + H_2O at high pressures and temperatures, J. Petrol. **17** 1–14.

Mysen, B.O., and Kushiro, I. 1977, Compositional variations of coexisting phases with degree of melting of peridotite in the upper mantle, Am. Mineral. **62** 845–865.

Mysen, B.O., Kushiro, I., and Fujii, T. 1977, Preliminary experimental data bearing on the mobility of H_2O in crystalline upper mantle, Carnegie Inst. Wash. Yearb. **77** 793–797.

Nicholls, I.A., and Ringwood, A.E. 1973, Effect of water on olivine stability in tholeiiotes and the production of silica-saturated magmas in the island arc environment, J. Geol. **81** 285–300.

Nicholls, I.A., and Whitford, D.J. 1976, Primary magmas associated with Quaternary volcanism in the western Sunda arc, Indonesia, *Volcanism in Australia* (Johnson, R.W. ed.), 77–90.

Nohda, S., and Tatsumi, Y. 1986, What happened in the mantle wedge during the Japan Sea opening ?, J. Geomag. Jeoelectr., in press.

Nohda, S., and Wasserburg, G.J. 1981, Nd and Sr isotopic study of volcanic rocks from Japan, Earth Planet. Sci. Lett. **52** 261–276.

Oba, T., 1978, Phase relationship of $Ca_2Mg_3Al_2Si_6Al_2O_{22}(OH)_2$–$Ca_2Mg_3Fe_4^{3+}Si_6Al_2O_{22}(OH)_2$ join at high temperatures and high pressures—the stability of tschermakite, J. Fac. Sci. Hokkaido Univ. Ser. 4 **18** 339–350.

O'Hara, M.J., 1965, Primary magmas and the origin of basalts, Scotish J. Geol. **1** 19–10.

O'Hara, 1968, The bearing of phase equilibria on synthetic and natural systems on the origin and evolution of basic and ultrabasic rocks, Earth Sci. Rev. **4** 69–133.

Okada, H., 1979, New evidences of the discontinuous structure of the descending lithosphere as revealed by ScSp phase, J. Phys. Earth **27** s53–s63.

Onuma, N., Hirano, M., and Issiki, N. 1981, Genesis of basalt magmas and their derivatives under Izu Islands, Japan, inferred from Sr/Ca–Ba/Ca systematics, J. Volcanol. Geotherm. Res. **18** 511–529.

Osborn, E.F., 1959, Role of oxygen pressure in the crystallization and differentiation of basaltic magmas, Am. L. Sci. **257** 609–647.

Perfit, M.R., Gust, D.A., Bence, A.E., Arculus, R.J., and Taylor, S.R. 1980, Chemical characteristics of island arc basalts: implication for mantle sources, Chem. Geol. **30** 227–256.

Pichler, H., and Zeil, W. 1969, Die quartre "Andesite"-Formation in der Hochkordillere Nord-Chiles, Geol. Rundsch **58** 866–903.

Pichler, H. and W. Zeil, 1972, The Cenozoic rhyolite-andesite association of the Chilean Andes, Bull. Volcanol. **35** 424–452.

Ramberg, H., 1972, Mantle diapirism and its tectonic and magmagenetic consequences, Phys. Earth Planet. Inter. **5** 45–60.

Ringwood, A.E., 1974, The petrological evolution of island arc system, J. Geol. Soc, London **130** 183–204.

Ringwood, A.E., 1975, *Composition and petrology of the Earth's mantle*, McGraw-Hill, New York.

Ryabchikov, I.D., and Boettcher, A.L. 1980, Experimental evidence at high pressure for potassic metasomatism in the upper mantle of the Earth, Am. Mineral. **65** 915–919.

Sakuyama, M., 1979, Lateral variations of H_2O contents in Quaternary magmas of northeastern Japan, Earth Planet. Sci. Lett. **43** 103–111.

Sakuyama, M., 1979, Evidence of magma mixing: petrological study of Shirouma-Oike calc-alkaline andesite volcano, Japan, J. Volcanol. Geotherm. Res. **5** 179–208.

Sakuyama, M., 1981, Petrological study of the Myoko and Kurohime volcanoes, Japan: crystallization sequence and evidence for magma mixing, J. Petrol. **22** 553–583.

Sakuyama, M., 1983, Petrology of arc volcanic rocks and their origin by mantle diapir, J. Volcanol. Geotherm Res. **18** 297–320.

Sakuyama, M., and Kushiro, I. 1979, Vesiculations of hydrous andesitic melt and transport of alkalis by separated vapor phase, Contrib. Mineral. Petrol. **71** 61–66.

Sakuyama, M. and R.W. Nesbitt, 1985, Trace element chemistry of Quaternary volcanic rocks of NE Japan arc: degree of partial melting and input from subduction zone, J. Volcanol. Geotherm. Res., in press.

Sato, H., 1977, Nickel content of basaltic magma: identification of primary magmas and measurement of the degree of olivine fractionation, Lithos, **10** 113–120.

Saunders, A.D., Tarney J., and Weaver, s.D. 1980, Transverse geochemical variations across the Antarctic Peninsula: implication for the genesis of calc-alkalic magmas, Earth Planet. Sci. Lett. **46** 341–360.

Sekine, T., Katsura, T., and Aramaki, S. 1979, Water saturated phase relations of some andesites with application to the estimation of the initial temperature and water pressure at the time of eruption, Geochim. Cosmochim. Acta. **43** 1367–1376.

Sekine, T., and Wyllie, P.J. 1982, Phase relationships in the system $KALSiO_4$–Mg_2SiO_4–SiO_2–H_2O as amodel for hybridization between hydrous siliceous melts and peridotite, Contrib. Mineral. Petrol. **79** 368–374.

Sekine, T., and Wyllie, P.J. 1982, Synthetic systems for modelling hybridization between hydrous siliceous magmas and peridotite in subduction zones, J. Geol. **90** 734–741.

Sekine, T., and Wyllie, P.J. 1982, The system granite-peridotite-H_2O at 30 kbar, with application to hybridization in subduction zone magmatism, Contrib. Mineral. Petrol. **81** 190–202.

Sekine, T., Wyllie, P.J., and Baker, D.R. 1981, Phase relationship at 30 kbar for quartz eclogite composition in CaO–MgO–Al_2–O_3–SiO_2–H_2O with implication for subduction zone magmas, Am. Mineral. **60** 935–950.

Stern, C.R., 1974, Melting product of olivine-tholeiite basalt in subduction zones, Geology **2** 227–230.

Stern, C.R., Huang, W.L. and Wyllie, P.J., Basalt-andesite-rhyolite-H_2O: Crystallization intervals with excess H_2O and H_2O-undersaturated liquidus surfaces to 35 kbars with inplications for magma genesis, Earth Planet. Sci. Lett. **28** 189–196.

Stern, C.R., and Wyllie, P.J. 1973, Melting relations of basalt-andesite-rhyolite-H_2O and a pelagic red clay at 30 kbar, Contrib. Mineral. Petrol. **42** 313–323.

Sudo, A., and Tutsumi, Y. 1990, Phlogopite & K-amphibole in the upper mantle: implication for magma genesis in subduction zones, Geophys. Res. Lett., **17** 29–32.

Sugimura, A., 1960, Zonal arrangemenet of some geophysical and petrological features in Japan and its environs, J. Fac. Sci. Univ. Tokyo, Sect II **12** 133–153.

Sugimura, A., Matsuda, T., Chinzei, K., and Nakamura, K. 1963, Quantitative distribution of late Cenozoic volcanic materials in Japan, Bull. Volcanol. **26** 125–140.

Suzuki, K., 1984, Personal communication

Takahashi, E. and I. Kushiro, 1983, Melting of a dry peridotite at high pressures and basalt magma genesis, Am. Mineral. **68** 859–879.

Tatsumi, Y., 1981, Melting experiments on a high magnesian andesite, Earth Planet. Sci. Lett. **54** 356–365.

Tatsumi, Y., 1982, Origin of high-magnesian andesites in the Setouchi volcanic belt, southwest Japan, II, melting phase relations at high pressures, Earth Planet. Sci. Lett. **60** 305–317.

Tatsumi, Y., 1983, High magnesian andesites in the Setouchi volcanic belt, southwest Japan and their possible relation to the evolutionary history of the Shikoku Inter-arc basin, *Geogynamics of the Western Pacific-Indonesian region, Geodynamics Series*, AGU **11** 331–341.

Tatsumi, Y., 1986, Formation of volcanic front, in subduction zones, Geophys. Res. Lett., **13**, 717–720.

Tatsumi, Y. (1989) Migration of fluid phases and genesis of basalt magmas in subduction zones, J. Geophys. Res., **94**, 4697–4707.

Tatsumi, Y., Hamilton, D.L., and Nesbitt, R.W. 1984, Transport of incompatible elements associated with the dehydration of serpentine in a down-going slab, *Prog. Experm. Petrol.*, **25**, 6–12.

Tatsumi, Y., Hamilton, D.L. and Nesbitt, R.W. 1986, Chemical characteristics of fluid phase from the subducted lithosphere and origin of arc magmas: evidence from high pressure experiments and natural rocks, J. Volcanol. Geothem. Res., **29**, 293–309.

Tatsumi, Y., and Ishizaka, K. 1981, Existence of andesitic primary magma: and example from southwest Japan, Earth Planet. Sci. Lett. **53** 124–130.

Tatsumi, Y., and Ishizaka, K. 1982, Origin of high-magnesian andesites in Setouchi volcanic belt, southwest Japan, I, petrographical and geochemical characteristics, Earth Planet. Sci. Lett. **60** 293–304.

Tatsumi, Y., and Ishizaka, K. 1982, High magnesian andesite and basalt from Shodo-Shima island, southwest Japan, and their bearing on the origin of calc-alkaline andesites, Lithos **15** 161–172.

Tatsumi, Y., and Isayamas H. 1988, Transportation of beryllium with H_2O at high pressures: implication for magma genesis in subduction zones, Geophys. Res. Lett., **15**, 180–183.

Tatsumi, Y., and Maruyama, S. 1989, Boninites and high -Mg andesites: tectonics and petrogenesis, *Boninite and Related Rocks* (Crawford, A.J. ed.), Unwin Hyman, 50–71.

Tatsumi, Y., Sakuyama, M., Fukuyama, H., and Kushiro, I. 1983, Generation of arc basalt magmas and thermal structure of the mantle wedge in subduction zones, J. Geophys. Res. **88** 5815–5825.

Tatsumi, Y., and Nakamura, N. 1986, Composition of aueous fluid from serpentinite in the subducted lithosphere, Geochem. J., **20**, 191–196.

Tatsumi, Y., and Nakano, S. 1984, Lateral variation of K/Hf ratios in Quaternary volcanic rocks of Northeastern Japan, Geochem. J. **18** 305–314.

Tatsumi, Y., Nohda, S., and Ishizaka, K. 1988, Secular variation of magma source compositions beneath the Northeast Japan arc, Chem. Geol., **68**, 309–316.

Tatsumi, Y., Orofuji, Y., Marsudu, T., and Nohda, S. 1988, Opening of the Sea of Japan back-arc basin by arthenosphere rejection, Tectonophys., **166**, 317–329.

Thompson, R.N., 1974, Primary basalts and magma genesis I. Skey, North-west Scotland, Contrib. Mineral. Petrol. **45** 317–341.

Toksoz, M.N., and Bird, P. 1977, Formation and evolution of marginal basins and continental plateaus, *Island Arcs Deep Sea trenches and Back-Arc Basins* (Talwani, M., and Pitman, W.C. eds.) Am. Geophys. Union, Washington, 379–393

Tohsoz, M.N., and Hsui, A.T. 1978, Numerical studies of back-arc convection and the formation of marginal basins, Tectonophys. **50** 177–196.

Toksoz, M.N., and Hsui, A.T. 1979, The evolution of thermal structures beneath a subduction zone, Tectonophys. **60** 43–60.

Toksoz, M.N., Minear, J.W., and Julian, B.R. 1971, Temperature field and geophysical effects of a downgoing slab, J. Geophys. Res. **76** 1113–1138.

Wilcox, R.E., 1954, Petrology of Paricutin Volcano, Mexico, US Geol. Surv. Bull. **965**–C.

Wood, D.A., Joron, J.L., and Treuil, M. 1979, A re-appraisal of the use of trace elements to classify and discriminate between magma series erupted in different tectonic setting, Earth Planet. Sci. Lett. **45** 326–336.

Wyllie, P.J., 1971, *The Dynamic Earth: Textbook in Geosiences*, Wiley, New York, 416.

Wyllie, P.J., 1979, Magmas and volatile components, Am. Mineral **64** 469–500.

Wyllie, P.J., and Sekine, T. 1982, The formation of mantle phlogopite in subduction zone hybridization, Contrib. Mineral. Petrol. **79** 375–380.

Yamamoto, K., and Akimoto, S.-I., 1977, The system $MgO-SiO_2-H_2O$ at high pressures and temperatures—stability field of for hydroxyl-chondrodite, hydroxyl-clinohumite and 10A-phase, Am. J. Sci. **277** 288–312.

Yoder, H.S.Jr., 1969, Calcalkalic andesites: experimental data bearing on the origin of their assumed characteristics, *Proc. Andesite Conf.* (A.R. McBirney ed), Oreg. Dept. Geol. Mineral. Ind. Bull. **65** 77–89.

Yoder, H.S.Jr., 1976, *Generation of Basaltic Magma*, Nat. Acad. Sci., Washington.

Yoder, H.S.Jr., and Kushiro, I. 1976, Melting of hydrous phase: phlogopite, Am. J. Sci. **267A** 558–582.

Yoder, H.S.Jr., and Tilley, C.E. 1962, Origin of basalt magmas: an experimenal study of natural and synthetic rock systems, J. Petrol. **3** 342–532.

Chapter 11
Effects of Fluid Composition on Melting Phase Relationships: The Application of Korzhinskii's Open Systems

Aleksey D. Kuznetsov and Mark B. Epel'baum

Introduction

Melting phase relationships in the alumino–silicate systems are important in explaining the diversity and the regularity governing the formation of magmatic rocks. An important thing about these natural processes is that they take place against the background of the ever-changing external conditions. Therefore, to build a quantitative theory of these processes, it is imperative to know how the various factors (parameters) of the environment influence the crystal–liquid equilibria in the magmatic systems.

Together with such "traditional" factors of the magmatic evolution as temperature and total pressure (T and P), the role of the fluid-melt interaction has been emphasized recently (e.g. Kuznetsov and Izoh, 1969; Korzhinskii, 1972; Ryabchikov, 1975; Letnikov et al., 1977; Burnham, 1979 and others). One of the aspects of this interaction is the so-called "effect of additional components" ("the effect of acid–base interaction" as it is known in Soviet literature), which represents the influence of fluid components on liquidus phase relations. However, unlike T and P, the influence of the various fluid components on melting relations has not been studied sufficiently. These influences have far-reaching consequences in igneous petrology (Korzhinskii, 1959, 1972; Zharikov, 1969; Kushiro, 1975; Fraser, 1977; Pichavant and Manning, 1984).

The purpose of this paper is to review the effect of the additional components on melting phenomena. This work includes: (1) a review of experimental data on the effect of various components on melting relations in granitic systems; (2) a brief review of theoretical models proposed to explain this effect; (3) a thermodynamic description of this effect offered by Korzhinskii's open systems that the authors are currently working on and (4) an attempt to use the resulting regularities for the analysis of fluid regimes during the formation of hypobissal multiphase granites in Central Kazakhstan (the USSR). In addition, the calculation of the albite liquidus surface in the T–P–μ_{H_2O} space where μ_{H_2O} is the chemical potential of the perfectly mobile H_2O is given in the Appendix.

Experimental Data

The authors have confined themselves to reviewing the available experimental data on the effects of various fluid components on shifts of eutectics (or minima) and cotectics in the granitic systems under moderate pressures.

Water

Tuttle and Bowen (1958), Shaw (1963), Luth, Jahns and Tuttle (1964) and Stewart (1967) determined that with an increase of P_{H_2O} in the binary quartz–feldspar systems, the eutectics shift first in the direction of quartz (up to $P_{H_2O} = 0.3$–0.5 kbar), and then towards feldspar. Similarly, in the ternary Qz–Ab–Or system the minimum melt composition (eutectic being above ~4 kbar) shifts with an increase of P_{H_2O} first in the direction of the quartz apex (in the range of 1–500 bar) and then toward the albite corner (Fig. 1). There is no unequivocal interpretation of such an extreme behavior of the eutectic curve[1] in these system; probably, it is connected with silica polymorphic transformations or with the change of the H_2O solubility mechanism (Dubrovskii, 1971; Epel'baum, 1980).

Unfortunately, there are so far only fragmentary data on the effect of P_{H_2O} on

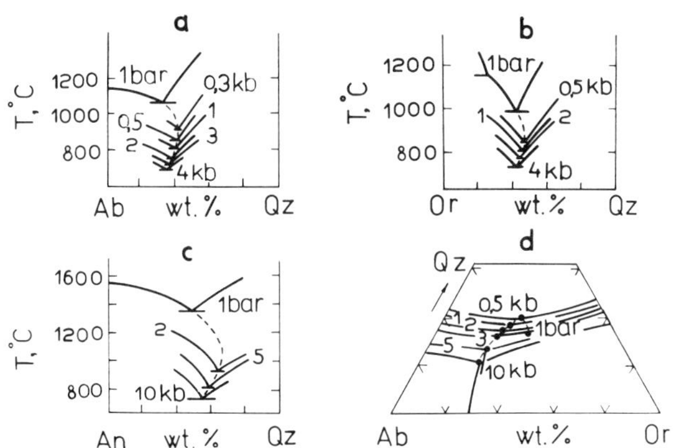

Fig. 1. Liquidus phase relationships for the Qz–Fsp systems as the functions of P_{H_2O} [After Tuttle and Bowen, 1958 (a), Shaw, 1963 (b), Stewart, 1967 (c) and Luth, Jahns and Tuttle, 1964 (d)].

[1] The eutectic curve is an univariant curve formed by invariant points of eutectics (or minima), relating to different contents of a fluid component in the system.

the liquidus phase equilibria involving plagioclase (e.g., James and Hamilton; 1969; Winkler and Lindemann, 1972; Johannes, 1984); therefore, a precise estimate of this effect is not available. It should also be noted that the eutectic shifts shown in Fig. 1, in fact, are caused by the summary effects of the additions of H_2O and of the total pressure (Manning et al., 1984). However P_{tot} and P_{H_2O} (above ~0.5–1 kbar) act in the same direction, i.e. they expand the quartz liquidus field at the expense of the feldspar field (Luth, 1969), but the effect of total pressure is less pronounced. So it may be assumed that the increase of P_{H_2O} as a whole reflects the influence of the H_2O additions.

Many experiments have been carried out to determine the phase stability fields in natural and synthetic granitic systems in the T–P_{H_2O} (or X_{H_2O} in the system) coordinates, both for H_2O-saturated and undersaturated conditions (Piwinskii and Wyllie, 1968; Steiner et al., 1975; Whitney, 1975; Naney and Swanson, 1980 and others). With an increase of P_{H_2O} (or X_{H_2O}), the relative position of the "mineral out curves" changes, which indicates a change in their liquidus fields and, consequently, in the eutectic and cotectic relations. But although the results of these studies are important for petrology (for example, Wyllie et al., 1976; Wyllie, 1979; Popov, 1981), they are of little use in quantitative interpretation of the effect in question because, as a rule, neither the proportions of solid phases and liquids nor the compositions of melts in equilibrium with crystals have been determined.

Chlorine

Von Platen (1965) studied the effect of the HCl aqueous solutions (0.0005–5 *N*) on the crystallization of natural obsidians and their mixtures with anortite and biotite under the total pressure of 2 kbar. Von Platen was the first to demonstrate that the presence of such additional components as HCl, HF and NH_3 had a considerable effect on eutectic relations in granitic systems. His results in reference to the HCl additions are shown in Fig. 2a. It should be noted, however, that the methods he used (a step-by-step decrease of temperature and a quantitative x-ray analysis of quenched products at each step) were outdated and the Cl solubility in melts were not determined. The authors have studied the effect of the HCl aqueous solutions (0.5–6*N*) on the position of eutectics in the Qz–Or–Bi system under P_{tot} = 1 kbar (Kuznetsov and Epel'baum, 1978), The stability of the system phases were also studied in run conditions and the results of the quench runs were checked by microprobe, including the Cl solubilities in eutectic and some other melts. The data obtained were then used for the calculation of eutectic shifts as functions of the changing of the melt basicity due to the Cl solubility (Kuznetsov and Epel'baum, 1979; Kuznetsov, 1982). The results of experiments in the binary Qz–Or system are shown in Fig. 2b.

Data on the influence of HCl on the liquidus phase relations in granitic systems involving biotite is of some interest. Using some original methods it was established that in the Qz–Or–Bi system a melt closed to cotectic minimum in

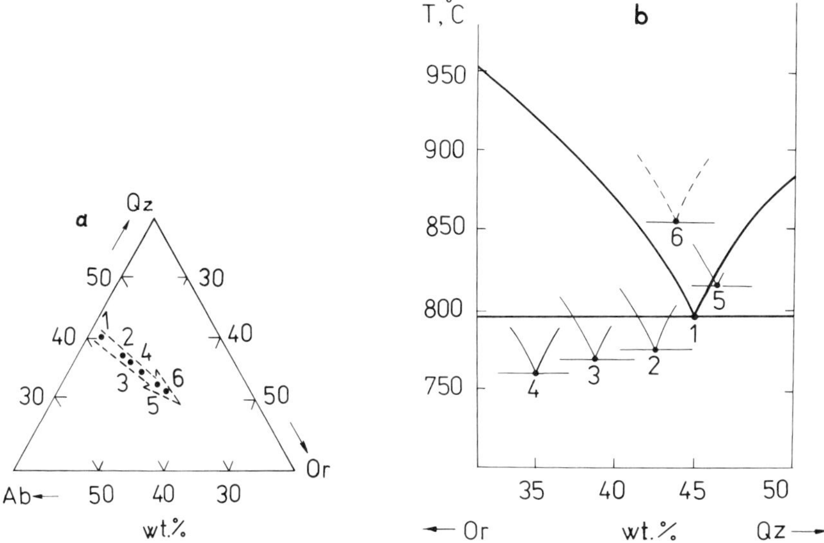

Fig. 2. (a) Normative compositions of the melt minima obtained during the crystallization of obsidian at $P_{tot} = 2$ kbar with excess H_2O (1) and with HCl aqueous solutions (2, 0.0005 *N*; 3, 0.005 *N*; 4, 0.05 *N*; 5, 0.5 *N*; 6, 5 *N*). (Drawn after Table 6 in von Platen, 1965.) (b) Shifts of the Qz + Or eutectic in the presence of various fluids at $P_{tot} = 1$ kbar (Kuznetsov and Epel'baum, 1978). 1, H_2O; 2, 0.5 *N* HCl; 3, 3 *N* HCl; 4, 6 *N* HCl; 5, 6.5 *N* NH_4OH; 6, 0.5 H_2O + 0.5 CO_2.

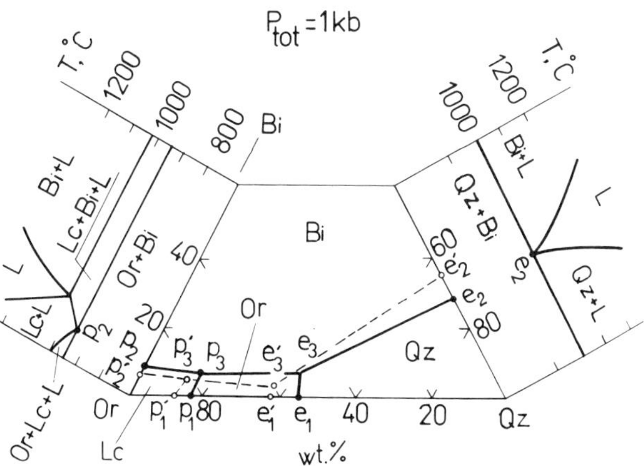

Fig. 3. Schematic liquidus diagram for the pseudoternary Qz–Or–Bi system in equilibrium with H_2O (solid lines) and with 3 *N* HCl (dashed lines) at $P_{tot} = 1$ kbar (Kuznetsov, 1982). The coordinates of points e_2, e_2' (only temperature), e_3, e_3', and P_2 were determined experimentally.

equilibrium with 3 N HCl solution is saturated with biotite at its content in the system being ~3 wt%, while in the presence of pure water the saturation occurs at ~7 wt% Bi (at P_{tot} = 1 kbar). Based on these experiments, the liquidus diagram of the Qz–Or–Bi system may be as shown in Fig. 3.

Fluorine

Detailed studies of the F additions were carried out by Anfilogov et al. (1973), Glyuk and Anfilogov (1973a,b), Kovalenko and Kovalenko (1976), Kovalenko (1979), and Manning (1981). As a result of studying such systems as granite–H_2O–HF, ongonite–H_2O–HF and granite–H_2O–MeF (Me = Li, Na, K, Cs) at ~0.5 and ~1 kbar it was found that aside from dramatically decreasing the solidus temperature, increasing F content causes a substantial expansion of the quartz liquidus field at the expense of the feldspar field, the appearance of F-containing phases (topaz and mica), simultaneous crystallization of albite and K-feldspar and, if present in sufficiently high amounts, F causes liquid immiscibility. Manning (1981 and others) determined the position of minima in the Qz–Ab–Or system with various F contents in the system and with excess H_2O at P_{tot} = 1 kbar (Fig. 4). With increasing F content the composition of the first easily melted liquid is progressively enriched with albite and the minimum temperature goes down substantially. In addition, with the F content in the system being above ~2 wt%, the solidus moves down onto the crest of feldspar solvus, i.e., cotectic minimum is replaced with true eutectic. Thus, the effect of F on the nature of changes of the liquidus phase relations in granitic system is rather similar to that of H_2O (above P_{H_2O} = 0.5–1 kbar) but is very much more significant.

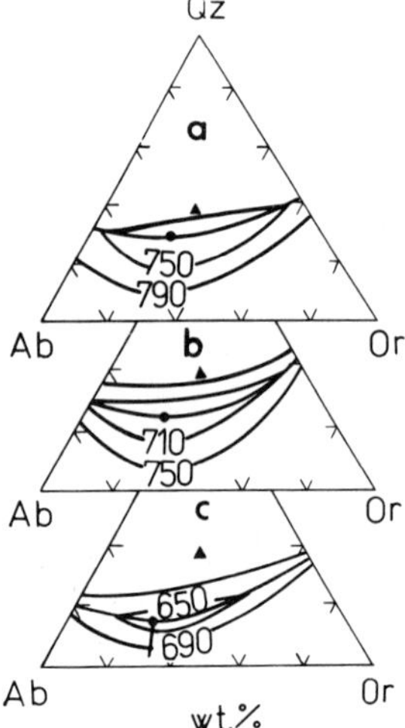

Fig. 4. Liquidus phase relationships for the Qz–Ab–Or system with excess H_2O and with 1 wt% F (a), 2 wt% F (b) and 4 wt% F (c) at P_{tot} = 1 kbar. (After Manning, 1981.) Solid triangles and solid circles are the minimum compositions for the F-free and the F-bearing systems correspondingly. Isotherm temperatures are in °C.

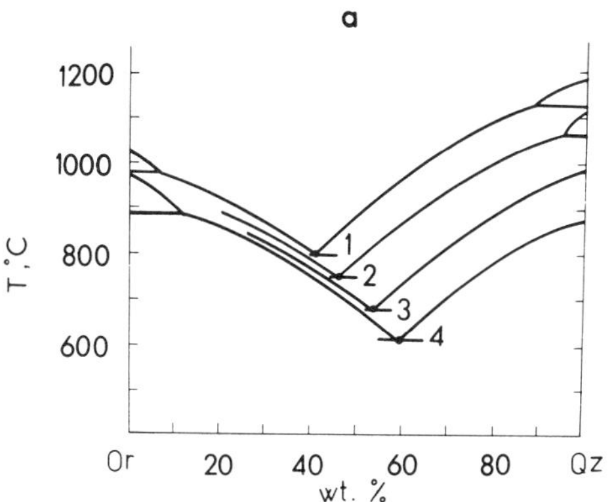

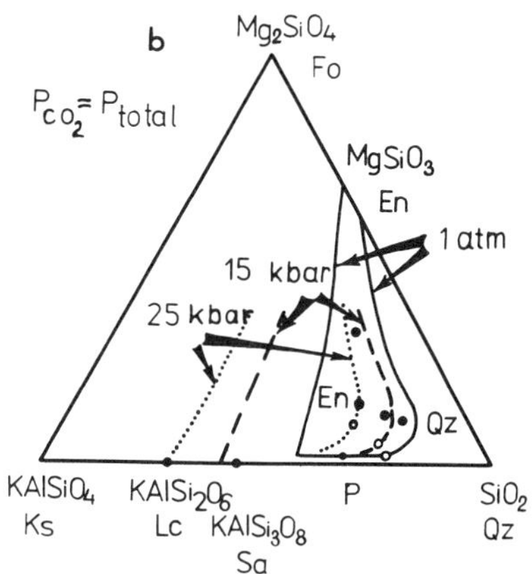

Fig. 5 (a) Shift of the Qz + Or eutectic as the function of the Na concentration in the system (1, 0%; 2, 3%; 3, 7%; 4, 12 wt% Na_2O) at $P_{H_2O} = 1000$ atm. (After Chelischev, 1967.) (b) Changes in liquidus phase relationships for the Mg_2SiO_4–$KAlSiO_4$–SiO_2 system at $P_{tot} = P_{CO_2}$. (After Wendlandt, 1981.) Solid circles are the starting composition and open circles are the liquid compositions in equilibrium with quartz and enstatite at 3, 15, and 25 kbar.

Boron

According to preliminary data of Pichavant, the minimum melt composition in the Qz–Ab–Or system shifts to a certain extent to the albite corner in the presence of 4.5 wt% B_2O_3 under water-saturated conditions at $P_{tot} = 1$ kbar (Manning and Pichavant, 1984; Manning et al., 1984). However, inasmuch as the boron oxide additions increase significantly the H_2O solubility in the melt (Pichavant, 1981, 1983), it is not clear to what extent the observing shift is caused by the adding of B_2O_3 alone.

Alkalies

Tuttle and Bowen (1958) showed that the adding of K_2O shifts the Qz + Or cotectic towards quartz. The same result for the Qz–Or system was obtained by Chelischev (1967) during studies of the crystallization sequence in the presence of NaOH aqueous solutions (Fig. 5a) and by the authors in the adding of 6.5 *N* NH_4OH (see Fig. 2b). All these data refer to the H_2O-saturated conditions and to $P_{tot} = 1$ kbar. The experiments of Carmichael and MacKenzie (1963) and Thompson and MacKenzie (1967) should be also mentioned in this connection. They discovered, that with the adding of 8.3 wt% acmite plus 8.3 wt% Na metasilicate the minimum for the Qz–Ab–Or system moves towards the Qz–Or sideline by ~13 wt% Ab ($P_{tot} = 1$ kbar). However, recently Martin and Henderson (1984) came to the coclusion that small amounts of Li_2O (1 and 2 wt% Li_2O, excess water, $P_{tot} = 1$ kbar) causes no considerable changes in the position of the haplogranitic minimum, it only essentially decreases its temperature.

Carbon Dioxide

The effect of CO_2 on the shift of liquidus boundaries was demonstrated by Wendlandt (1981) in the framework of a more basic Mg_2SiO_4–$KAlSiO_4$–SiO_2 system (Fig. 5b). From a review of a part of the system of the author's interest, it can be seen that CO_2 expands the quartz liquidus field at the expense of the K-feldspar field, i.e., it acts as an acid component. A similar result was obtained by the authors (see Fig. 2b), but because of a lower pressure and, consequently, CO_2 solubility in the melt, the shift is less pronounced.

The effects of other components on liquidus relations for granitic system have not been given special attention because the sparse data available are difficult to interpret in terms of the effect under review here.

Thus, the experimental data demonstrate the importance of the effect of liquidus boundary shifts due to the additional components in granitic systems, but a lot of problems are still unclear In particular, the effects of various factors should be separated and, vice versa, the effect of the conjoint efficacy of the added components should also be estimated (Manning et al., 1984). Apparently, this work could be done only on the basis of the clear theoretical generalizations.

Theoretical Models

Several models have been proposed to explain the effect of additional components on liquidus relations. Some of the basic ones are considered here.

Korzhinskii (1959) was the first to notice the effect. The regularities in shifts of cotectics in ternary liquidus diagrams from a certain "ideal" position were explained by him as being the result of acid-base interaction in melt between a "third" component and the components that make up cotectic phases. The degree of the shifts is determined by the differences in acid-base properties of the system components. With the help of his theory of systems with perfectly mobile components (PMC), Korzhinskii formulated the equations for this effect:

$$(\partial \ln \gamma^S_{MeO}/\partial \ln a_{O^{2-}})_{T,P,n_i} = \alpha, \quad (1)$$

$$(\partial \ln \gamma^S_{SiO_2}/\partial \ln a_{O^{2-}})_{T,P,n_i} = -\beta, \quad (2)$$

where γ_i^S is the bulk activity coefficient of the inert component i in melt (i = MeO, ..., SiO_2), i.e. related to the total content of i in dissociated and undissociated species; $a_{O^{2-}}$ is the activity of O^{2-} ion in the melt which indicates the basicity of silicate melts (Lux, 1939; Flood and Förland, 1947) and reflects the total effect of all additional (perfectly mobile) components on the system state; and α and β are the degrees of dissociation of the basic (MeO) and acid (SiO_2) inert components in the melt respectively.

Thus, in Korzhinskii's model, dissolving of the additional components in the melt changes its basicity and through (1) and (2)—the bulk activity coefficients and, hence, the activities of the melt components. In turn, this effect should lead

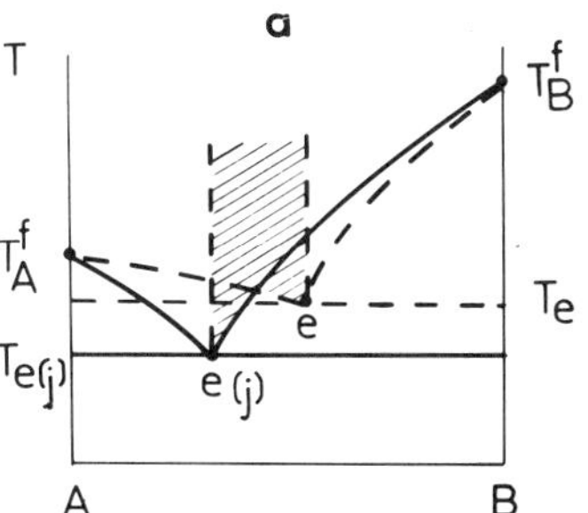

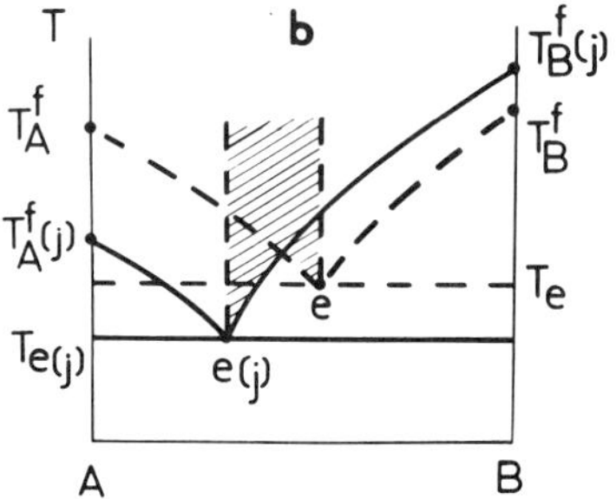

Fig. 6. Schematic diagram for a binary system A–B showing the changes in eutectic relationships in the presence of an additional (perfectly mobile) component j: *a*, due to acid–base interaction of components *a*, *b*, and j in melt phase; *b*, due to changes in fusion temperatures of phases A and B caused by various solubility of j in these phases. Striped area shows the composition range with the changing crystallization sequence.

to a change in the liquidus temperatures of the phases built up by these components and, therefore, their liquidus fields (Fig. 6a).

It should be noted that Korzhinskii's principle of acid base interaction reflects the general principle of the maximum polarity of chemical bonds (Ramberg, 1952). According to this principle, exchange reactions "are always shifted toward the combination of more acid (electronegative) anions with more basic (leas electronegative) cations and vice versa, less acid anions with less basic cations" (Kogarko and Ryabchikov, 1980, pp. 505–506).

Korzhinskii's approach was further developed by Zharikov (1969, 1976). He showed that in systems with PMC it was necessary to take into consideration the dependence of fusion temperatures of phases consisting of inert components on the chemical potentials of PMC. True, the equation of equilibrium melting of a certain phase A, constituted by inert component a, in the presence of PMC j

$$-\Delta S_A\, dT + \Delta V_A\, dP - \Delta n_j\, d\mu_j = 0 \tag{3}$$

leads to

$$dT/d\mu_j = \Delta n_j/\Delta S_A, \tag{4}$$

where $\Delta S_A = S_A^l - S_A^S$ is the entropy of fusion of the phase A and $\Delta n_j = n_j^l - n_j^A$ is the difference between the mole numbers ("solubilities") of PMC j in molten and crystalline phase A. As $S_A^l > S_A^S$, and the solubilities of PMC j are, in general, different in various phases of the system, one and the same change of μ_j will cause unequal changes in fusion temperatures of the phases. Hence, the liquidus fields of the phases and. in a certain composition range, the sequences of their crystallization, should change accordingly (Fig. 6b).

It should be emphasized that the dissolution of a component *j* in a "pure" melt phase may also be viewed in terms of acid–base interaction. On the other hand, it is obvious that the dissolution of a *j* in a multicomponent melt will affect not only the fusion temperatures of "pure" phases but also the mixing behavior of the melt components. Thus, in reality these two sides of the effect of *j* addition are inseparable from each other and should be taken into account together in any general thermodynamic description of the effect (see below).

An alternative approach was offered by Kushiro (1975) and later was developed by Mysen (Mysen, 1976; Mysen et al., 1980a,b) and other researchers. Analysing a vast amount of liquidus diagrams and his own experimental data, Kushiro found that the addition of oxides of monovalent and, to a lesser extent, divalent cations shifts cotectics towards the SiO_2 phase whereas the oxides of polyvalent elements produce an opposite effect (Fig. 7a). These regularities were explained by Kushiro through structural changes in the melt network due to the dissolution of additional components. So it was concluded that the Me_2O- and MeO-type oxides facilitate the depolymerization of Si–O tetrahedra and, therefore, expand the liquidus fields of the less polymerized mineral (fayalite, forsterite, and pseudowollastonite). The addition of such oxides as P_2O_5 and TiO_2 produces a reverse effect due to the polymerization of the network.

These deviations of cotectics in closed systems could also be explained by the salt-forming effect in melts (Epel'baum and Kuznetsov, 1980a), which illustrates the principle of maximum polarity of chemical bonds through the formation of corresponding compounds with the cotectics naturally leading to their figurative points on the liquidus diagrams. Thus, a "fan" of tie-triangles will correspond to the Kushiro's "fan" of cotectics (Fig. 7b).

Kushiro's approach has become widely-accepted because of the possibility of proving it by direct physical methods. Thus, it was through changes in the degree of polymerization of the melt network that Mysen et al. (1980a,b) came to describe the observed shifts in the liquidus boundaries between phases in a number of systems with the addition of H_2O and TiO_2. Manning et al. (1980) linked changes in the liquidus phase relations in granitic systems having increasing F (H_2O) contents with the bonding of part of Al with F (H_2O) and its transformation from the tetrahedral to octahedral coordination in melts,[2] and so on.

An interesting combination of the structural and thermodynamic approaches was introduced by Fraser (1977). He assumed that silicate polyanions of the melt can consist of five structural units ("structons"): ^{ij}Si, where *i* stands for the number of non-bridging (O^-) and *j* is the number of bridging (O°) oxygen species.

[2] According to Kogarko and Krigman (1981) the increase of SiO_2 activity in acid melts containing F is caused by the formation of the bonds between F with the network-modifying ions (alkalies), not with the glass-forming ions (Al).

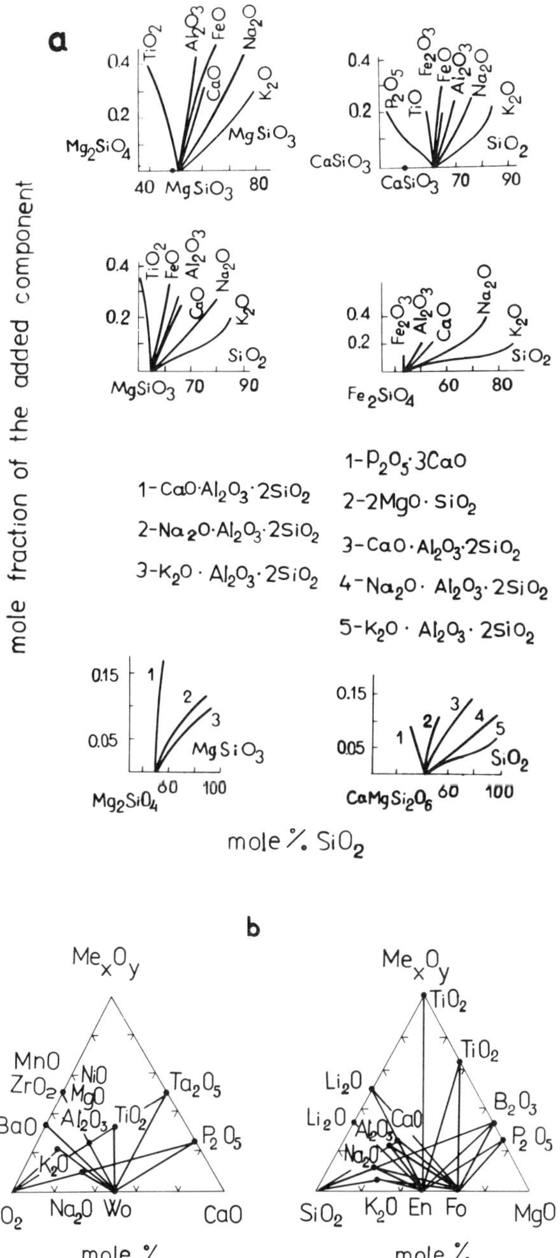

Fig. 7. The effect of additional components in "close" systems: (a) "Fans" of cotectics (Kushiro, 1975). (b) Corresponding elementary phase triangles (Epel'baum, 1980) in the SiO_2–MgO–Me_xO_y and SiO_2–CaO–Me_xO_y systems.

The crystallizing mineral uses only part (one, as a rule) of the structons: pyroxenes–^{22}Si, olivine–^{40}Si and so on. Regarding a melt as an ideal mixture of structons and cations and introducing the notion of "the solubility product of a mineral in the melt" ($K_{s.p.}$), the activities of the mineral components (enstatite and forsterite, for example) on liquidus could be described as

$$K^{En}_{s.p.} = a_{MgSiO_3} = x_{Mg^{2+}} \cdot x_{22_{Si}}, \tag{5}$$

$$K^{Fo}_{s.p.} = a_{Mg_2SiO_4} = x^2_{Mg^{2+}} \cdot x_{40_{Si}}, \tag{6}$$

The values of $x_{ij_{Si}}$ can be calculated from the equilibrium proportions of O^- and O°. Data on the dependence of the equilibrium polymerization constant on temperature being available, in principle it is possible to calculate the activities of these mineral components on the liquidus.

The added component (for example, K_2O which dissociates in the melt according to base-type: $K_2O \rightleftarrows 2K^+ + O^{2-}$- will shift the polymerization equilibrium:

$$O^\circ + O^{2-} \rightleftarrows 2O^- \tag{7}$$

leading to a relative increase of $x_{40_{Si}}$ $(=x^4_{O^-})$ compared with $x_{22_{Si}}$ $(=6x^2_{O^\circ} \cdot x^2_{O^-})$. Therefore, the "activity of $MgSiO_3$ will decrease. This decrease will cause enstatite to dissolve and forsterite to crystallize. To maintain both minerals in equilibrium with the melt, the cotectic must move to more SiO_2-rich compositions or otherwise add O° to the system" (Fraser, 1977, p. 323).

Thus, according to Fraser's model, the dissolution of additional components in the melt is regarded as a kind of "perturbation" of the melt structure which the melt-crystals system seeks to put down. In terms of acid-base formalism this effect means that the system has a tendency to preserve along the cotectic the inherent basicity corresponding to the basicity of the eutectic in the absence of added components.

It should be noted that change in the liquidus phase relations with increasing volatile contents evidently reflects the change of melt component activities due to their interaction with the dissolving components. Therefore, any thermodynamic model of silicate liquids (e.g. Bottinga et al., 1981; Ghiorso et al., 1983; Burnham, 1975, 1979) can undoubtedly be applied to estimation of the liquidus boundary shifts if the activities of the melt components are known as well as the concentration functions of the dissolving additional components. However, this aspect of these models has not been thoroughly worked out. Furthermore, these models are not available for the thermodynamic description of the alterations of the liquidus relations caused by independent changes of chemical potentials of some components in open fluid-magma system, because it is Korzhinskii's potential, but not Gibbs', that is the characteristic state function of such systems. So the development of the thermodynamic model, suitable for the quantitative description of the liquidus phase relations in open systems was the main purpose of the present paper.

Eutectic Relationships in the Systems with Perfectly Mobile Components

General Thermodynamics

To develop this model the thermodynamic formalization of Korzhinskii's open system is very convenient because it treats the parameters of state of a number of components ("perfectly mobile") as being their chemical potentials[3], that is, intensive parameters. The parameters of state of other components ("inert") as in Gibbs' systems, are their masses or mole numbers in the system, i.e. extensive parameters (Korzhinskii, 1973; Zharikov, 1976). The essence of Korzhinskii's systems is that the total sum of phases, singled out as a system, is considered to be in a certain "power chemical field" and its intensity is set by the PMC chemical potentials from outside (Karpov and Kiselev, 1979). In terms of the thermodynamics of irreversible processes this proposal means that local equilibrium in the system with PMC is determined not only by the gradients of T and P but also by the gradients of the PMC chemical potentials.

Thus, assuming that inert components are the mineral components that make up eutectic and cotectic phases and that PMC are the additional (fluid) components, it is possible to describe in the framework of the equilibrium thermodynamics the changes in parameters of eutectic and cotectic equilibria as functions of the independently set values of PMC chemical potentials. Before doing so, it is necessary to discuss some essential features of a thermodynamic description of systems with PMC.

Korzhinskii's free energy, G_m, and its total differential are given by equations:

$$G_m = G_m(T, P, \mu_f, \ldots, \mu_k, n_a, \ldots, n_e), \tag{8}$$

$$G_m = G - \sum_{j=f}^{k} \mu_j n_j = \sum_{i=a}^{e} \mu_i n_i, \tag{9}$$

$$dG_m = -S_m\,dT + V_m\,dP - \sum_{j=f}^{k} n_j\,d\mu_j + \sum_{i=a}^{e} \mu_i\,dn_i, \tag{10}$$

where G is the Gibbs free energy, T is the temperature, P is the total pressure, μ_j and n_j are the chemical potential and the mole number of the PMC j ($j = f, \ldots, h, \ldots, k$), μ_i and n_i are the chemical potential and the mole number of the inert component i ($i = a, \ldots, l, \ldots, e$), S_m is the total entropy and V_m is the total volume of Korzhinskii's system. The index m hereafter refers to the thermodynamic functions in the system with PMC.

As may be seem from Eq. (10), the effect of chemical potential of PMC j on

[3] In the canonical-type systems. We shall discuss only systems of this type.

the chemical potential of inert component i is expressed through a partial derivative:

$$(\partial\mu_i/\partial\mu_j)_{T,P,\mu_{h\neq j},n_i} = -(\partial n_j/\partial n_i)_{T,P,\mu_j,n_{l\neq i}} = -N_j^i. \quad (11)$$

This partial derivative, N_j^i, we called "the partial molar capacity of the inert component i with respect to the perfectly mobile component j" (Kuznetsov and Epel'baum, 1981). This value bears a clear physical meaning (the mole number of j in the system at a given T, P, and μ of PMC related to 1 mol of i) and may be calculated from the common solubility data. In particular, the partial molar capacity N_j^i is equal simply to the n_j/n_i molar ratio in the system consisting of one inert component i and, hence, is connected with the mole fraction of j in a closed system by the apparent equation: $N_j^i = x_j/(1 - x_j)$. It characterizes the individual property of inert component i with respect to its interaction with PMC j and should possess all the properties of partial molar values. For instance,

$$n_j = \sum_{i=a}^{e} n_i N_j^i, \quad (12)$$

i.e., the mole number of PMC j in the system, which is included in the (9)–(10) equations for Korzhinskii's free energy, is composed of the partial molar capacities of all inert components of the system with respect to PMC j.

So, describing the chemical potential of inert component i[4] as

$$\mu_i = \mu_i^{m(o)} + RT\ln(\gamma_i x_i^m) \quad (13)$$

and putting (13) in the expression for the partial derivative (11), we shall get (differentiation constants are ommitted here for the sake of simplicity):

$$\partial\mu_i/\partial\mu_j = \partial\mu_i^{m(o)}/\partial\mu_j + \partial RT\ln\gamma_i/\partial\mu_j =$$

$$\partial\mu_i^{m(o)}/\partial\mu_j + \partial\mu_i^{m(ex)}/\partial\mu_j = -N_j^{i(o)} - N_j^{i(ex)} = -N_j^i \quad (14)$$

inasmuch as $(\partial RT\ln x_i^m/\partial\mu_j)_{T,P,\mu_{h\neq j},n_j} = 0$.[5]

In these and subsequent equations the mole fraction of the inert component i is determined as

$$x_i^m = n_i/\sum_{i=a}^{e} n_i, \qquad \sum_{i=a}^{e-1} x_i^m = 1 \quad (15)$$

and the standard state for the chemical potential of the inert component i in the system with PMC is the state of i in the phase composed of the sole inert component i at a given T and P, and the vector of chemical potentials of all

[4] The chemical potential of the inert component i is regarded as the partial molar Korzhinskii's free energy, i.e. $\mu_i^m \equiv (\partial G_m/\partial n_i)_{T,P,\mu_j,n_{l\neq i}}$; thus, $\mu_i^m = \mu_i$.

[5] The upper indices o, id, and ex refer to the thermodynamic functions for pure substances, to the functions in an ideal system and to the excess functions, respectively.

perfectly mobile components (Ryabchikov, 1965; Kuznetsov and Epel'baum, 1981).

As may be seen from (14), partial molar capacity, N_j^i, includes two members: $N_j^{i(o)} = N_j^{i(id)}$ that characterize the interaction of "pure" (or existing in an ideal mixture with other inert components) inert component i with PMC j, and $N_j^{i(ex)}$ that reflects the effect of PMC j on the activity coefficient of inert component i in a real system. Thus, the first member refers to the effect of μ PMC on fusion temperatures of "pure" mineral components (phases), and the second member relates to the change in their mixing behavior in the melt (i.e. relates to the effect of acid-based interaction proper, see Fig. 6a and b).

On the basis of (14) it is possible to write the following equation for the total differential of the chemical potential of inert component i in a system with PMC:

$$d\mu_i = -S_i^{m(id)}dT + V_i^{m(id)}dP - \sum_{j=f}^{k} N_j^{i(id)}d\mu_j + RTd\ln x_i^m + d\mu_i^{m(ex)}, \quad (16)$$

where

$$S_i^{m(id)} = S_i^{m(o)} - R\ln x_i^m, \quad (17a)$$

$$V_i^{m(id)} = V_i^{m(o)}, \quad (17b)$$

$$N_j^{i(id)} = N_j^{i(o)}, \quad (17c)$$

$$d\mu_i^{m(ex)} = -S_i^{m(ex)}dT + V_i^{m(ex)}dP - \sum_{j=f}^{k} N_j^{i(ex)}d\mu_j + \sum_{i=a}^{e-1} (\partial\mu_i^{m(ex)}/\partial x_i^m)_{T,P,\mu_j,x_{l\neq i}^m} dx_i^m. \quad (18)$$

It should be pointed out that the formal extension of the principal relations of equilibrium thermodynamics, expressed in generalized thermodynamic values, to Korzhinskii's systems (Kuznetsov and Epel'baum, 1981), has shown that the total entropy and the total volume in these systems are determined by the equations:

$$S_m = \sum_{i=a}^{e} n_i S_i^m \quad (19a)$$

and

$$V_m = \sum_{i=a}^{e} n_i V_i^m \quad (19b)$$

in which the partial molar entropies and the partial molar volumes of PMC are absent. However, by way of differentiation of S_m and V_m with respect to n_i we can have the following:

$$S_i^m = S_i + \sum_{j=f}^{k} N_j^i S_j \quad (20a)$$

and

$$V_i^m = V_i + \sum_{j=f}^{k} N_j^i V_j \tag{20b}$$

where $S_{i,j}$ and $V_{i,j}$ are the partial molar entropies and the partial molar volumes of components i, and j in the materially analogous Gibbs' system (in this instance $S_m = S$ and $V_m = V$). Thus, the partial molar entropies and the partial molar volumes of PMC are, in fact, present in the total entropy and in the total volume of a system with PMC but not obviously.

In conclusion it should be emphasized that, as in Korzhinskii's systems, the chemical potentials of PMC are a priori regarded as the parameters of state, and these values should be set independently of T and P. This independence is the key difference (not always taken into account) of chemical potentials of PMC from chemical potentials of the same components in Gibbs' systems that are taken as the functions of T and P and the bulk composition of a system.

Topology and Thermodynamics of T–P–μ_j–x Liquidus Diagrams

Let us consider the liquidus phase relations in a simple system consisting of two inert components a and b and of one perfectly mobile ("additional") component j with now be considered. The components a and b denote neutral chemical units corresponding to mineral phases A and B. For simplicity it is assumed that A and B do not form solid solutions or compounds with one another but mix completely in the melt phase, and PMC j can dissolve in both melt and crystalline phases A and B. Therefore, this system belongs to the eutectic type and its parameters of state are temperature, total pressure, chemical potential of PMC j and mole fraction of one of the inert components in the melt.

Interest will be focused on the change in parameters of the three-phase (eutectic) equilibrium

$$A + B \rightleftarrows L \tag{21}$$

in response to the change of intensive parameters, i.e., of the conditions of the system's existence. At equilibrium, the chemical potentials of a and b in the coexisting melt and solid phases are equal:

$$\begin{cases} \mu_a^L = \mu_a^A, \\ \mu_b^L = \mu_b^B \end{cases} \tag{22}$$

and to maintain this equilibrium it is necessary and enough to keep to

$$\begin{cases} d\mu_a^L = d\mu_a^A, \\ d\mu_b^L = d\mu_b^B. \end{cases} \tag{23}$$

According to the phase rule of Gibbs–Korzhinskii, in general equilibrium (21) is divariant:

$$n = k_{\text{intert}} + k_{\text{mobile}} + 2 - r = 2 + 1 + 2 - 3 = 2$$

and therefore will be depicted by some surface in the four-dimensional T–P–μ_j–x_i^m space marking the intersection of liquidus phase volumes of A and B.

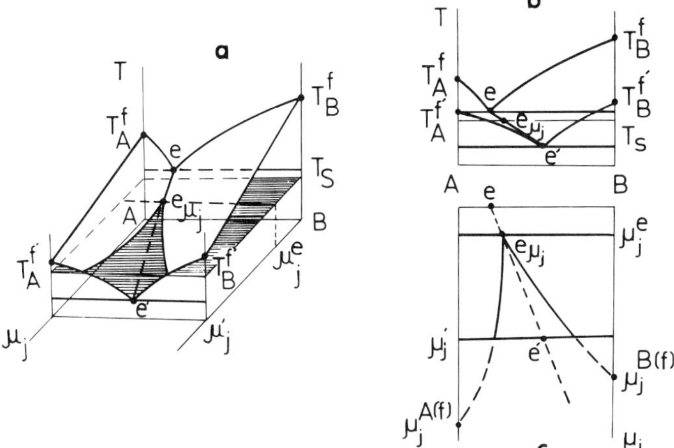

Fig. 8. Schematic diagrams for a binary system A–B showing the changes in eutectic relationships in the presence of perfectly mobile component j at constant P. (a) General T–μ_j–x liquidus diagram of the system. (b) Eutectics related to temperature (T–x sections of T–μ_j–x diagram in the absence of j in the system and at $\mu_j = \mu_j'$). (c) Eutectic related to μ_j(μ_j–x *section of* T–μ_j–x diagram at $T = T_s$). e–e′ is the univariant line in the system, $\mu_j^{A(f)}$ and $\mu_j^{B(f)}$ are the chemical potentials of perfectly mobile component j corresponding to the fusion of "pure" phases A and B at a given T and P.

If one of the intensive parameters is considered as constant (i.e. in the section on this parameter), there will be divariant surfaces for liquidus of A and B and univariant (eutectic) line for equilibrium (21) in three-dimensional space (Fig. 8a).

Finally, if the system is considered with two intensials being constant, a corresponding geometrical image of equilibrium (21) will be an invariant point (an eutectic) serving as the intersection for univariant liquidus lines of A and B (Fig. 8b and c). An eutectic point therefore, is characterized by constant proportion of solid phases, by constant melt composition and, as follows from general considerations (e.g. Zharikov, 1976), by the extreme value of variable intensive parameter at which the melt phase appears (or disappears).

It is obvious that the beginning of melting or the end of crystallization should correspond to any of the intensials (T, P, or μ_j), reaching this extreme value in the environment. Depending on the purpose of the investigation and real conditions for the system's existance, three types of eutectics can be distinguished: related to temperature, to pressure and to the chemical potential of PMC. From the point of view of thermodynamics, these types of eutectics are quite equipollent because the corresponding intensials have been a priori chosen as independent variables. However, the physico-chemical and geological literature traditionally interprets the term "eutectic" as being related only to temperature. The necessity to consider total pressure in an analogous manner has led to the definition "eutectic related to pressure" (Zharikov, 1976). Having introduced the notion of "eutectic related to the chemical potential of PMC" (Epel'baum and Kuznetsov, 1980b), which reflects the effect of additional components on the equilibrium melting (crystallization), the authors have followed what appears to be logical extension of this approach to complex fluid–magma system.

Thermodynamic relations describing eutectic relationships for the given system will next be considered. Substituting

$$\mu_i^L = \mu_i(T, P, \mu_j, x_i^m) \tag{24a}$$

and

$$\mu_i^S = \mu_i(T, P, \mu_j) \tag{24b}$$

into (23) taking into consideration (15) and (16), gives for the eutectic surface in the T–P–μ_j–x_i^m space:

$$\begin{cases} RTd\ln x_a^m = \Delta S_a^{m(id)}dT - \Delta V_a^{m(id)}dP + \Delta N_j^{a(id)}d\mu_j - d\mu_a^{m(ex)}, \\ RTd\ln x_b^m = \Delta S_b^{m(id)}dT - \Delta V_b^{m(id)}dP + \Delta N_j^{b(id)}d\mu_j - d\mu_b^{m(ex)}, \\ x_a^m + x_b^m = 1, \end{cases} \tag{25}$$

where

$$S_i^{m(id)} = S_i^{mL(id)} - S_i^{mS}, \tag{26a}$$

$$V_i^{m(id)} = V_i^{mL(id)} - V_i^{mS}, \tag{26b}$$

$$N_j^{i(id)} = N_j^{iL(id)} - N_j^{iS} \tag{26c}$$

are the changes in the corresponding partial molar values of inert component i during melting of the solid phase and $\mu_i^{m(ex)}$ in general is determined by equation (18).

Assuming that one or two intensive parameters are constant, it is logical to derive equations from (25) that describe univariant eutectic lines and invariant eutectic points for this system respectively. For example, at constant T and P the result is, for eutectic related to μ_j:

$$\begin{cases} RTd\ln x_a^m = \Delta N_j^{a(id)}d\mu_j - d\mu_a^{m(ex)}, \\ RTd\ln x_b^m = \Delta N_j^{b(id)}d\mu_j - d\mu_b^{m(ex)}, \\ x_a^m + x_b^m = 1. \end{cases} \tag{27}$$

By simple transformation these equations can also be presented as:

$$\begin{cases} d\ln x_a^m = (\Delta N_j^{a(o)}/RT)d\mu_j + (\partial \ln \gamma_a/\partial \mu_j)d\mu_j - (\partial \ln \gamma_a/\partial x_a^m)dx_a^m, \\ d\ln x_b^m = (\Delta N_j^{b(o)}/RT)d\mu_j + (\partial \ln \gamma_b/\partial \mu_j)d\mu_j - (\partial \ln \gamma_b/\partial x_b^m)dx_b^m, \\ x_a^m + x_b^m = 1, \end{cases} \tag{28}$$

where γ_i is the activity coefficient of the inert component i in the melt.

Thus, at a given T and P the location of eutectic related to μ_j is determined by: (1) the differences of the molar capacities of inert components a and b with

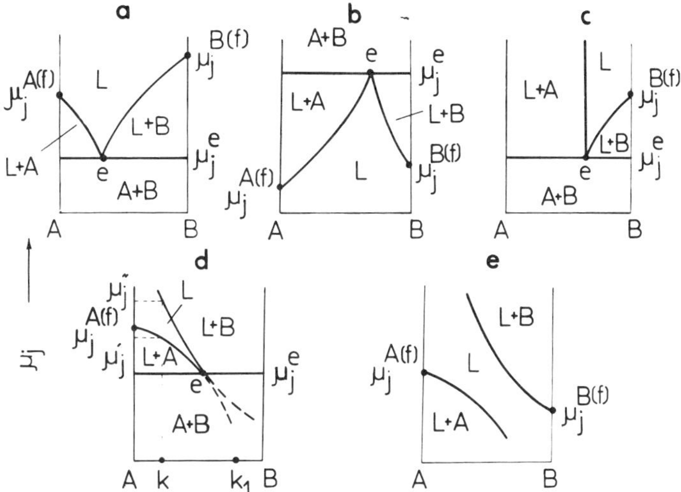

Fig. 9. Topological types of μ_j–x liquidus diagram for an ideal binary system A–B in the presence of perfectly mobile component j (see the text).

respect to PMC j in phases A and B in their molten and solid states; (2) the effect of μ_j on the mixing behavior of a and b in the melt phase; and (3) the "proper" unideality of mixture a and b in the melt at a given T and P. The first factor reflects the effect of PMC j on the fusion temperatures of phases A and B and the second and third factors—Korzhinskii's effect of acid–base interaction of components a, b, and j in the melt (see Fig. 6).

In conclusion, consider the topology of μ_j–x liquidus diagrams. For simplicity, discussion is here confined to the ideal approximation of the system A–B (Epel'baum and Kuznetsov, 1980b). As follows from (28), in this instance ($\gamma_i = 1$) the sign of $(d \ln x_i^m / d\mu_j)$, i.e. the incline of the liquidus curves in μ_j–x coordinates, will depend only on the correlation of molar capacities $N_j^{i(o)}$ in molten and solid phases A and B. Five variants are possible (Fig. 9).

1. For both inert components $N_j^{iL} > N_j^{iS}$. Then $(d \ln x_i^m / d\mu_j) > 0$ and the liquidus diagram will be the same as ordinary T–x diagrams of silicate systems (Fig. 9a).

2. For both inert components $N_j^{iL} < N_j^{iS}$. The liquidus diagram is turned "upside down" and resembles typical liquidus diagrams of eutectic related to pressure (Zharikov, 1976). With the increase of μ_j there will be progressive crystallization ending at the eutectic point (Fig. 9b).

3. For one of the components $N_j^{iL} = N_j^{iS}$. In this instance, the initial mixture with the higher content of this component than in the eutectic can not be melted irrespective of the value of μ_j (Fig. 9c).

4–5. When $N_j^{aL} > N_j^{aA}$ and $N_j^{bL} < N_j^{bB}$. In general, the curves of two-phase equilibrium will intersect (Fig. 9d). Stability will apply to the part of the diagram from the intersection point to where the eutectic temperature goes down while μ_j

increases. In a specific example, the liquidus curves of phases A and B may not intersect (Fig. 9e), so that at these T–μ_j conditions three-phase equilibrium (21) cannot be achieved.

The situation shown in Fig. 9d is of special interest as far as the problem of the formation of antimonomineral magmatic rocks (anortosites, for example) is concerned. True, if the melting of arbitrary initial mixtures k and k_1 is traced with the increase of μ_j, it is easy to see that the production of melt of "pure" phase A will be the result of this process while in the common T–x systems, with the increase of temperature, the melt composition tends to the composition of the initial mixture.

Thus, the nature of the interaction of a perfectly mobile ("additional") component with the mineral components making up a system's phases, even in the simple example of a binary eutectic system, could result in several varieties of melting (or crystallization) behavior. Of course, the reviewed types of μ_j–x liquidus diagrams are only an illustration of the main principles of the approach. For fluid–magma systems it would be more appropriate to study systems with solid solutions, with the formation of compounds, with several PMC, etc. For example, Fig. 10 shows schematic T–μ_j–x liquidus diagrams of binary systems with gap miscibility of solid solution in subsolidus (a), with incongruent melting (b, c) and with liquid immiscibility (d). As can be seen, a progressive accumulation of an additional component j, due to its influx from outside or due to magmatic differentiation (i.e. the increase of μ_j) can lead to cardinal changes in the configuration of the diagrams and, therefore, in the behavior of the system at equilibrium melting or crystallization. The description of the "movement" of the elements of these diagrams as μ_j increases is possible on the basis of the aforementioned thermodynamic relations.

It should be also noted that the thermodynamics of Korzhinskii's systems at the appropriate redetermination of such notions as "system" and "medium" can be applied even to the "internal" buffering of the chemical potentials (activities) of the fluid components. Such an application enables one to obtain the uniform thermodynamic description of the liquidus equilibria, depending on the fluid regimen, both in open and closed systems, because it allows the investigator to single out this effect and to study it pure and simple, without taking into consideration the other factors (temperature, total pressure, etc.).

The calculation peculiarities of the liquidus relation in systems with PMC are discussed in the Appendix using the Ab–H_2O system as an example.

Fig. 10. T–μ_j–x liquidus diagrams for binary systems A–B of different types: a, with the solid solutions and subsolidus gap miscibility; b, c, with the incongruent melting of intermediate compound C, and d, with the metastable (subliquidus) liquid immiscibility. In all instances the type of system transforms into another type with the increase of μ_j.

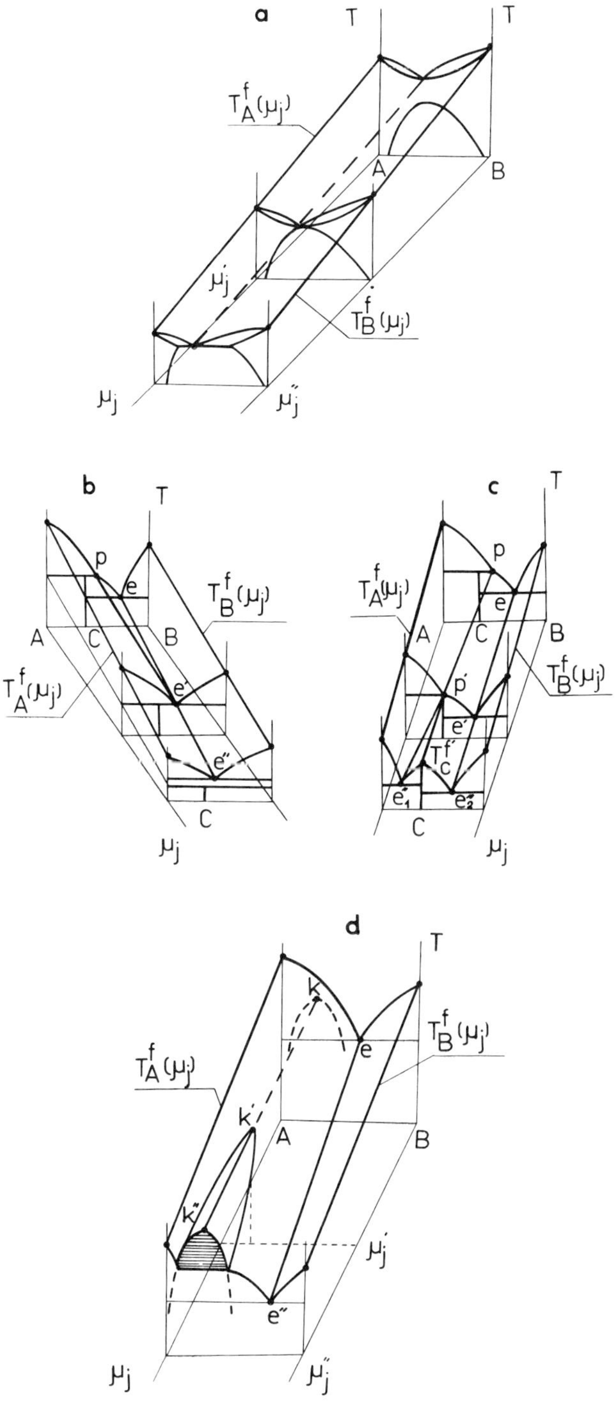
a
T
T
$T_A^f(\mu_j)$
A
B
μ_j'
$T_B^f(\mu_j)$
μ_j
μ_j''
b
T
p
e
$T_B^f(\mu_j)$
A
C
B
$T_A^f(\mu_j)$
e'
e''
C
μ_j
c
T
$T_A^f(\mu_j)$
p
e
A
C
B
$T_B^f(\mu_j)$
p'
e'
$T_C^{f'}$
e_1''
e_2''
C
μ_j
d
T
k
e
$T_B^f(\mu_j)$
$T_A^f(\mu_j)$
k'
A
B
k''
μ_j'
e''
μ_j
μ_j''

Geological Implications

On the whole the behavior of magmatic systems is rather well governed by the regularities that have been established in the relatively simple model systems. In some instances this kind of approximation proves inadequate, dictating the necessity to take into account the effect of additional components (Korzhinskii, 1972; Kushiro, 1975; Kogarko and Ryabchikov, 1980; Pichavant and Manning, 1984, and others). It would also be interesting to apply this effect to the reconstruction of the fluid regime during the formation of specific objects, granite in particular, because it is the fluid regimen that determines their potential ore-bearing capacity (e.g. Kovalenko and Kovalenko, 1976; Marin, 1976; Tauson, 1977).

To solve this problem, the late-stage high-level granites that are well developed in the area of the Tokrau synclinorium (Central Kazakhstan, the USSR) have been chosen for study. These granites form laccolith-type plutons and fall into three complexes: Kaldyrma biotite granites, Akchatau two-feldspar granites and Kizilray leucogranites (alaskites) (Negrey, 1983). All these high-silica granites are of similar chemical compositions but are different in terms of their modal composition, texture and structure features, nature of late- and postmagmatic alterations, etc. (Table 1). It should be emphasized that all these granites were formed during a comparatively short time interval (290–310 m.y. ago) in similar geotectonic conditions and at about the same depth level (ca. 1–2 km).

These plutons are usually the composite ones. The greater part of a pluton is composed of coarse-grained rocks of major intrusive phases in which sheet- or lens-shaped bodies of sharply porphiritic ("additional") granites and dykes and veins of fine-grained granites occur. Commonly, within a pluton, several intrusive phases are distinguished each of which in general consists of the major, additional and veined granites (Serykh et al., 1976).

Available chemical data (Serykh et al., 1976) from 12 plutons (4 most differentiated ones for each complex) were recalculated to norm compositions and were then plotted on Qz–Ab–Or diagrams separately for (1) the sequences of major intrusive granites of various phases and for (2) the sequences of major—additional—veined granites within each phase. The results allow determination of the following regularities: (1) the quartz and the albite "eutectic" trends, i.e. shifts of granitic minimum towards the quartz- and the albite-apexes of the Qz–Ab–Or diagram respectively, are the most typical of these granites; (2) the first one is present in all granites and connects the compositions of major intrusive granites of various phases; (3) the second one is marked by a shift in compositions of granites within one and the same phase towards the more differentiated varieties (Fig. 11). The albite trend is typical of granites of the Akchatau complex, not typical of alaskites and unusual for granites of Kaldyrma complex. Statistical analysis of the analytical data has shown the satisfactory significance of these trends.

The authors assume that these granites in the first approximation are the

Table 1. Chemical compositions and some essential features of Upper-Paleozoic granites, in the area to the north-west of Balkhash Lake, Central Kazakhstan, the USSR

Major elements, in wt% (mean and range)	1 Biotite granites of the Kaldyrma complex	2 Two-feldspar granites of the Akchatau complex	3 Leucogranites of the Kizilray complex
SiO_2	73.18 (71.50–75.00)	75.23 (74.25–76.65)	75.87 (75.50–76.61)
TiO_2	0.25 (0.16–0.33)	0.17 (0.13–0.41)	0.19 (0.13–0.22)
Al_2O_3	13.37 (12.60–14.00)	12.87 (12.25–14.08)	12.39 (11.87–12.95)
Fe_2O_3	1.03 (0.19–1.95)	0.95 (0.49–1.48)	0.98 (0.60–1.25)
FeO	1.37 (0.30–3.20)	0.92 (0.41–2.25)	0.79 (0.35–1.53)
MnO	0.05 (0.02–0.08)	0.04 (0.02–0.09)	0.06 (0.02–0.12)
MgO	0.61 (0.27–1.34)	0.37 (0.09–0.95)	0.15 (0.06–0.21)
CaO	1.46 (0.96–3.08)	0.84 (0.44–1.80)	0.55 (0.40–0.80)
Na_2O	3.61 (3.41–4.39)	3.60 (3.45–4.62)	4.12 (4.07–4.18)
K_2O	4.52 (4.00–4.81)	4.63 (3.70–4.93)	4.83 (4.67–5.00)
Average norms (Qz + Ab + Or recast into 100%)	Qz 34.8, Ab 34.7, Or 30.5	Qz 37.1, Ab 33.1, Or 29.8	Qz 34.4, Ab 36.2, Or 29.4
Modal composition, in volume %	Qz 27–33, Fsp 36–46, Pl 21–31, (Bi + Amph) 1.5–2.5	Qz 33–35, Fsp 43–48, Pl 16–22, Bi 0.6–1.3	Qz 30–35, Fsp 50–65, Bi < 1
Normative composition of Fsp	Or 58–61, Ab 37–40, An 0.5–1.0	Or 55–61, Ab 39–44, An 1.0–1.4	Or ~44, Ab ~55, An ~1
Composition of Pl	Oligoclase–andesine	Albite–oligoclase	(Albite)
Composition of Bi	In fields, IV, III, rarely in II, I (on the Si/Al-(Mg + Fe)/Al diagram of Marakushev and Tararin, 1965)		In field V, differ a high F content (up to 5.4%)
Structure	Enriched with basic xenolithes	Common (massive)	Miarolitic varieties are typical in apical parts of plutons
Late- and post-magmatic alterations	Not typical	K-feldspathization, muscovitization, extensive greizenization	Albitization, riebeckitization
Associated ore mineralization	No mineralization	Rare-metal (W, Mo, Bi) mineralization and deposits of Qz–greizen formation	Poor rare-metal mineralization, Qz–fluorite pegmatites (without mica)

Compiled after Negrey, 1983.

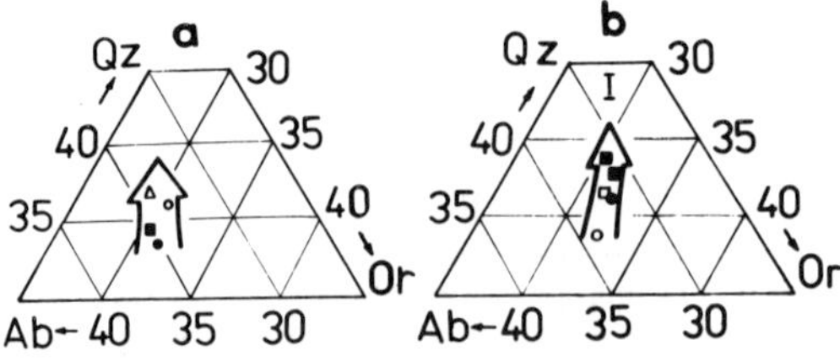

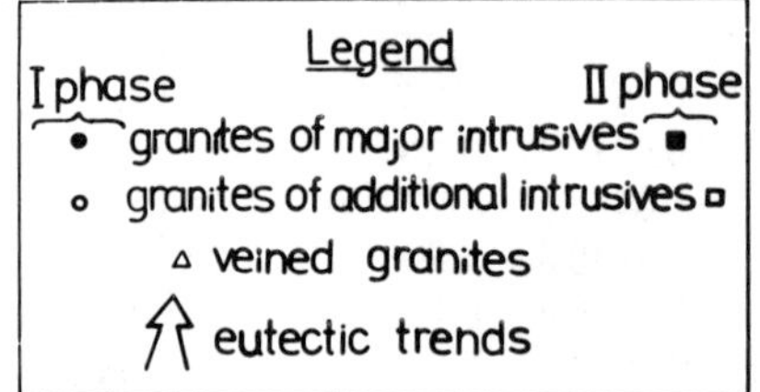

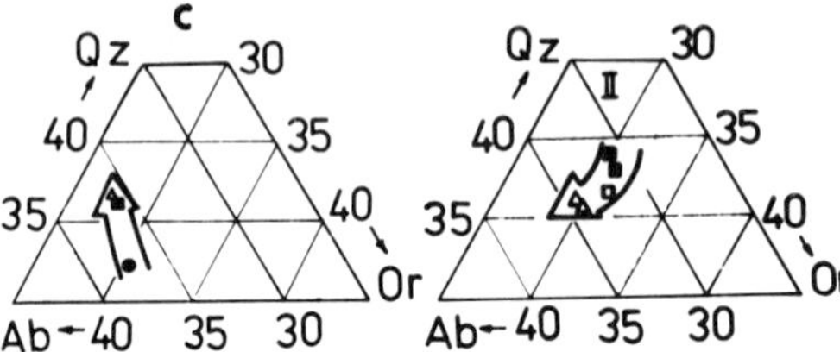

Fig. 11. Eutectic crystallization trends for granites of Central Kazakhstan, the USSR. (a) The Kaldyrma complex, Shaltas pluton. (b) The Akchatau complex, Akchatau pluton (*I*, the granitic series of major intrusives (two facies for the II phase); II, the granitic series of II phase). (c) The Kizilray complex, Kizilray pluton.

products of the eutectic crystallization[6] of the corresponding magmas in open systems at about the same depth. Therefore, the revealed compositional trends can be attributed to the fluid–magma differentiation in the chamber and be interpreted with the help of the T–μ_j–x liquidus diagrams. The crystal fractionation for granitic intrusives of this type seems to have a subordinate role (Marin, 1976; Tauson, 1977).

The principal scheme of the influence of the complex fluid phase upon the crystallization behavior, using the T–μ_j–x diagram of the binary eutectic system as an example (Fig. 12), will next be considered, where quartz (phase A) and alkali feldspar (phase B) are taken as the inert mineral components and an acid component j, characterizing the summary effect of a complex fluid is taken as a perfectly mobile one. It is assumed that the melt in the magmatic chamber was of the "m" composition, corresponding to the eutectic due to the partial melting at the depth, but it differs from the eutectic, corresponding to the consolidation level. Then there could be three generalized crystallization paths, depending on

[6] In reality, the composition of minimum (eutectic) should correspond only to the composition of the fine-grained matrix excluding the phenocrysts and over-cotectic minerals, but thus far such data are not available.

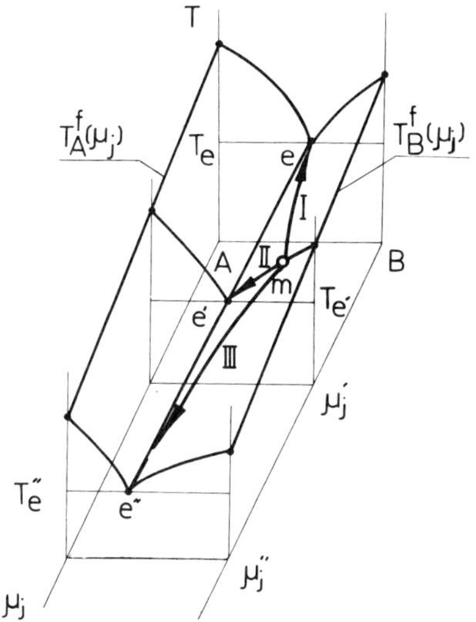

Fig. 12. Schematic T–μ_j–x liquidus diagram for granitic system demonstrating various possible crystallization paths (see the text).

both the external conditions and the nature of the fluid–magma system itself. The first one is realized at the constant temperature, when the melt crystallization is caused by the volatile separation. The curve I (eutectic with respect to μ_j) corresponds to this path in Fig. 12. The second crystallization path occurs, when the volatile separation is hampered in one or another way and μ_j in the crystallized system is maintained on one and the same level. The curve II (eutectic with respect to T) corresponds to this path in Fig. 12. Finally, there is quite a common situation when the volatiles, escaping under the solidification of the main magma volume or influxing by the transmagmatic fluids, accumulate in the residual melts. With the decreasing of T this accumulation will result in the increase of μ_j (curve III in Fig. 12).

It is quite obvious that this or that type of magma crystallization will lead to the different compositions of the granitic eutectic (minimum). In reality, these processes are complicated by the other effects. Thus for instance, the fluid regime is conditioned by both the magmatic and the depth origin flows, by the degree of the volatiles saturation of the parent magma, by the mechanical properties of the country rocks, by the character of the thermal field around of the intrusion, by the possible participation of the meteoric waters, etc. However, these generalized crystallization paths characterize even some extreme cases and they could be applied to the explanation of the observed "eutectic" trends of the granitic plutons.

The quartz trend, being present in all plutons and corresponding to the regular changes of the granite compositions from the major intrusive phases, could be connected with the evolution of fluid-magma systems at the constant T and under the decrease of μ_j (the path *I* in Fig. 12). This decrease of μ_j (i.e. the relative increase of alkalinity) could have a number of causes: by the "washing" of magma by the flows of the transmagmatic fluids enriched with alkalies; by the accumulation of alkalies in residual fluids and melts during the crystallization of the main magma volume; by the preferential separation into the vapor phase of the acid volatiles, etc. All these processes promote increasing fluid–magma system alkalinity and, therefore, can shift the granitic eutectic (minimum) towards the quartz corner[7].

The nature of the albite trend is well explained by the increase of chemical potentials of such acid components as H_2O, F and B in the fluid phase and, as crystallization of the major intrusive granites proceeds, in the residual melts (the path III in Fig. 12). This interpretation is based on the available experimental data on the effect of H_2O, F, and B additions (see above). It was found out that the increase both of fluorine contents and modal plagioclase amounts in the granitic series depends on the presence of the albite trend. These findings serve as additional evidence of this interpretation, because the crystallization of one or two alkali feldspar from granitic melts depends not only on the temperature but also on the value of chemical potentials of these components in a coexisting fluid phase (see Fig. 10a, where *A*, *B*, and *j* denote Ab, Or and H_2O, F and B).

Thus, two general types of fluid regime can be distinguished under the crystallization of composite granitic plutons of Central Kazakhstan: one is towards higher alkalinity and the other towards higher acidity. The first one is typical of all granitic plutons while the second is typical only of ore-bearing granites of Akchatau complex. The validity of these two types of fluid evolution is also proved by the geochemical and petro-chemical data (Marin et al., 1983). Naturally, two questions are bound to arise: (1) what are the reasons for the relationship between the specific acidity-type of fluid regimen and the ore-bearing capacity of these granites and (2) what are the reasons directing the evolution of fluid–magma system along the "ore-bearing" way?

To answer these questions, the effect of eutectic shift can again be applied, but on the level of magma generation. It is first assumed that the parent magmas corresponding to these granitic complexes represent the eutectics that were generated due to the ascending fluids (Kuznetsov and Izoh, 1969; Korzhinskii, 1972) at about the same low depth and the same source composition. Then the distinctions in the normative compositions of granites from the major intrusives of the *I* phase, which are less differentiated and compose the main volume of the plutons, could indicate the different fluid regimes at their generation stage.

It is established that the alaskites of the *I* intrusive phase of Kizilray complex are characterized by the relative enrichment with normative albite, which could indicate the generation of the corresponding parent magma at the increased values of the chemical potentials of the acid volatiles (probably, F to the great extent) in the magma-generating fluids. The minimum biotite content in the alaskites as compared with the other granites, its very high F content and some other features (see the Table) prove this assumption. Due to the partitioning of such acid components as H_2O, F and B essentially in favor of the melt phase (e.g. Kogarko and Krigman, 1981; Manning, 1981; Pichavant, 1983) the alaskitic

[7] The increase of P_{H_2O} in the range of low P_{H_2O} (up to about 1 kbar) from the earlier intrusive phases towards the later ones can be the alternative reason for such a shift (see Fig. 1d).

magma should be highly enriched with these volatiles and, therefore, it couldn't evolve towards their accumulation in residual melts at the consolidation level. So, the crystallization of this magma proceeds within a sufficiently narrow temperature interval, accompanied by the "dispersed" separation of the volatiles on the crystallization front. The local accumulation and the redistribution of volatiles could lead only to the formation of poor mineralization and of camera pegmatites.

At the same time the relative enrichment (as compared with the former) of the granites of the *I* phase of Akchatau complex with the normative quartz could indicate that the corresponding magma-generating fluids were poorer in such components as H_2O, F, and B and richer in alkalies (probably, in $K > Na$). Therefore, while this magma ascends and emplaces on the shallow level there should still be a tendency toward the accumulation of volatiles in residual melts during crystallization of major intrusive granites. For this reason, the crystallization interval of this magma was extended with respect to temperature, and the melt could have served as collector and conductor for the fluids for a long time. So, the specific fluid-magma open system, which is evolved through a salt-melt one into a hydrothermal system (Manning, 1981; Pichavant and Manning, 1984) could have formed. It seems that the generation of this system favors the formation of large-scale ore deposits, if only of the investigated rare-metal type[8].

In contrast, certain features (the weaker differentiation of the plutons, a great amount of basic xenolithes, more basic plagioclase, etc.; see the Table) typical of the barren granites of Kaldyrma complex are evidence of the insignificant role of the fluid phase during their formation. Perhaps the corresponding magma are generated in relatively "dry" conditions and/or at the lower depth and from the more mafic source.

Thus, the fluid regime on the magma generation level could play the role of a "trigger", determining the evolution of a fluid–magma system and their crystallization behavior at the consolidation level. This model of course, is schematic because many other factors of magmatic evolution were not taken into consideration. However, a special attempt has been made to choose a geological situation in which the role of the other factors in the granites compared could be regarded as about the same and, therefore, all the distinctions between these granites could be attributed mainly to the fluid–magma interaction.

Conclusion

The effect of additional components (including the fluid components) on the shift of eutectics and cotectics is widely spread in nature. Apparently it is impossible to reproduce all the behavior varieties of multicomponent magmatic systems in laboratory investigations. Therefore, along with the experimental studies it

[8] It is remarkable that the albite trend in Akchatau pluton is just in the II intrusive phase with which the industrial ore mineralization is associated (see Fig. 11b).

is necessary to develop a quantitative model of fluid-magma interaction that would take this effect into account. One of the possible approaches in this direction, demonstrated in this paper, is the use of thermodynamics of Korzhinskii's open systems. Further progress in this approach consists mainly of calculations of T–P–μ_j–x liquidus diagrams modelled on the open fluid–magma systems. On the other hand, as has been shown, it is possible to shed light on some petrological problems, using the already established regularities.

Appendix: Calculation of Albite Liquidus in the Presence of Perfectly Mobile H_2O

The purpose of this section is to demonstrate the peculiarities of quantitative calculation of the liquidus relations in open systems on the example of a concrete system with the help of thermodynamics, as discussed earlier. As a model, the albite–H_2O system is selected, for which there are available experimental and thermodynamic data (Burnham and Davis, 1974; Boettcher et al., 1982; Hervig and Navrotsky, 1984, and others). Albite ($NaAlSi_3O_8$) ia considered as an inert component, and H_2O as a perfectly mobile one.

The focus of interest is the changes of parameters of crystal melt equilibrium for albite in Korzhinskii's open system. Therefore, temperature (T), total pressure (P) and chemical potential of the perfectly mobile H_2O (μ_{H_2O}) are taken as the parameters of state. This equilibrium is divariant according to the phase rule:

$$n = k_{\text{inert}} + k_{\text{mobile}} + 2 - r = 1 + 1 + 2 - 2 = 2,$$

which means that only two parameters could be set up independently. So, the corresponding geometrical image of this equilibrium will be a certain surface (liquidus) in T–P–μ_{H_2O} space.

Using the above mentioned equations (25, for example) for only for one component (phase) and omitting the insignificant members, for this surface gives:

$$\Delta S_{\text{Ab}}^{f(m)} dT - \Delta V_{\text{Ab}}^{f(m)} dP + \Delta N_{\text{H}_2\text{O}}^{\text{Ab}} d\mu_{\text{H}_2\text{O}} = 0, \tag{A1}$$

where according to (26)

$$\Delta S_{\text{Ab}}^{f(m)} = S_{\text{Ab}}^{l(m)} - S_{\text{Ab}}^{s}, \tag{A2a}$$

$$\Delta V_{\text{Ab}}^{f(m)} = V_{\text{Ab}}^{l(m)} - V_{\text{Ab}}^{s}, \tag{A2b}$$

$$\Delta N_{\text{H}_2\text{O}}^{\text{Ab}} = N_{\text{H}_2\text{O}}^{\text{Ab}(l)} - N_{\text{H}_2\text{O}}^{\text{Ab}(s)}. \tag{A2c}$$

are the changes of the partial molar entropy, volume and capacity of the albite component with respect to H_2O, correspondingly, during albite fusion in Korzhinskii's system.

For simplicity, the liquidus surface will be considered as a series of univariant isobaric curves in T–μ_{H_2O} space. Then these curves will be described by the following equation:

$$\Delta S_{Ab}^{f(m)} dT + \Delta N_{H_2O}^{Ab} d\mu_{H_2O} = 0, \tag{A3}$$

where the values of $\Delta S_{Ab}^{f(m)}$ and $\Delta N_{H_2O}^{Ab}$ correspond to a given P.

Apparently, the values of $\Delta S_{Ab}^{f(m)}$ and $\Delta N_{H_2O}^{Ab}$ should be preliminarily estimated as functions of T, P and μ_{H_2O}. For this purpose Burnham's and Davis' (1974) thermodynamic data on the Ab–H_2O closed system and the correlationship (20) between the partial molar values in Gibbs' and Korzhinskii's systems. In addition, calculations will be held within the moderate pressure range, where H_2O solubility in crystalline albite and albite solubility in water vapor could be neglected (i.e. $N_{H_2O}^{Ab(s)} \approx 0$ and on the H_2O saturation surface $\mu_{H_2O}^{l} \approx G_{H_2O}^{0}$). Then we shall have from (20) and (A3):

$$(x_{Ab}^{l} \Delta S_{Ab}^{f} + x_{H_2O}^{l} S_{H_2O}^{*l}) dT + x_{H_2O}^{l} d\mu_{H_2O} = 0, \tag{A4}$$

where

$$\Delta S_{Ab}^{f} = S_{Ab}^{l} - S_{Ab}^{s} = \Delta S_{Ab}^{f(o)} - R \ln x_{Ab}^{l} + S_{Ab}^{l(ex)} \tag{A5}$$

and

$$x_{H_2O}^{l} = N_{H_2O}^{Ab(l)}/(1 + N_{H_2O}^{Ab(l)}). \tag{A6}$$

In the equations (A4)–(A6) $S_{Ab}^{f(o)}$ is the entropy of fusion of anhydrous albite, $S_{Ab}^{l(ex)}$ is the partial molar excess entropy of Ab-component in the melt, $S_{H_2O}^{*l}$ is the partial molar entropy of H_2O in the melt which is standardized with respect to the hypothetical state with $x_{H_2O}^{l} = 1$ (see below), x_{Ab}^{l} and $x_{H_2O}^{l}$ are the mole fractions of Ab-component and of H_2O in the melt correspondingly. Naturally, all the functions in equation (A4) refer to a given P.

The peculiarity of this approach is connected not only with the usage of new thermodynamic functions ($S_{Ab}^{f(m)}$ and $N_{H_2O}^{Ab}$) but also with the choice of the fusion temperatures of anhydrous albite at any P as initial boundary conditions for the calculations of the liquidus surface of albite according to (A4), whereas the coordinates of the H_2O-saturated liquidus surface of albite serve as the final boundary conditions. The procedure is made necessary by the fact that μ_{H_2O} a priori, was taken as the independent intensive parameter, inducing the melting of the albite at a fixed P (cf. Fig. 9). So, the main difficulties of this calculation were connected with the restandardization of the thermodynamic functions for H_2O in the melt and with their correct description in the range of a very dilute solution of H_2O in the albite melt.

This question will now be considered in more detail. The chemical potential of H_2O in water–albite melt at a given T and P could be presented as:

$$\mu_{H_2O}^{l} = \mu_{H_2O}^{\ominus l} + RT \ln a_{H_2O}^{*l} = \mu_{H_2O}^{\ominus l} + RT \ln x_{H_2O}^{l} + RT \ln \gamma_{H_2O}^{*l}, \tag{A7}$$

where $a_{H_2O}^{*l}$ and $\gamma_{H_2O}^{*l}$ are the activity and the activity coefficient and $\mu_{H_2O}^{\ominus l}$ is the standard chemical potential of H_2O in albite melt referring to the hypothetical state of H_2O in the melt at the fictitious mole fraction of H_2O in the melt equal to one (at a given T and P). The condition $\gamma_{H_2O}^{*l} \to 1$ when $x_{H_2O}^{l} \to 0$ corresponds to this standard state (e.g. Münster, 1971; Robinson and Stokes, 1963).

According to (A7):

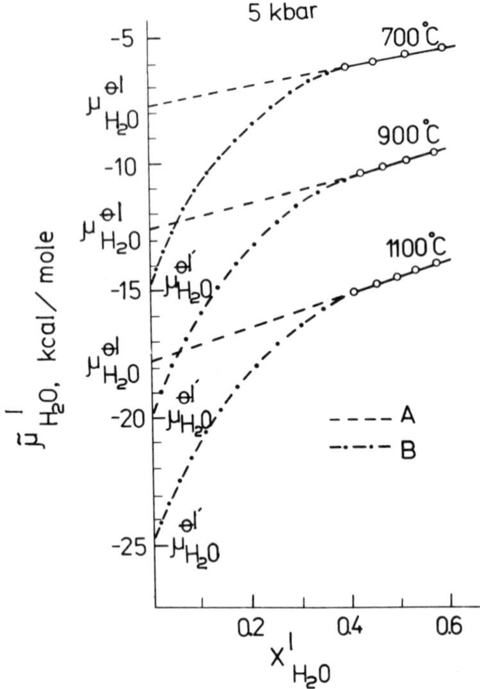

Fig. A1. The dependence of "reduced" chemical potential of H_2O in albite melt ($\tilde{\mu}^l_{H_2O}$) on the H_2O mole fraction in the melt and the determination of the standard chemical potential of H_2O in albite melt ($\mu^{\ominus l}_{H_2O}$) at various temperatures and $P_{tot} = 5$ kbar. *A* and *B* stand for different models (see the text).

$$\mu^{\ominus l}_{H_2O} = \lim_{x^l_{H_2O} \to 0} (\mu^l_{H_2O} - RT \ln x^l_{H_2O}) = \lim_{x^l_{H_2O} \to 0} \tilde{\mu}^l_{H_2O}, \quad \text{(A8)}$$

where $\tilde{\mu}^l_{H_2O}$ is the "reduced" chemical potential of H_2O in an albite melt, characterizing the state of H_2O in a hypothetical melt at the same T, P, and $\gamma^{*l}_{H_2O}$, but extrapolated to the fictitious $x^l_{H_2O} = 1$.

So, having $\mu^l_{H_2O}$ dependencies upon $x^l_{H_2O}$ for water–albite melt in the appropriate range of T (700–1100°C), P (1–10 kbar) and $x^l_{H_2O}$ (0.4–0.7) (Burnham and Davis, 1974, Fig. 2–6), $\tilde{\mu}^l_{H_2O}$ values could be determined as the function of real $x^l_{H_2O}$ (in this work the chemical potential of H_2O in the melt on the H_2O saturation surface is taken as the standard state of $\mu^l_{H_2O}$). It was found that in all T–P–x_{H_2O} ranges, these dependencies are well approximated by straight lines (Fig. A1). However, the extrapolation of these dependencies to $x^l_{H_2O} = 0$ is a complicated task, because the behavior of $\mu^l_{H_2O}$ as a function of $x^l_{H_2O}$ within the range of the low $x^l_{H_2O}$ is unknown a priori. This range is determined by the nature of the inter-action between the dissolving water and the melt (by its dissociation degree in the melt in particular). It was recently shown (Stolper, 1982; Epel'baum et al., 1984), that both the dissociated and nondissociated (molecular) forms of H_2O are present in the albite melt, practically within the whole range of total pressure, and the former prevails at the very beginning, but with increasing H_2O content in the melt, the proportion of the molecular H_2O increases also.

For the sake of approbation of the general calculation scheme the authors have determined the $\mu_{H_2O}^{\ominus l}$ values (as the initial approximation) by means of the linear extrapolation of the $\tilde{\mu}_{H_2O}^{l}$ on $x_{H_2O}^{l}$ dependencies to the $x_{H_2O}^{l} = 0$ (model A). So, it is assumed that the proportion between the dissociated and nondissociated forma of dissolved water within the range of the low $x_{H_2O}^{l}$ is the same as at $x_{H_2O}^{l} = 0.4$–0.6. Therefore, the $\mu_{H_2O}^{\ominus l}$ value thus obtained can be considered as a somewhat conditional standard state chemical potential of H_2O in albite melt, corresponding to the concentration range of the water–albite melt.

In this model, apparently, the excess chemical potential of H_2O in the melt at a given T and P ia proportional to the slope of the straight lines, i.e.

$$\mu_{H_2O}^{*l(ex)} = RT \ln \gamma_{H_2O}^{*l} = \tilde{\mu}_{H_2O}^{l} - \mu_{H_2O}^{\ominus l} = A x_{H_2O}^{l}, \tag{A9}$$

where the A parameter is the function of T and P.

Substituting the $\mu_{H_2O}^{\ominus l}$ and $\gamma_{H_2O}^{*l}$ values thus obtained into equation (A7), allows evaluation of total $\mu_{H_2O}^{l}$ values at any T, P, and $x_{H_2O}^{l}$ including the range of a very dilute solution of H_2O in albite melt. Then, taking the partial derivative from the total $\mu_{H_2O}^{l}$ values with respect to T gives the partial molar entropy of H_2O in albite melt ($S_{H_2O}^{*l}$) as the function of T, P, and $x_{H_2O}^{l}$ in the whole concentration range, which is standardized with respect to the same hypothetical reference state. By means of the Gibbs–Duhem equation also it is possible to determine the value of the partial molar excess entropy of the albite component in the melt. All the other data necessary for the calculation of the albite liquidus surface according to (A4) were taken: the entropy of fusion of anhydrous albite from Hervig and Navrotsky (1984) and T–P coordinates of the "dry" albite liquidus from Boettcher et al. (1982).

The solution of the differential equations (A4) at $P = 3$, 5 and 7 kbar was computed numerically using Eiler's method. Due to the fact that the $dT/d\mu_{H_2O}$ derivative in the starting point happens to be equal to zero, the curves of the T_{Ab}^{f} dependence on μ_{H_2O} have an initial horizontal part, so the numerical solutions, beginning with the sufficiently small, but finite μ_{H_2O} values, essentially tally (a beginning was with $-50{,}000$ cal/mol). So, the initial condition of $\mu_{H_2O} \to -\infty$ causes no calculation difficulties.

The albite isobaric liquidus curves "cut off" at the values of $\mu_{H_2O}^{l}$ equal to Gibbs' free energy of water ($G_{H_2O}^{0}$), when water vapor in equilibrium with albite crystals and H_2O-albite melt appears. The $G_{H_2O}^{0}$ values were taken from Burnham et al. (1969). This calculated H_2O-saturated liquidus of albite is in agreement with the experimental data (summed up in Burnham and Davis, 1974, Fig. 1) with respect to the H_2O solubility, but lies at about 50–70°C above it (Table 2, model *A*). Apparently, the difference results from ignorence of the well-known fact of H_2O dissociation in albite melt into two particles (Burnham and Davis, 1974; Epel'baum, 1980 and others) in this model.

For this reason: the later calculations have been carried out in the approximation of the subregular solution of H_2O in the albite melt within the low $x_{H_2O}^{l}$ range, in which both molecular and dissociated forms of H_2O are present (model *B*). Unfortunately, experimental data on H_2O solubility in this region are practically not available, and the H_2O dissociation constant as a function of T, P, and

Table 2. The experimental and calculated temperatures and H_2O solubilities in the melt at the water-saturated liquidus of the albite–H_2O system

	T, °C			$x^1_{H_2O}$		
	Calculation		Experiment	Calculation		Experiment
P_{tot}, kbar	Model A	Model B	(±10°C)	Model A	Model B	(±0.02)
3	865	790	800	0.545	0.550	0.552
5	825	765	755	0.615	0.617	0.616
7	775	715	725	0.675	0.675	0.675

The experimental data were taken from Burham and Davis (1974, Fig. 1).

$x^l_{H_2O}$ is unknown. So, in the model B has been assigned only a certain type of the function of the "reduced" chemical potential of H_2O in albite melt:

$$\mu^l_{H_2O} = \mu^{\ominus l'}_{H_2O} + \mu^{*l(ex)'}_{H_2O}, \tag{A10}$$

where

$$\mu^{\ominus l'}_{H_2O} = \mu^{\ominus l}_{H_2O} - A_0, \tag{A11}$$

$$\mu^{*l(ex)'}_{H_2O} = k_1 x^l_{H_2O} + k_2 (x^l_{H_2O})^2 = \left(A + \frac{2A_0}{x_0}\right) x^l_{H_2O} - \frac{A_0}{x_0^2}(x^l_{H_2O})^2. \tag{A12}$$

In equations (A11) and (A12) the parameter $A(T, P)$ is the same as in (A9), x_0 is the mole fraction of H_2O in the melt at which $\mu^l_{H_2O}$ begins deviating from linear dependence on $x^l_{H_2O}$ (see Fig. 11) and A_0 is a parameter, forming the new value of $\mu^{\ominus l}_{H_2O}$ and depending only on the choice of the x_0 value. It should be noted that (A12) corresponds to the general type of dependence of the activity coefficient of a solute on the concentration, being typical of the strong electrolyte solutions, whose properties, even at low concentrations, differ greatly from the properties of the infinitely dilute solution that serves as a reference state (Shakhparonov, 1956).

The value of the A_0 parameter was adjusted from experimental data referring to $P = 5$ kbar, and it turns out to be 7 kcal/mol at $x_0 = 0.4$. The other thermodynamic functions, included in (A4) were determined the same way as in the A model. The general technique of computing also remained the same.

As is shown in Table 2, one could reach a good agreement between the calculated and experimental coordinates of the water-saturated albite liquidus within the whole range of investigated pressures (3–7 kbar), using the constant A_0 parameter, which has been chosen for a single pressure.

This calculation results in the construction of the albite liquidus surface in T–P–μ_{H_2O} space. The liquidus isobars at 3; 5 and 7 kbar and H_2O solubility in the albite melt along these isobars are shown in Fig. A2. The μ_{H_2O} values on this surface are the $\mu^{Ab(f)}_{H_2O}$, i.e. the values of the chemical potential of the perfectly mobile H_2O, corresponding to the melting of the "pure" albite at a given T and

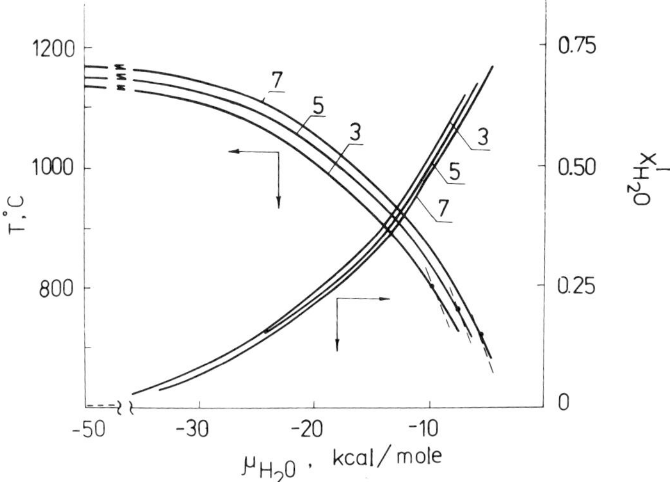

Fig. A2. The calculated isobaric liquidus curves of albite and isobaric H_2O solubility curves as function of the chemical potential of perfectly mobile H_2O. The intersections of calculated liquidus curves with the $G^0_{H_2O}$ isobars (dashed lines) correspond to the points of the three-phase equilibrium (albite crystals–melt–water vapor).

P (analogous to T^f_{Ab} in the common T–x liquidus diagrams). These $\mu^{Ab(f)}_{H_2O}$ values are necessary for calculation of the liquidus relations for the more complex systems including albite in the presence of the perfectly mobile H_2O.

In conclusion, the calculations undertaken which are of an illustrative character to some extent, demonstrate the practical use of the peculiarities of the thermodynamic description of the systems with PMC discussed above: i.e. the standardization of the μ of the inert component (albite) with respect to the μ of a perfectly mobile one (H_2O) and the use of the new thermodynamic values ($N^{Ab(l)}_{H_2O}$ and $S^{m(l)}_{Ab}$). The use of thermodynamic data on the "closed" albite–water system serves only for calibration and for finding out the necessary functions and could be considered as auxiliary operations. In general, these calculations have shown the validity of our approach to the quantitative thermodynamic description of melting phenomena in Korzhinskii's open systems.

Acknowledgments

The authors wish to express their sincere gratitude to Prof. V.A. Zharikov for his encouragement in connection with this work. They would also like to thank Dr. Yu.V. Alekhin for the helpful discussions on thermodynamics. The authors are grateful to A.G. Simakin for his help in the calculations of albite–H_2O system. They thank K.I. Shmulovich for his constructive comments and remarks.

References

Anfilogov, V.N., Glyuk, D.S., and Trufanova, L.G. (1973) Phase relations in interaction between granite and sodium fluoride at a water vapor pressure of 1000 kg/cm^2. *Geochem. Int.*, **10**, 30–33.

Boettcher, A.N., Burnham, C.W., Windom, K.E., and Bohlen, S.R. (1982) Liquids, glasses, and the melting of silicates at high pressures, *J. Geol.*, **90**, 127–138.

Bottinga, Y., Weill, D.F., and Richet, P. (1981) Thermodynamic modelling of silicate melts. In: *Thermodynamics of Minerals and Melts, Advances in Physical Geochemistry, Vol. 1*, edited by S.K. Saxena, pp. 207–245, Springer-Verlag, New York.

Burnham, C.W. (1975) Water and magmas: a mixing model. *Geochim. Cosmochim. Acta*, **39**, 1077–1084.

Burnham, C.W. (1979) The importance of volatile constituents. In: *The Evolution of the Igneous Rocks, Fiftieth Anniversary Perspectives*, edited by H.S. Yoder, Jr., pp. 439–478. Princeton University Press, Princeton, N.J.

Burnham, C.W., and Davis, N.F. (1974) The role of H_2O in silicate melts: II. Thermodynamic and phase relations in the system $NaAlSi_3O_8$–H_2O to 10 kilobars, 700° to 1100°C, *Amer. J. Sci.*, **274**, 902–940.

Burnham, C.W., Holloway, J.R., and Davis, N.F. (1969) Thermodynamic properties of water to 1000°C and 10,000 bars, *Geol. Soc. Amer. Spec. Paper* **132**, 96 p.

Carmichael, I.S.E., and MacKenzie, W.S. (1963) Feldspar–liquid equilibria in pantellerites: an experimental study, *Amer. J. Sci.*, **261**, 382–396.

Chelischev, N.F. (1967) An experimental study of the crystallization sequence of feldspars under the conditions of varying alkalinity of granitic melts. In: *Experimental Investigations in Mineralogy and Geochemistry of Rare Elements*, pp. 47–52. Nauka, Moscow (in Russian).

Dubrovskii, M.I. (1971) Granitic eutectics, minima and magmas. In: *Problems in Petrology and Geochemistry of Granitoids*, edited by D.S. Shteinberg and G.B. Fershtater, pp. 54–68. Sverdlovsk (in Russian).

Epel'baum, M.B. (1980) *Silicate Melts with Volatile Components*, p. 255, Nauka, Moscow (in Russian).

Epel'baum, M.B., and Kuznetsov, A.D. (1980a) On cotectic relations in closed three-component system and the effect of acid-base interaction of components in melts. *Geochimia*, 513–520 (in Russian).

Epel'baum, M.B., and Kuznetsov, A.D. (1980b) Eutectic related to chemical potential of a perfectly mobile component. *Dokl. USSR AS* **254**, 200–204 (in Russian).

Epel'baum, M.B., Persikov, E.S., and Zhigun, I.G. (1984) On the ratios of various water species in water-albite glass. In: *Contrib. Physico-Chemical Petrology, Vol. 12*, edited by V.A. Zharikov and V.V. Fedkin, pp. 72–78, Nauka, Moscow (in Russian).

Flood, H., and Förland, T. (1947) The acidic and basic properties of oxides. *Acta Chem. Scand.* **1**, 592–604.

Fraser, D.G. (1977) Thermodynamic properties of silicate melts. In: *Thermodynamics in Geology*, edited by D.G. Fraser, pp. 301–325. Dordrecht-Holland.

Ghiorso, H.S., Carmichael, I.S.E., Rivers, M.L., and Sack, R.O. (1983) The Gibbs free energy of mixing of natural silicate liquids; an expanded regular solution approximation for the calculation of magmatic intensive variables. *Contrib. Mineral. Petrol.*, **84**, 107–145.

Glyuk, D.S., and Anfilogov, V.N. (1973a) Phase equilibria in the system granite–H_2O–HF at a pressure of 1000 kg/cm^2, *Geochem. Int.*, **10**, 321–325.

Hervig, R.L., and Navrotsky, A. (1984) Thermochemical study of glasses in the system

$NaAlSi_3O_8$–$KAlSi_3O_8$–Si_4O_8 and the join $Na_{1.6}Al_{1.6}Si_{2.4}O_8$–$K_{1.6}Al_{1.6}Si_{2.4}O_8$. *Cosmochim. Acta*, **48**, 513–522.

James, R.S., and Hamilton, D.L. (1969) Phase relations in the system $NaAlSi_3O_8$–$KAlSi_3O_8$–$CaAl_2Si_2O_8$–SiO_2 at 1 kilobar water vapor pressure. *Contrib. Mineral. Petrol.*, **21**, 111–141.

Johannes, W. (1984) Beginning of melting in the granitic system Qz–Or–Ab–An–H_2O. *Contrib. Mineral. Petrol.*, **86**, 264–273.

Karpov, I.K., and Kiselev, A.I. (1979) Some general questions of theory of open systems with perfectly mobile components. In: *Physical-Chemistry of Endogenic Processes*, edited by F.A. Letnikov and Yu.V. Komarov, pp. 24–44. Nauka, Novosibirsk (in Russian).

Kogarko, L.N., and Krigman, L.D. (1981) *Fluorine in Silicate Melts and Magmas*, p. 126. Nauka, Moscow (in Russian).

Kogarko, L.N., and Ryabchikov, I.D. (1980) Polarity of magmatic equilibria. *Zapiski Vses. Mineral. Obsch.*, **109**, 505–516 (in Russian).

Korzhinskii, D.S. (1959) Acid–base interaction of components in silicate melts and directions of cotectic lines. *Dokl. USSR AS*, **228**, 383–386 (in Russian).

Korzhinskii, D.S. (1972) Flows of transmagmatic fluids and processes of granitization. In: *Magmatism, Formation of Crystalline Rocks and Depths of the Earth*, edited by A.K. Symon, pp. 144–153. Nauka, Moscow (in Russian).

Korzhinskii, D.S. (1973) *The Theoretical Foundations of Paragenetic Analysis of Minerals*, p. 288, Nauka, Moscow (in Russian).

Kovalenko, V.I., and Kovalenko, N.I. (1976) Ongonites (topaz-bearing quartz keratophyre)—subvolcanic analogues of rare-metal Li–F granites. *Trans. Joint Soviet-Mongolian Sci. Rea. Geol. Exped.*, *Vol.* **15**, p. 128. Nauka, Moscow (in Russian).

Kovalenko, N.I. (1979) *Experimental Investigation of Formation of Rare-Metal Li–F Granites*, p. 152. Nauka, Moscow (in Russian).

Kushiro, I. (1975) On the nature of silicate melt and its significance in magma genesis: regularities in the shift of the liquidus boundaries involving olivine, pyroxene, and silica minerals. *Amer. J. Sci.* **275**, 411–431.

Kuznetsov, A.D. (1982) The effect of fluid composition on eutectic relations in granitic systems. Unpublished Ph. D. Diss., *Inst. Exper. Miner., Chernogolovka*, p. 240 (in Russian).

Kuznetsov, A.D., and Epel'baum, M.B. (1978) An experimental study of the effect of acid-base interaction in acid melts. II. The influence of acidity of the equilibrium fluid on the shift of eutectics quartz + orthoclase and quartz + orthoclase + biotite. In: *Contrib. Physico-Chemical Petrology*, *Vol.* **8**, edited by V.A, Zharikov, pp. 62–75, Nauka, Moscow (in Russian).

Kuznetsov, A.D., and Epel'baum, M.B. (1979) On the quantitative estimation of the effect of acid–base interaction in melts, In: *Problems of Physico-Chemical Petrology. Metamorphism, Magmatism*, *Vol.* **1**, edited by V.A. Zharikov, pp. 242–255. Nauka, Moscow (in Russian).

Kuznetsov, A.D., and Epel'baum, M.B. (1981) Some peculiarities of thermodynamic descriptions of systems with perfectly mobile components. *Geochimia*, 820–835 (in Russian).

Kuznetsov, Yu.A., and Izoh, E.P. (1969) Geological evidences of intratelluric heat- and mass-flows as agents of metamorphism and magma genesis. In: *Problems of Petrology and Genetical Mineralogy*, *Vol.* **1**, edited by Yu. A. Kuznetsov, pp. 7–20. Nauka, Moscow (in Russian).

Letnikov, F.A., Karpov, I.K., Kiselev, A.I., and Shkandrii, B.O. (1977) *Fluid Regime of the Earth's Crust and Upper Mantle*, p. 216. Nauka, Moscow (in Russian).

Luth, W.C. (1969) The systems $NaAlSi_3O_8$–SiO_2 and $KAlSi_3O_8$–SiO_2 to 20 kb and the relationship between H_2O content, P_{H_2O}, and P_{total} in granitic magmas. *Amer. J. Sci.*, **267–A**, 325–341.

Luth, W.C., Jahns, R.H., and Tuttle, O.F. (1964) The granite system at pressures of 4 to 10 kilobars, *J. Geophys. Res.*, **69**, 759–773.

Lux, H. (1939) "Sauren" und "Basen" im Schmelzfluss: die Bestimmung der Sauerstoffionenkonzentration. *Z. Electrochem.*, **45**, 303–309.

Manning, D.A.C. (1981) The effect of fluorine on liquidus phase relationships in the system Qz–Ab–Or with excess water at 1 kb. *Contrib. Mineral. Petrol.*, **76**, 206–215.

Manning, D.A.C., and Pichavant, M. (1984) Experimental studies of the role of fluorine and boron in the formation of late-stage granitic rocks and associated mineralization, in *27th Inter. Geol. Congr., Moscow*, 4–14 Aug., 1984. *Dokl. Vol. 9. Section C.09. Petrology*, pp. 166–174, Moscow (in Russian).

Manning, D.A.C., Martin, J.S., Pichavant, M., and Henderson, C.M.B. (1984) The effect of F, B and Li on melt structure in the granitic system: different mechanisms? In: *Progress in Experimental Petrology, N.E.R.C., Publ. Ser. D No. 25*, 36–41.

Manning, D.A.C., Hamilton, D.L., Henderson, C.M.B., and Dempsey, M.J. (1980) The probable occurence of interstitial Al in hydrous, F-bearing and F-free aluminosilicate melts. *Contrib. Mineral. Petrol.*, **75**, 257–262.

Marakushev, A.A., and Tararin, I.A. (1965) On mineralogical criteria of alkalinity of granitoids. *Izvestia USSR AS, Ser. Geol.*, 20–37 (in Russian).

Marin, Yu.B. (1976) *Granitoidic Formations of Shallow and Moderate Depths*, p. 144, Leningrad State University, Leningrad (in Russian).

Marin, Yu.B., Skublov, G.T., and Vanshtein, B.G. (1983) *The Petrochemical Evolution of Phanerozoic Granitoidic Formations*, p. 152, Nedra, Leningrad (in Russian).

Martin, J.S., and Henderson, C.M.B. (1984) An experimental study of the effects of small amounts of lithium on the granitic system. In: *Progress in Experimental Petrology, N.E. R.C., Publ. Ser. D No.* **25**, 30–35.

Münster, A. (1971) *Chemical Thermodynamics*, p. 296, Mir, Moscow (in Russian).

Mysen, B.O. (1976) The role of volatilies in silicate melts: solubility of carbon dioxide and water in feldspar, pyroxene, and feldapathoid melts to 30 kb and 1625°C. *Amer. J. Sci.*, **276**, 969–996.

Mysen, B.O., Ryerson, F.J, and Virgo, D. (1980a) The influence of TiO_2 on the structure and derivative properties of silicate melts. *Amer. Mineral.* **65**, 1150–1165.

Mysen, B.O., Virgo, D., Harrison, W.J., and Scarfe, C.M. (1980b) Solubility mechanisms of H_2O in silicate melts at high pressures and temperatures: a Raman spectroscopic study, *Amer. Mineral.* **65**, 900–914.

Naney, M.T., and S.E. Swanson (1980) The effect of Fe and Mg on crystallization in granitic systems, *Amer. Mineral.* **65**, 639–653.

Negrey, E.V. (1983) *Petrology of the Late-Paleozoic Granitoids of the Central Kazakhstan*, p. 168, Nauka, Moscow (in Russian)

Pichavant, M. (1981) An experimental study of the effect of boron on a water saturated haplogranite at 1 kbar vapour pressure. Geological applications. *Contrib. Mineral. Petrol.*, **76**, 430–439.

Pichavant, M. (1983) Melt–fluid interaction deduced from studies of silicate–B_2O_3–H_2O systems at 1 kbar. *Bull. Mineral.*, **106**, 201–211.

Pichavant, M., and Manning, D.A.C. (1984) Petrogenesis of tourmaline granites and topaz granites; the contribution of experimental data. *Phys. Earth Planet. Inter.* **35**, 31–50.

Piwinskii, A.J., and Wyllie, P.J. (1968) Experimental studies of igneous rock series: a zoned pluton in the Wallowa batholith, Oregon. *J. Geol.* **76**, 205–234.

Platen, H. von (1965) Kristallisation granitischer Schmelzen. *Beitr. Mineral. Petrogr.*, **11**, 334–381.

Popov, V.S. (1981) The crystallization sequence of calc-alkali magmas and their petrological significance. *Geochimia*, 1665–1676 (in Russian).

Ramberg, H.S. (1952) Chemical bonds and distribution of cations in silicates. *J. Geol.*, **60**, 331–355.

Robinson, R., and Stokes, R. (1963) *Electrolyte Solutions*, p. 646. IL, Moscow (in Russian).

Ryabchikov, I.D. (1965) On the methods of thermodynamic analysis of equilibria in open systems. *Izvestia USSR AS Ser. Geol.*, 144–149 (in Russian).

Ryabchikov, I.D. (1975) *Thermodynamics of Fluid Phase of Granitoidic Magmas*, p. 232. Nauka, Moscow (in Russian).

Serykh, V.I., Gabov, Yu. A., Novichkova, A.P., Samoilova, V.A., and Nazarova, K.M. (1976) *The Mineral and Chemical Composition of Ultraacid Granitoids of the Central Kazakhstan*, p. 194, Nauka, Alma-Ata (in Russian).

Shakhparonov, M.I. (1956) *Introduction to the Molecular Theory of Solutions*, p. 507, GITTL, Moscow (in Russian).

Shaw, H.R. (1963) The four-phase curve sanidine—quartz—liquid—gas between 500 and 4,000 bars. *Amer. Mineral.*, **48**, 883–896.

Steiner, J.C., Jahns, R.H., and Luth, W.C. (1975) Crystallization of alkali feldspar and quartz in the haplogranite system $NaAlSi_3O_8$–$KAlSi_3O_8$–SiO_2–H_2O at 4 kb. *Geol. Soc. Amer. Bull.*, **86**, 83–98.

Stewart, D.B. (1967) Four-phase curve in the system $CaAl_2Si_2O_8$–SiO_2–H_2O between 1 and 10 kilobars, *Schweiz. mineral. Petrogr. Mitt.*, **47**, 35–59.

Stolper, E. (1982) Water in silicate glasses: an infrared spectroscopic study. *Contrib. Mineral. Petrol.*, **81**, 1–17.

Tauson, L.V. (1977) *The Geochemical Types and Potential Ore-Bearing Capacity of Granitoids*, p. 279, Nauka, Moscow (in Russian).

Thompson, R.N., and MacKenzie, W.S. (1967) Feldspar-liquid equilibria in peralkaline acid liquids: an experimental study. *Amer. J. Sci.*, **265**, 714–734.

Tuttle, O.F., and Bowen, N.L. (1958) Origin of granite in the light of experimental studies in the system $NaAlSi_3O_8$–$KAlSi_3O_8$–SiO_2–H_2O. *Geol. Soc. Amer. Mem.*, **74**, 153.

Wendlandt, R.F. (1981) Influence of CO_2 on melting of model granulite facies assemblages: a model for the genesis of charnockites. *Amer. Mineral.* **66**, 1164–1174.

Whitney, J.A. (1975) The effect of pressure, temperature, and X_{H_2O} on phase assemblege in four synthetic rock compositions, *J. Geol.*, **83**, 1–31,

Winkler, H.G.F., and Lindemann, W. (1972) The system Qz–Or–An–H_2O within the granitic system Qz–Or–Ab–An–H_2O. Application to granitic magma formation. *N. Jb. Mineral. Mh.*, 49–61.

Wyllie, P.J. (1979) Magmas and volatile components. *Amer. Mineral.*, **64**, 469–500.

Wyllie, P.J., Huang, W.L., Stern, C.R., and Maaløe, S. (1976) Granitic magmas: possible and impossible sources, water contents, and crystallization sequences. *Can. J. Earth Sci.*, **13**, 1007–1019.

Zharikov, V.A. (1969) Regime of components in melts and magmatic replacement. In: *Problem of Petrology and Genetical Mineralogy, Vol. 1*, edited by Yu. A. Kuznetsov, pp. 62–79, Nauka, Moscow (in Russian).

Zharikov, V.A. (1976) *Foundations of Physico-Chemical Petrology*, p. 420, Moscow State University, Moscow (in Russian).

Index